NONIONIC
SURFACTANTS

SURFACTANT SCIENCE SERIES

CONSULTING EDITOR

MARTIN J. SCHICK
Consultant
New York, New York

ADDITIONAL VOLUMES IN PREPARATION

NONIONIC SURFACTANTS

Organic Chemistry

edited by

Nico M. van Os

Shell International Chemicals B.V.
Amsterdam, The Netherlands

CRC Press
Taylor & Francis Group
Boca Raton London New York

CRC Press is an imprint of the
Taylor & Francis Group, an **informa** business

CRC Press
Taylor & Francis Group
6000 Broken Sound Parkway NW, Suite 300
Boca Raton, FL 33487-2742

First issued in paperback 2019

© 1998 by Taylor & Francis Group, LLC
CRC Press is an imprint of Taylor & Francis Group, an Informa business

No claim to original U.S. Government works

ISBN-13: 978-0-8247-9997-7 (hbk)
ISBN-13: 978-0-367-40078-1 (pbk)

Visit the Taylor & Francis Web site at
http://www.taylorandfrancis.com

and the CRC Press Web site at
http://www.crcpress.com

Nico M. van Os
(1944–1996)

Foreword

Nonionic surfactants were the subject of the first volume of the Surfactant Science series published by Marcel Dekker, Inc. Because this volume was published 30 years ago, a thorough reappraisal of this class of surfactant, which has attained ever-increasing importance, has become necessary. An initial step in this direction was Volume 23 of the series *Nonionic Surfactants: Physical Chemistry*, which appeared in 1987. An only slightly more recent volume was concerned with the analytical chemistry of nonionic surfactants. The present volume, *Nonionic Surfactants: Organic Chemistry*, brings the presentation of this field up to date.

Through his many years of experience as a research chemist in the field of surfactant chemistry at the Shell Research and Technology Center, Amsterdam, Nico van Os had a profound knowledge of this subject matter. After completing the manuscript for this book, Nico van Os' untimely death denied him the opportunity to witness its publication. He had succeeded in recruiting fifteen well-versed specialists with many years of practical experience in the industrial sector to present their expert knowledge on synthesis and application of the various types of nonionic surfactants with the appropriate conciseness.

The rapid growth in the field of nonionic surfactants continues unabated today, with new classes of nonionic surfactants being introduced and new applications being discovered for these compounds. The area of detergents, cleaning agents, and personal care products will remain the primary domain for the application of nonionic surfactants in the future. Other applications for these compounds must also be taken into account, however. Recent studies on the phase behavior of functionalized nonionic surfactants indicate interesting prospects with regard to the homogeneous catalysis of chemical processes:

nonionic surfactant phosphines control the temperature of a process that takes place in aqueous solution as "intelligent" complex ligands, by separating the catalyst complex from the reaction mixture when a desired temperature identical to the cloud point of the nonionic phosphine ligands is exceeded. Nonionic surfactant phosphines are able to play a similarly "intelligent" role in two-phase catalysis, by transferring the homogeneous catalyst complex with these surfactant phosphine ligands from the aqueous phase to the organic substrate phase at the reaction temperature, where it is able to act like a normal homogeneous catalyst. When the temperature drops below the cloud point, it returns to the aqueous phase and is thus separated from the reaction product. Such examples clearly demonstrate the chemical potential harbored by the nonionic surfactants and their wide range of potential applications in areas other than detergents and cleaning agents.

Bernhard Fell
Rheinisch-Westfälische Technische Hochschule
Aachen, Germany

Preface

The main use for nonionic surfactants has been, and in the foreseeable future will be, in the area of the detergent and personal care formulations in combination with anionic surfactants. Nonionic surfactants have also begun to find applications in new areas such as environmental protection systems, paper processing, and thin films.

In detergent and personal care formulation, nonionic surfactants are represented mostly by linear alcohol ethoxylates, with the alcohols being derived from either petrochemical raw materials or natural feedstock. Their share of the total use of surfactants has increased from 4% in 1970 to 35% in the 1990s. The reasons for this are their excellent detergency properties and favorable environmental properties such as rapid biodegradability and low toxicity. In addition, detergent manufacturers use nonionic surfactants because of their convenience in formulating concentrated powders and liquids.

Not surprisingly, this increase in market share is also reflected in the literature on nonionic surfactants. Since the early 1960s, over 50,000 articles, patents, and books have treated the subject of nonionic surfactants in one way or another. The need for review is therefore great, and this undoubtedly was the main motive of Martin Schick in publishing in 1967 his *Nonionic Surfactants* as Volume 1 in the Surfactant Science Series. This volume covered the following disciplines: organic, physical, and analytical chemistry and biology. The book was well received by the scientific and technical community, and is a much referenced introductory text for many scientists in industry and academia who have since become acquainted with nonionic surfactants

With the passage of time and with the enormous growth in fundamental and applied research on nonionic surfactants, it was decided to split the second edition of Dr. Schick's book into separate volumes. These treat the chemical analysis of nonionic surfactants, the physical chemistry of nonionic surfactants, the polyalkylene oxide block copolymers, the alkylglucosides, the biology of nonionic surfactants, and in the present volume, organic chemistry.

The polyoxyethylene surfactants, constitute a significant class of nonionic surfactants and therefore are discussed extensively in the book. The materials are commercially produced by reaction of ethylene with an active hydrogen-containing compound RXH (initiator). A polyoxyethylene surfactant can be represented by the general formula $RX(CH_2CH_2O)xH$, where R can represent H (for polyethylene glycols) or a hydrophobic group (usually C_1-C_{20}) and X represents a hetero atom, such as O (giving polyoxyethylene alcohols), N (giving polyoxyethylene alkylamines or polyoxyethylene alkylamides), S (giving polyoxyethylene mercaptans), or phenol (giving polyoxyethylene alkylphenols).

The number of oxyethylene groups in the product x represents the average degree of polymerization and is therefore not necessarily an integer. The product is thus a mixture of polyoxyethylene surfactants, and properties of the surfactants depend on the details of this distribution. In Chapter 1, "Distribution of the Polyethylene Chain," Edwards discusses the details of the mechanisms of the reaction of ethylene oxide with various initiators. Of particular importance is the development of a novel class of catalysts that are able to produce surfactants with a much narrower distribution than conventional catalysts.

Over the last 50 years, a substantial portion of the nonionic surfactant market has been represented by the polyoxyethylene alkylphenols. In Chapter 2, "Polyoxyethylene Alkylphenols," Weinheimer and Varineau discuss the manufacturing of these type of surfactants. They are used in a wide range of industries from cleaning systems to paints and coatings to agricultural chemistry. Because polyoxyethylene alkylphenols are produced by a large number of different companies and have been on the market for many years many environmental studies have been performed.

Another important class of nonionic surfactants is the "Polyoxyethylene Alcohols," which are discussed in Chapter 3 by Edwards. These materials are produced in an excess of a billion pounds a year and are used in applications ranging from household and institutional laundry, textile scouring, pulp and paper manufacturing, oil field surfactants, and agricultural spray adjuvants. With an increasing emphasis on environmental issues, soil remediation could become an important new area for applications for polyoxyethylene alcohols.

Among the first nonionic surfactants to be commercialized are the "Polyoxyethylene Esters of Fatty Acids." Kosswig describes various synthesis methods of these materials in Chapter 4. The applications of these surfactants are numerous. For example, they are used as emulsifiers, softeners, antistatic agents, or mold lubricants. Because polyoxyethylene esters of fatty acids have in general a low toxicity and irritancy to the skin and mucous membranes, they are particularly suited for cosmetic and pharmaceutical applications.

Mercaptans (or thiols) can also be reacted with ethylene oxides to produce surfactants. In Chapter 5, "Polyoxyethylene Mercaptans," Edwards discusses their preparation. Compared to their oxygen or nitrogen counterparts, there are fewer applications for these polyoxyethylene mercaptans because of the potential for odor development from impurities in the product. However, the sulfur atom does give these surfactants some unique properties, utilizing, for example, the reactivity of the sulfur atom in the chain, which results in their use in industrial and institutional applications that do not have stringent odor requirements.

In Chapter 6, "Polyoxyethylene Alkylamines," Hoey and Gadberry describe the synthesis of these surfactants. Most of the applications utilize their affinity for surfaces. For example, polyoxyethylene alkylamines are used as corrosion inhibitors, to coat films or glass, or as wettability aids, and as additives in gasoline, drilling fluids, and plastics. Hoey and Gadberry show that more applications of these type of surfactants could be foreseen if such problems as the dark color and the formation of side products during the reactions could be solved.

Nonionic surfactants containing an amide group are a small part of the total volume of surfactants. Lif and Hellsten show in Chapter 7, "Nonionic Surfactants Containing an Amide Group," that because of their simple manufacturing process combined with their good chemical stability and their strong surface activity, it is likely that the importance of these type of surfactants will increase.

The "Polyol Ester Surfactants," discussed in Chapter 8 by Lewis, are a different class of nonionic surfactants. These materials are made from reactions of fatty acids with polyols such as glycols, sorbitol, and sugars. Because these surfactants are made from natural materials, they are often used as food additives. The surfactants made from glycols, sorbitol, and sugars are often too "oily," which limits their value as useful surfactants. To enhance the hydrophilicity of these materials the residual hydroxyl functionality in these compounds can be used as the basis for reaction with ethylene oxide. Recently, there have been interesting developments in the use of enzyme catalysis to add a degree of product specificity.

The surfactant properties of nonionic surfactants can be improved by introducing an ionic group. In Chapter 9, "Nonionics as Intermediates for Ionic Surfactants," Behler et al. discuss the synthesis of these type of surfactants. It is shown that (alkyl) polyoxyethylene alcohols can be used as a feedstock to build up sulfates, sulfonates, phosphates, and carboxylates. The most important group is the polyoxyethylene alkyl sulfates, which are used in dishwashing liquids and cosmetic formulations. Phosphates and carboxylates are mainly used in technical applications in the fiber and textile industries.

The nine chapters of this book represented an update of the organic chemistry described in the first edition of Volume 1. Each of the chapters contains discussions on laboratory and in particular technical syntheses of nonionic surfactants. The authors represent a cross-section of researchers working in industry who are specialists in the discipline of organic chemistry. Taken together, the chapters of this book constitute an exhaustive, yet concise view of all aspects of the organic chemistry of nonionic surfactants. The researcher who is new to the field will find this volume a valuable introduction; at the other end of the arc, the specialist in the area will benefit from the review material contained in the present volume.

Nico M. van Os

Contents

Contributors

Ansgar Behler   Henkel KGaA, Düsseldorf, Germany

Charles L. Edwards   Westhollow Technology Center, Shell Chemical Company, Houston, Texas

James F. Gadberry   Surfactants America Research and Development, Akzo Nobel Central Research, Dobbs Ferry, New York

Martin Hellsten*   Central Research and Development, Akzo Nobel Surface Chemistry AB, Stenungsund, Sweden

Karlheinz Hill   Henkel KGaA, Düsseldorf, Germany

Michael D. Hoey   Surfactants America Research and Development, Akzo Nobel Central Research, Dobbs Ferry, New York

Kurt Kosswig*   Hüls AG, Marl, Germany

Andreas Kusch†   Henkel KGaA, Düsseldorf, Germany

Jeremy J. Lewis   Research and Technology, ICI Surfactants, Everbe Belgium

---

*Retired.

†*Current affiliation*: Boehringer Ingelheim KG, Ingelheim au Rhein, Germany.

Anna Lif   Central Research and Development, Akzo Nobel Surface Chemistry AB, Stenungsund, Sweden

Stefan Podubrin   Henkel KGaA, Düsseldorf, Germany

Hans-Christian Raths   Henkel KGaA, Düsseldorf, Germany

Günter Uphues   Henkel KGaA, Düsseldorf, Germany

Pierre T. Varineau   Research and Development, Union Carbide Corporation, South Charleston, West Virginia

Robert M. Weinheimer   Research and Development, Union Carbide Corporation, South Charleston, West Virginia

rg,

1

Distribution of the Polyoxyethylene Chain

CHARLES L. EDWARDS Westhollow Technology Center,
Shell Chemical Company, Houston, Texas

I. INTRODUCTION

Nonionic polyoxyethylene surfactants are used extensively in the chemical industry, e.g., in such areas as detergents, health and personal care, coatings, and polymers. These materials are commercially produced by reaction of ethylene oxide with an active hydrogen-containing compound using a basic catalyst. The reaction can be represented by the following general equation:

$$RXH \ + \ x \ CH_2{-}CH_2 \ \xrightarrow{\hspace{2cm}} \ RX(CH_2CH_2O)_xH$$

For commercially available surfactants R represents either H (as for polyethylene glycols) or a hydrophobic group (usually C_1 to C_{20}), and X represents a heteroatom such as O, S, or N. The number of moles of ethylene oxide reacting with the initiator (RXH) need not be an integer and represents the number average degree of polymerization of the ethylene oxide in the product. However, the

1

product is a mixture of polyoxyethylene adducts of the initiator with each adduct having the formula of $RX(CH_2CH_2O)_xH$ with x being an integer. This product distribution is often referred to as the ethoxylate distribution (EOD) and is defined by the nature of R, X, and, importantly, the type of catalyst used in its preparation. The physical characteristics of the nonionic polyoxyethylene-based surfactant and its ultimate utility is dependent on the nature of R, X, the value of x, and the adduct or oligomeric distribution of the species present in the mixture.

This chapter addresses the mechanism of reaction of ethylene oxide with various initiators using both conventional homogeneous catalysts and the recently developed narrow-range polyoxyethylation catalysts. The latter catalyst systems are capable of producing surfactants containing much narrower polyoxyethylene adduct distributions compared with those produced using conventional catalysts. These materials are reported to have improved physical and/or performance properties [1] and will be discussed at the end of this chapter. A discussion of chain length distribution equations that best fits experimentally derived product distributions obtained using the various catalyst systems will follow each discussion of reaction mechanism.

This chapter will only address the reaction of initiators with ethylene oxide. General information of the chemistry of higher molecular weight epoxides (i.e., for epoxides containing more than two carbons) are described in other chapters in this book and elsewhere [2,3]. Information concerning performance features of polyoxyethylene alcohols with different values of R, X, and x will be found in other chapters in this book.

II. CHEMISTRY AND HANDLING OF ETHYLENE OXIDE

A. Structure and Reactivity of Ethylene Oxide

Ethylene oxide is the most reactive member of the family of cylic ethers. It can form adducts with a variety of nucleophiles under the mildest of conditions. In contrast with larger cyclic ethers, it can react not only under acidic or basic conditions, but also under essentially neutral conditions [4]. This marked difference in reactivity is attributed to the release of the ring strain of the three-membered compound [5]. This ring strain has been determined indirectly by several methods [6] with an average value of approximately 54 kJ/mole. The release of energy during reaction with initiators accounts for the high heat release (ΔH_f) that occurs during the polyoxyethylation reaction and that has been reported to be 92kJ/mole [7]. General physical properties for the molecule can be found in information concerning the safe handling of ethylene oxide [8], which is described further in the next section.

The structure of ethylene oxide has been reported [9,10] with attendant bond lengths and angles. As is the case for cyclopropane the three-membered ethy-

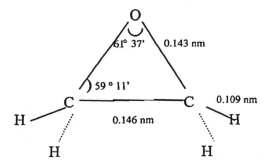

FIG. 1 Bond lengths and angles for ethylene oxide [8].

lene oxide ring has considerable double-bond character. For example, the C–C bond length as shown in Fig. 1 has been measured to be 0.146 nm [11], which is midway between a normal C–C single bond (0.154 nm) and C=C double bond (0.135 nm). The strained ring prevents C–O–C bond rotation and thus the oxygen had directed lone pairs of electrons, which accounts for the increased nucleophilicity and basicity of the ether compared with other ethers, such as diethyl ether. Thus ethylene oxide is a much stronger ligand for coordination with metallic compounds than other ethers. It is clear from the reports that the high reactivity of ethylene oxide is due primarily to the ring strain and double-bond character of the molecule.

B. Handling, Storage, and Transfer of Ethylene Oxide

Ethylene oxide is a particularly hazardous chemical. When using this material, care must be taken to minimize personal exposure to ethylene oxide due to its extreme toxicity and potential for explosion [8]. Ethylene oxide has been reported to be a carcinogen [12,13] as well as a fetotoxin [14–16]. The current acceptable Occupational Safety and Health Administration (OSHA) limits of exposure are <1 ppm per 8-h day (time-weighted average) [17]. When handling, transporting, or transferring ethylene oxide, the use of respiratory equipment has been recommended by safety organizations.

Unlike other members of the epoxide family and ethers, ethylene oxide has an explosion limit of 2.6–100% in air [18]. The upper explosion limit of 100% means that pure ethylene oxide can decompose even in the absence of air or oxygen. Flammability limits of ethylene oxide have been reported as mixtures with various inert gases, such as with H_2O [19], N_2 [20], CO_2 [21], and CH_4 [22]. These reports strongly recommend that ethylene oxide be stored and used under an inert gas atmosphere (most commonly N_2). Most commercial operations use N_2 pressure in the vapor phase to ensure that the vapor phase ethylene oxide concentration never reaches the explosion region. Explosion limit tables should always be consulted prior to use.

III. POLYOXYETHYLENE CHAIN LENGTH DISTRIBUTIONS

The reaction of active hydrogen-containing substrates with ethylene oxide normally results in an oligomeric distribution of products even when using less than 1 mole of ethylene oxide per mole of initiator [23]. Exceptions to this general rule will be discussed later in this chapter. The mixture of adducts results from the fact that the inititally formed adduct ($RXCH_2CH_2OH$) is also an active hydrogen-containing species and can compete with the initiator for reaction with more ethylene oxide. This can be represented by the following general equations:

$$RXH \ + \ H_2C \overset{O}{\diagdown} CH_2 \longrightarrow RXCH_2CH_2OH$$

$$RXCH_2CH_2OH \ + \ x \ H_2C \overset{O}{\diagup} CH_2 \longrightarrow RX(CH_2CH_2O)_{x+1}H$$

The polyoxyethylene product distribution is determined by the choice of initiator (RXH), the number of moles of ethylene oxide reacted per mole of initiator (x), and the choice of catalyst used. Recent work in the area of new polyoxyethylation catalysts has resulted in the development of catalysts capable of producing unusually narrow or peaked polyoxyethylene distributions. Prior to the discovery of these new catalysts, the narrowest distributions attainable followed a Poisson type of distribution [23]. However, discovery of these new heterogeneous catalysts allowed distributions to be attained that permitted the production of ethoxylates in which the major single adduct approached 25 wt% of the total product mixture [1].

The polyoxyethylene distributions will be divided into two sections—conventional distributions obtained using homogeneous basic or acidic catalysts and narrow-range distributions obtained using the new heterogeneous narrow-range ethoxylate (NRE) catalysts. In each case the proposed mechanism of reaction will be discussed followed by distribution equations that best describe each system.

A. Distributions Obtained Using Conventional Homogeneous Catalysts

Early work involved the use of acidic and basic catalysts for the ring opening reactions of ethylene oxide [3,23–26]. This section will discuss the effect of the use of conventional homogeneous catalysts for the reaction of substrates with ethylene oxide. Conventional distributions fall into two categories—broader distributions obtained using basic catalysts, and narrower distributions obtained using acidic catalysts. Although the latter produce narrower ethoxylate distributions, they are rarely used commercially because they result in significant production of undesirable side products that would have to be removed prior to use.

From a mechanistic perspective, the main difference between conventional base-catalyzed and acid-catalyzed ring opening of ethylene oxide is which species, RXH or EO, is primarily associated with the catalyst species. For base-catalyzed reactions (e.g., when using KOH as catalyst), the metal is primarily associated with RXH (as RX$^-$), whereas for acid-catalyzed reactions the catalyst is associated with EO. These differences are a major factor in determining the polyoxyethylene chain distribution and will be discussed in each section.

1. Base-Catalyzed Ring Opening of Ethylene Oxide

(a) Mechanism of Reaction. The vast majority of commercially available polyoxyethylene-based surfactants are produced using basic catalysts. Although the most commonly used catalyst is potassium hydroxide, many others have been reported. These include catalysts derived from alkali and alkaline earth metal hydroxides, oxides, or alkoxides [27], as well as amines [28]. Catalyst activity is directly related to the size of the metallic cation within a specific family in the periodic table, the larger the ionic radius, the more active the catalyst (e.g., Li < Na < K < Rb < Cs for group I and Mg < Ca < Sr < Ba for group II metals). Lanthanide hydroxide or alkoxides have also been reported to be capable of catalyzing this reaction [29]. The most commonly employed basic catalyst is potassium hydroxide conveniently introduced to the reactor by using 50% aqueous potassium hydroxide solutions. There are several reasons for using this catalyst system: ease of handling, low cost, high activity, high selectivity to desired products, and low levels of byproducts.

The mechanism of the ring opening reaction using basic catalysts can be summarized in the following steps for the case of potassium hydroxide as catalyst:

Activation $\qquad$ RXH $\quad + \quad$ KOH $\quad\longrightarrow\quad$ RX$^-$ K$^+$ $\quad + \quad$ H$_2$O $\uparrow$ $\qquad$ (1)

Initiation $\qquad$ RX$^-$ $\quad + \quad$ $\underset{\displaystyle H_2C \text{——} CH_2}{\overset{\displaystyle O}{\diagup\diagdown}}$ $\quad\longrightarrow\quad$ RXCH$_2$CH$_2$O$^-$ $\qquad$ (2)

Chain Transfer $\quad$ RXH $\;+\;$ RXCH$_2$CH$_2$O$^-$ $\;\underset{\displaystyle \longleftarrow}{\overset{\text{fast}}{\longrightarrow}}\;$ RX$^-$ $\;+\;$ RXCH$_2$CH$_2$OH $\qquad$ (3)

Propagation
$$RXCH_2CH_2O^- + x\ H_2C\overset{O}{\overset{\diagup\diagdown}{-\!\!-}}CH_2 \longrightarrow RX(CH_2CH_2O)_x^- \qquad (4)$$

Termination
$$RX(CH_2CH_2O)_x^- + HX \longrightarrow RX(CH_2CH_2O)_xH + X^-$$
$$\text{where } HX = \text{acid} \qquad (5a)$$

$$RX(CH_2CH_2O)_x^- + RX(CH_2CH_2O)_yH \rightarrow RX(CH_2CH_2O)_2R + {}^-O(CH_2CH_2O)nH \quad (5b)$$

In this mechanism the metallic catalyst (in this case potassium cation) is primarily associated with the RXH or $RX(OCH_2CH_2OH)$ and not EO. This significantly controls the polyoxyethylene oxide chain distribution.

Activation of RXH. The activation step refers to the reaction of RXH with base to produce the potassium salt of the initiator and water [see Eq. (1)]. This is an equilibrium reaction, and the water must be remove to achieve high conversion to the potassium salt. The water can be removed either by N_2 sparging or under vacuum during heating of the mixture. Azeotropic distillation using aromatic hydrocarbons is also employed under special conditions. Unreacted base or residual water will produce polyethylene glycol during reaction, so significant effort must be made to drive the reaction to completion. While most commercial ethoxylation processes use potassium hydroxide as catalyst, potassium metal or low molecular weight potassium alkoxides (such as the ethoxides or *t*-butoxides) can also be used. When potassium metal is used, water is not formed and thus eliminates the need for a drying step. However, when potassium alkoxides are used as catalyst, the corresponding low molecular weight alcohol is produced via transalcoholysis, which must be removed to prevent competitive polyoxyethylation of this alcohol.

Initiation. The initiation step [Eq. (2)] involves reaction of the deprotonated initiator (RX^-) with ethylene oxide, which falls under the category of nucleophilic substitution. Because the rate has been shown to be dependent on both the concentration of deprotonated substrate and EO that is upon $[RX^-][C_2H_4O]$, it is a second-order reaction known as an S_N2 reaction (30). The actual rate of reaction of RX^- with ethylene oxide for a given metallic catalyst depends on (1) the nucleophilicity of the X^- group [31] and (2) the steric hindrance of the R group listed below [32]:

$\cdots > RO^-$ (alkoxide) $> RO^-$ (phenoxide) $> RCO_2^-$

primary $>$ secondary $\gg$ tertiary

Further, the order of reactivity of RX^- within a series of metallic catalyst is dependent on the size and valency of the metallic cation.

Chain transfer. The chain transfer reaction [Eq. (3)] is the fastest step of the overall reaction because it involves an acid–base proton transfer. Because this is an equilibrium reaction, then the concentration of the active (anionic) species in solution will be determined by the relative acidities of RXH and $RXCH_2CH_2OH$. This dramatically affects the outcome of the reaction, and this effect will be discussed under the following three situations:

   i.   Where the acidity of RXH >> $RXCH_2CH_2OH$
  ii.   Where the acidity of RXH = $RXCH_2CH_2OH$
 iii.   Where the acidity of RXH >> $RXCH_2CH_2OH$

   (i) Acidity of RXH >> $RXCH_2CH_2OH$: conditions for phenols, mercaptans, and carboxylic acids. When the acidity of RXH is much greater than $RXCH_2CH_2OH$, as for the cases with phenols, mercaptans, and carboxylic acids, the equilibrium constant for the proton exchange reaction will be very large.

$$RXH + RXCH_2CH_2O^- \; \rightleftharpoons \; RX^- + RXCH_2CH_2OH$$

$$K_{eq} = \frac{[RX^-]\,[RXCH_2CH_2OH]}{[RXH]\,[RXCH_2CH_2O^-]} >> 1$$

Under these conditions essentially all of the initiator is converted to the first polyoxyethylene adduct prior to further reaction of the monoadduct to higher adducts. Therefore, chain propagation does not occur even in the presence of excess ethylene oxide until all RXH has been consumed. Therefore, it is possible to isolate extremely high yields (in some cases >98 mol%) of $RXCH_2CH_2OH$ if equimolar concentrations of RXH and ethylene oxide are reacted using basic catalysts. This is true for the ethoxylation of phenols [33], mercaptans [34], and carboxylic acids [35].

   An interesting rate effect is observed during the polyoxyethylation of phenols and carboxylic acids. Because phenolate and carboxylate anions are less nucleophilic than the polyoxyethylene anion [31], ethylene oxide adds to the initiator much more slowly than the ethylene oxide adds to the monoadducts. This results in the observance of a sharp increase in rate of ethylene oxide consumption after an equimolar amount of ethylene oxide has reacted with phenols [33,36] or carboxylic acids [35], as shown in Figs. 2 and 3.

   The reaction between phenols and ethylene oxide has been studied extensively and the kinetics reported [33,37]. The reaction of carboxylic acids with ethylene oxide has also been reported [35], and under the basic reaction conditions, ester interchange occurs rapidly leading to a mixture of products, including polyethylene glycol. This leads to products that have a relatively broad molecular weight distribution relative to that of products obtained from the polyoxyethylation of phenols.

$$2\ \overset{\displaystyle O}{\underset{\displaystyle \|}{R}}CO(CH_2CH_2O)_xH \; \underset{\longleftarrow}{\longrightarrow} \; \overset{\displaystyle O}{\underset{\displaystyle \|}{R}}CO(CH_2CH_2O)_x\overset{\displaystyle O}{\underset{\displaystyle \|}{C}}R \; + \; HO(CH_2CH_2O)_xH$$

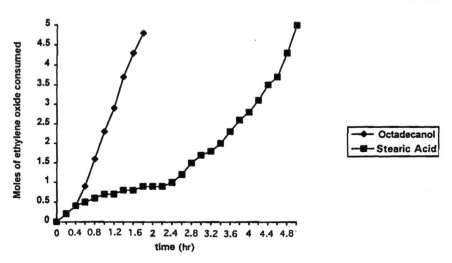

FIG. 2 Rate of polyoxyethylation of stearic acid and octadecanol at 185°C using basic catalysts [35].

The reaction of mercaptans with ethylene oxide differs in one respect from reactions of phenols and carboxylic acids with ethylene oxide. Mercaptide anions are much more nucleophilic than either phenolate, carboxylates, or polyoxyethylene anions [31]. Therefore, the rate of addition of RS$^-$ to ethylene oxide should be much greater than that of $RS(CH_2CH_2O)_x^-$ to ethylene oxide. This is true only if the RS$^-$K$^+$, which is usually insoluble in the mercaptan, is

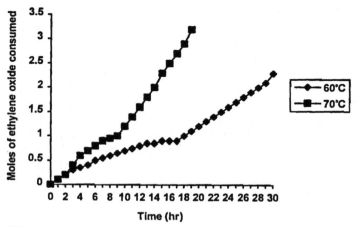

FIG. 3 Rate of addition of ethylene oxide to phenol in excess ethylene oxide using sodium phenoxide catalyst [33].

solubilized using an additive such as an alcohol or the monoadduct of the mercaptan to be used [34]. Early reports that mercaptans were less reactive than alcohols were presumably describing insoluble anhydrous RS^-K^+ systems [38]. In this case, the apparent slow reaction with ethylene oxide would be due to the insolubility of the catalyst at the beginning of the reaction. Induction periods would be observed under these circumstances. Using solubilized catalyst systems (e.g., of molar ratios of $RS^-K^+/RSCH_2CH_2OH = 1$), temperatures as low as 80–100°C can be used during the addition of the first mole of ethylene oxide per mole of mercaptan. The reaction temperature is then allowed to increase during the addition of the first mole of ethylene oxide to the initiator to the final polymerization temperatures (e.g., 140–160°C) [34]. As with phenols and carboxylic acids, mercaptans react with ethylene oxide to produce the monoethoxylate adduct prior to polymerization even in using an excess of ethylene oxide. Consequently, $RSCH_2CH_2OH$ can be produced in >99 mol% yield.

(ii) Acidity of RXH comparable to RXCH$_2$CH$_2$OH: conditions for alcohols, amides, and water. When the acidity of RXH and $RXCH_2CH_2OH$ is comparable, as for the cases with alcohols, amides, and water, the equilibrium constant for the proton exchange reaction will be ~1.

$$K_{eq} = \frac{[RX^-]\,[RXCH_2CH_2OH]}{[RXH]\,[RXCH_2CH_2O^-]} = \sim 1$$

Under these conditions, chain growth (chain propagation) is competitive with chain initiation. This means that there will be substantial chain growth prior to complete conversion of initiator to monoadduct formation. The overall product distribution will be determined not only by the equilibrium reaction (chain transfer), but by the relative rates of addition of RX^- and $RXCH_2CH_2O^-$ to ethylene oxide. This will be described in more detail in the chain propagation section.

The most important class of compounds in this category is fatty alcohols. These materials are used extensively in the detergents industry, and consequently much has been published concerning their reaction with ethylene oxide. The relative acidities of various alcohols have been studied and compared with that of the comparable monoethyleneoxy adduct [39,40]. These are illustrated in Table 1. The rate constants of the reaction of ethylene oxide with various alcohols have also been reported [41] and are presented in Table 2. From the pK_a values and relative reaction rates, the distribution equations can be calculated. This will be addressed later in this chapter.

The reaction rate for polyoxyethylation of alcohols increases with an increase in the basicity of the catalyst. Despite a report [24] that Na, $NaOCH_3$, $NaOC_2H_5$, and KOH showed comparable activity for the oxyethylation of tridecanol, close inspection of the data shows that KOH is actually more active than

TABLE 1 Relative Acidities of Some Alcohols in Isopropyl Alcohol Solution

Compound	K_e[a]	Ref.
Isopropyl alcohol	0.076[b]	39
t-Butanol	c	39
s-Butanol	c	39
s-Butanol	<0.8	40
n-Propanol	~0.5	39
i-Butanol	~0.5	39
n-Butanol	~0.6	39
n-Butanol	0.8	40
Ethanol	0.95	39
Water	1.2	40
Water	1.2	39
Methanol	4.0	39
Ethylene glycol mono-s-butyl ether	6.3	40
Ethylene glycol mono-n-butyl ether	6.4	40
Ethylene glycol monomethyl ether	8.0	39
Diethylene glycol mono-s-butyl ether	10.5	40
Diethylene glycol mono-n-butyl ether	11	40
Ethylene glycol monophenyl ether	11	39
Ethylene glycol monoethyl ether	12	39
Trimethylene glycol	12	39
Tetramethylene glycol	13	39
Propylene glycol	26	39
Ethylene glycol	43	39
Ethylene glycol	44	40
Diethylene glycol	66	40

[a] $K_e = [A^-]/[HA][i\text{-}PrO^-]$, where HA is the compound studied.
[b] This value is fixed by the definition of K_e.
[c] These acids were too weak to measure; their K_e values are probably <0.2.

NaOH. This is consistent with the observation that catalyst activity increases with the size of the metallic cation (i.e, K > Na >> Li). The catalytic activity of Li+ is quite low and is very seldom used as an oxyethylation catalyst. Both the hydroxides and alkoxides of the group I metals have been shown to be useful as catalysts.

Several studies have also shown that the initial reaction rate of polyoxyethylation of an alcohol decreases as the carbon number of the alcohol increases and that primary alcohols react more readily than secondary alcohols, which in turn are more reactive than tertiary alcohols [42,43]. In fact, successful polyoxyethylation of secondary and tertiary alcohols can only be achieved using either acidic catalysts or a two-step procedure initially developed by Carter [43] using first an acidic catalyst, followed by continued polyoxyethylation using basic catalysts. These observations may be explained by a combination of a

TABLE 2 Rate Constants for Addition of Ethylene Oxide to Various Alcohols at 30°C[a] [41]

Alcohol	10[NaOR] (M)	$10^5 k_1$[b] (s^{-1})	$10^5 k_2$[b] (Liter mol^{-1} s^{-1})
CH_3OH	2.63	5.63	21.4
C_2H_5OH	0.68	3.74	55.0
n-C_3H_7OH	1.25	7.30	58.5
n-C_4H_9OH	0.97	4.57	47.1
$HO(CH_2)_2OH$	1.76	3.71	21.1
$HO(CH_2)_3OH$	0.62	3.94	63.6
$CH_3OC_2H_4OH$	3.54	6.62	18.7
$C_2H_5OC_2H_4OH$	1.17	1.77	15.1
$CH_3(OC_2H_4)_2OH$	2.67	3.15	11.8
$C_2H_5(OC_2H_4)_2OH$	2.72	4.51	16.6
$C_2H_5(OC_2H_4)_3OH$	2.09	2.36	11.3

[a] The initial concentrations of ethylene oxide were 1–2 M. At least a fivefold excess of alcohol, which also served as the solvent, was present.
[b] $k_1 = -d \ln [C_2H_4O]/dt$; $k_2 = k_1/[NaOR]$.

decrease in the dielectric constant of the media, a decrease in the acidity of and an increase in steric bulk of secondary and tertiary alcohols compared to primary alcohols.

 (iii) *Acidity of RXH >> RXCH$_2$CH$_2$OH: conditions for amines and activated hydrocarbons.* When the acidity of RXH is much less than that of RXCH$_2$CH$_2$OH, as for the case of amines or some activated hydrocarbons, conventional basic catalysts such as KOH will be ineffective for deprotonation of RXH [44]. Consequently, there will be little or no reaction of the initiator, RXH,

RXH + KOH $\rightleftarrows$ RX$^-$K$^+$ + H$_2$O

with ethylene oxide. If this reaction is attempted using KOH as catalyst, only the catalyst will react with ethylene oxide, leading to the production of polyethylene glycol. If a very strong base is used in catalytic concentrations to produce RX$^-$M$^+$, this species will react stoichiometrically with ethylene oxide with the rest of RXH acting only as a solvent. This condition can be exploited synthetically to produce hydroxyethyl derivatives of desired compounds. Grignard or organolithium reagents can react with excess ethylene oxide to produce the monoadduct in near-quantitative yields [45].

RLi$^+$ + H$_2$C—CH$_2$ $\xrightarrow[\text{2 H}_2\text{O}]{\text{1 solvent}}$ R-CH$_2$CH$_2$OH

Amines are sufficiently nucleophilic to react with ethylene oxide without the aid of a basic catalyst [46–48]. The reaction of primary amines with ethylene oxide occurs in two step, with the first step reacting more slowly than the second. Consequently, it is not possible to isolate the monoethoxylated product in high purity using this reaction scheme. Water or acids have been shown to accelerate the uncatalyzed reaction of amines with ethylene oxide to form diethanolamines either by facilitation of proton transfer or mild activation of the ethylene oxide via an acid-catalyzed reaction mixture [47,49].

$$RNH_2 \;+\; CH_2\!\!\overset{\displaystyle O}{\overset{/\backslash}{\rule{0pt}{0pt}}}\!\!CH_2 \;\xrightarrow{\;\text{slow}\;}\; RNHCH_2CH_2OH$$

$$RNHCH_2CH_2OH \;+\; CH_2\!\!\overset{\displaystyle O}{\overset{/\backslash}{\rule{0pt}{0pt}}}\!\!CH_2 \;\xrightarrow{\;\text{fast}\;}\; RN(CH_2CH_2OH)_2$$

Polyoxyethylene amines are produced commercially in two independent steps—by the formation of diethanolamine in an uncatalyzed reaction, followed by polyoxyethylation using basic catalysts. Polyoxyethylene amine surfactants will be discussed in more detail in Chapter 6.

The previous discussion concerning the relative acidities and thus relative reactivities of various hydrogen-containing initiators is particularly useful for predicting the results of polyoxyethylation of initiators containing more than one type of functional group. If more than one type of active hydrogen-containing functional group is present in an initiator, the relative acidities and thus competitive reaction rates will determine which group will react with the ethylene oxide. Often, both hydroxyl and acid groups or hydroxyl and amine groups are present in an initiator. In these cases, the first mole of ethylene oxide might react exclusively with one group, leaving the less reactive group underivatized. Such is the case for the reaction of hydroxy acids, such as 12-hydroxystearic acid, with 1 mole of ethylene oxide per mole of initiator. For 12-hydroxystearic acid, the ethylene oxide reacts exclusively with the acid moiety when only 1 mole of ethylene oxide is added [50,51]. Reaction of 2-hydroxyethylamine with 2 moles of ethylene oxide results in the production of tris(2-hydroxyethyl) amine. In this case, the primary amine reacts with the ethylene oxide without any significant reaction of the original alcohol with ethylene oxide. Therefore, differences in functional group reactivities can be exploited to prepare more complicated molecules.

Chain propagation. The propagation of the polyoxyethylene chain under basic conditions proceeds by kinetically nearly identical steps, since the structure of the propagating polyoxyethylene anions are essentially the same. The acidities of the polyoxyethylene alcohols, $RX(CH_2CH_2O)_xH$, are nearly identical when $x > 3$ [36,40,52] and therefore independent of chain length. Consequently, proton exchange between species should not only be fast but should also occur statistically with no species being favored over another. At low degrees of polymerization there are small differences between the nucleophilicity and acidity of the adducts that lead to small differences in the reaction kinetics. As the degree of polymerization increases the effect of the initiator on the polyoxyethylene chain length rapidly diminishes, with the chain length distribution controlled by a statistical addition of ethylene oxide to the chain. Consequently, the chain length distribution will follow the Poisson distribution equation [53,54] at high degrees of polymerization. The Poison distribution equation will be discussed in the section on chain distribution equations.

$$RXCH_2CH_2OH \; + \; H_2C \overset{O}{\underset{}{\triangle}} CH_2 \;\; \xrightarrow{k_1} \;\; RX(CH_2CH_2O)_2H$$

$$RX(CH_2CH_2O)_2H \; + \; H_2C \overset{O}{\underset{}{\triangle}} CH_2 \;\; \xrightarrow{k_2} \;\; RX(CH_2CH_2O)_3H$$

$$RX(CH_2CH_2O)_nH \; + \; H_2C \overset{O}{\underset{}{\triangle}} CH_2 \;\; \xrightarrow{k_n} \;\; RX(CH_2CH_2O)_{n+1}H$$

where $k_1 \sim k_2 = k_3 = k_4 \; .... = k_n$

During chain length propagation the dielectric constant of the medium increases, leading to an increase in the absolute rate of ethylene oxide consumption. The use of solvents for polyoxyethylation has been investigated and found to influence the absolute reaction rate, i.e., the overall rate of ethylene oxide consumption, presumably by affecting the dissociation of the metal alkoxide propagating species. In nonpolar solvent media the metal alkoxide will be less reactive because it exists as a tight ion pair. In polar solvents the metallic cation is substantially dissociated producing a free anion pair. Other methods of producing free anion pairs or solvated anions, such as the use of crown ethers or polyamines, have been shown to substantially increase the overall reaction rate (ethylene oxide consumption), but not to significantly affect the chain length distributions [55,56]. A study has shown that reaction parameters such as temperature, ethylene oxide pressure, and catalyst concentration (at relatively low catalyst concentrations) do not affect the polyoxyethylene chain length distribution over a 15-fold change in absolute polyoxyethylation rates [42]. However, as will be discussed later in the section on distributions obtained using basic catalyst, use of extremely large catalyst concentrations (resulting in high concentrations of RX^-K^+ relative to RXH) dramatically affects product distributions.

Chain termination. Chain growth normally continues until all of the ethylene oxide has been consumed and the basic chains are neutralized with an acid. Therefore, polyoxyethylation is an example of a living polymerization. There have been reports that chain termination occurs at very high degrees of polymerization, which explains the difficulty in obtaining polyethylene glycols with molecular weights greater than 1 million [57] using basic catalysts. It is proposed that the growing chains can undergo chain cleavage either via S_N2 reactions [see Eq. 5] or hydrogen abstraction reactions. Special catalyst systems have been developed that do not suffer from chain termination via chain transfer in order to obtain extremely high molecular weight polyethylene glycols (58).

(b) Distributions Obtained Using Group I Basic Catalysts. Various distribution equations for the base-catalyzed polymerization of an initiator have been derived from an understanding of the reaction mechanism. In general, these equations have been based on various assumptions of the relative rates of the stepwise addition of ethylene oxide to the deprotonated species (i.e., RX^- or $RX(CH_2CH_2O)_x^-$). Inherent in these assumptions is the understanding that the concentration of the various species will be dependent on the relative acidities of the hydrogen-containing species. Therefore, the equilibrium equations [see Eq. (3) above] must be considered in any distribution equation, especially at low orders of polymerization and when using relatively high catalyst concentrations.

These general polymer distribution equations initially developed by Flory [53,54], Gold [59], Natta [60], Maget [61], and Weilbull and Nycander [62] vary in complexity from simple to very complicated. Some of these early distribution equations have been modified to improve the agreement between predicted and experimentally obtained distributions, e.g., the Weibull–Nycander–Gold [63] and Natta-Mantica [64] distribution equations. As expected, the simplest model leads to the simplest equations. The simplest model is one based on the assumption that all reaction steps are kinetically identical, which can be represented by the familiar Poisson distribution as given by Flory [53,54]:

$$[RX(EO)_nH] = [RXH]_0\, e^{-u}(u^n/n!)$$

In this equation, n is an integer representing each polyoxyethylene adduct, u is the average molar ratio of ethylene oxide to initiator used, and o refers to the initial concentration. The product distribution is completely defined by the term u.

The Poisson equation adequately describes product mixtures obtained using base under two types of conditions:

1. Reactions where the acidity of RXH is the same as or greater than $RXCH_2CH_2OH$ (i.e., for phenols, mecaptans, and carboxylic acids)

2. For reactions of ethylene oxide with alcohol where base is used in stoichiometric concentrations (i.e., molar ratio of KOH/ROH = 1).

Under both of these circumstances, the equilibria described in Eq. (3), do not influence the course of reaction, especially for condition 2 where there is no free RXH present. Therefore, the relative reaction rates of each stepwise addition of ethylene oxide to anionic species is virtually equivalent (i.e., $k_1 \approx k_2 \approx k_3 \ldots k_n$). The second condition is a special case for the reaction of ethylene oxide with alcohols, with the product mixture being narrow range [65] as opposed to the normally broad-range distributions obtained when base is used in catalytic concentrations. For the base-catalyzed reaction of ethylene oxide with alcohols, equilibrium considerations [Eq. (3)] have considerable influence, the differences in nucleophilicity between RO^- and $RO(CH_2CH_2O)_n^-$ is significant and thus the rate of reaction for the first few steps could not be identical (i.e., $k_1 \neq k_2 \neq k_3$). Therefore, for the polyoxyethylation of alcohols using low concentrations of basic catalyst, other more complicated distribution equations must be used.

The most complex equation was derived by Natta [58] who made the assumption that all steps were kinetically different. This leads to the following equation, the derivation of which is described in detail elsewhere [23]:

$$[RX(EO)_nH] = [RXH]_0 \, (-1)^n \prod_{j=1}^{n-1} r_j \sum_{j=0}^{n} \left[\prod_{\substack{i=0 \\ i \neq j}}^{n} (r_j - r_i) \right]^{-1} \left[\frac{[RXH]}{[RXH]_0} \right]^{r_j}$$

where   RXH is the initiator

          EO is ethylene oxide

          [ ] indicates instantaneous concentration of species enclosed; subscript zero is for initial concentrations

          k_i is the second-order rate constant of the initial reaction

          k_p is the second-order rate constant for the further propagation steps

          n is the number of oxyethylene units in a polymer molecule

$r = k_p/k_i$ is the ratio of the propagation rate constant to the initiation rate constant

          i,j are integral numbers

A less complicated model developed by Weibull and Nycander [62] is based on a simplification of the Natta equation which assumes that the chain propagation steps are kinetically identical and differ only from the chain initiation step (i.e., $k_1 \neq k_2 = k_3 = k_4 \ldots$ etc). This leads to the Weibull–Nycander equation, which has been shown to fit reasonably well with the observed distributions of the base-catalyzed polyoxyethylation of most initiators.

$$[RX(EO)_nH] = -[RXH]^r \left\{ \frac{r^{n-1}}{S[RXH]_0^{\ s}} \sum_{i=0}^{n-1} \frac{1}{s^{n-1-i}} \frac{z^i}{i!} \right\} + \left\{ \frac{r^{n-1}}{s^n} \right\} [RXH]$$

where $\quad s = r - 1$ for $r \neq 1$

$\qquad z = \ln[RXH]_0 / [RXH]$

Early work comparing experimentally obtained distributions with theoretical distributions were complicated due to the difficulty of accurately separating the oxyethylene adducts. However, advances made in the analysis of the oxyethylene adducts of most initiators [42] have led to a high degree of confidence in comparing experimental data with theoretical curves. One can compare the product mixture obtained from the reaction of various alcohols with ethylene oxide using sodium methoxide as catalyst and ethylene oxide/alcohol molar ratios of 6.0 with the Poisson and Weibull–Nycander equations [66]. As seen in Fig. 4, the Poisson equation does not adequately describe the product distribution, confirming that the relative reaction rates are indeed not equivalent. However, the Weibull–Nycander equation, which assumes that $k_1 < k_2 = k_3 = k_4$, etc., fits the product distributions quite well. When the Weibull–Nycander or Weibull–Nycander–Gold distribution equations do not adequately predict the product distributions, the Natta–Mantica equations [64], which assume that each addition of ethylene oxide to an activated species is kinetically different, can be used [67].

A key parameter often used in the Weibull–Nycander type of equations is the term r where $r = k_p/k_i$, the ratio of the rate constants of the propagating species to that of the initiator species. The r value can be taken as a measure of how

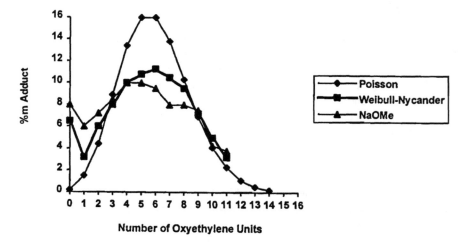

FIG. 4 A comparison of the product distribution of an NaOMe-catalyzed reaction of dodecanol with 6.0 moles of ethylene oxide with a Poisson and Weibull–Nycander distribution.

narrow or broad the polyoxyethylene adduct distribution will be [65]. When $r = 1$, then each addition step is kinetically identical, and the Poisson equation will predict the experimental results. However, the larger the value of r, the broader will be the adduct distribution. Typical values for primary alcohols have been reported to be 3, whereas secondary and tertiary alcohols (which as initiators react much more slowly with ethylene oxide compared with their polyoxyethylene adducts) have values of ~8–22 and 33–64, respectively [68]. In these cases, the product mixtures contain significant amounts of unreacted secondary or tertiary alcohols even when moderately high ethylene oxide/alcohol molar ratios are used. Typical values derived for r for various alcohols are shown in Table 3.

Work continues on the refinement of methods to determine the best r values or distribution coefficients to use with the Weibull–Nycander–Gold and Natta–Mantica distribution equations. In many studies using homogeneous group I basic catalysts, the product distribution coefficients, or oligomer reactivity coefficients (ORCs) as defined by Geissler and Johnson [42], are observed to change with a change in the degree of polyoxyethylation. This phenomenon is known as the Weibull–Tornquist effect [69] and results in an increase in the reactivity of oligomers with ethylene oxide up to a maximum, which occurs around the sixth polyoxyethylene adduct $[RO(EO)_6^-]$. This increased reactivity of the polyoxyethylene adducts from c_1 to c_6 leads to a broadening of the product distribution. In one study using sodium hydroxide as catalyst the distribution coefficients (c_i) were determined using the Natta–Mantica relationship and found to increase to a maximum at c_6 and then begin to decrease again [67].

The Weibull–Tornquist effect is explained by the intramolecular complexation of the cation (e.g., K^+) by the polyoxyethylene chain, which leads to an

TABLE 3 Various r Values for the Base-Catalyzed Addition of Ethylene Oxide to Alcohols [68]

Compound	Ethylene oxide/ROH[a] molar ratio	r[b]	Range[c]
Methyl alcohol	1.3	0.7	0.6–0.8
Ethyl alcohol	1.3	2.5	2.2–2.6
n-Propyl alcohol	1.3	2.4	2.2–2.5
i-Propyl alcohol	1.3	8.8	7.9–9.7
n-Butyl alcohol	1.3	3.3	2.9–3.5
i-Butyl alcohol	1.3	4.7	4.5–5.1
s-Butyl alcohol	1.3	22	19–23
t-Butyl alcohol	1	33	31–36
t-Butyl alcohol	3	64	64–65

[a] Reactions run at 100°, 120°C using 1 wt% sodium hydroxide catalyst.
[b] r value from the Weibull–Nycander equation.
[c] Range of r over four experiments.

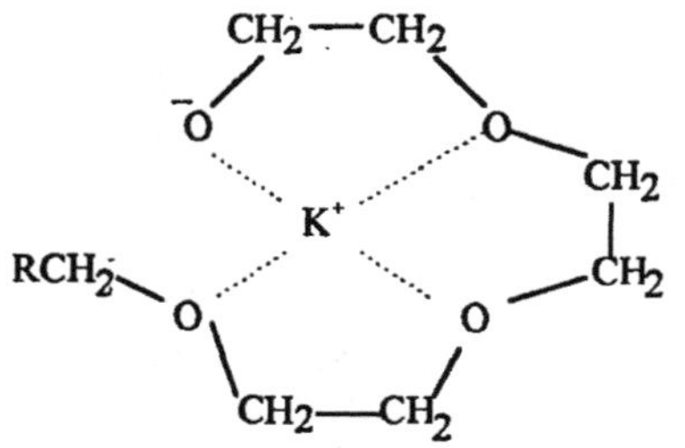

FIG. 5 Intramolecular complexation of the metal cation in base-catalyzed poly-oxyethylation.

increase in oligomer reactivity due to the activation of the metal alkoxide ion pair; see Fig. 5. Sallay et al. [70] suggested that the sixth polyoxyethylene adduct represents the highest stability of the series of oligomers, $RO(EO)_x^-$, with stability increasing for $x = 1$–6 and decreasing with $x > 6$. This effect has been seen in studies involving the use of soluble group I metal hydroxides [67,42,27,70,71] as well as aliphatic tertiary amines [27,72].

Weibull [71] recently investigated the reaction of ethylene oxide with 3-phenyl-1-propanol using lithium, sodium, and potassium as the counterion for the alkoxide. Using modified Natta equations, Weibull reported in this comprehensive study that the parameters used in the formulas, often referred to as distribution constants, are not true constants because they depend on the molar ratio of ethylene oxide to alcohol used (i.e., the degree of polyoxyethylation). This dependence was most pronounced for sodium and potassium but was not observed when using lithium. Weibull suggests that this phenomenon is caused by the complexation of the cation by the higher molecular weight polyoxy-ethylene adducts. His results were confirmed by experiments using 18-crown-6 as an effective complexing agent for the metal cations for polyoxyethylation reactions.

It should be expected that if this complexation between the polyoxyethylene adduct and the metal cation cannot take place, then the Weibull–Tornquist effect would not be observed. Such seems to be the case as reported in experiments using bridgehead nitrogen-containing compounds [72] or group II metal hydroxides (alkoxides) [67] as catalysts for polyoxyethylation reactions. In each case it has been proposed that the metal cation is effectively shielded and cannot participate in intramolecular complexation.

Although r values and oligomer distribution coefficients are often employed to compare various product distributions obtained using basic catalysts, the effect of the equilibrium reaction [Eq. (3)] on the resulting product distribution should be emphasized [36]. In fact, for the case of polyoxyethylation of secondary alcohols, the equilibrium expression significantly impacts the final

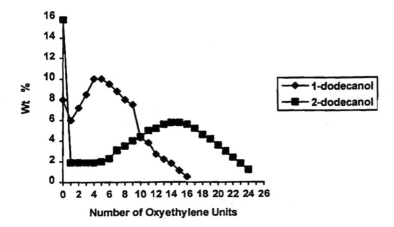

FIG. 6 Products of the reaction of 1-dodecanol and 2-dodecanol with 6 moles of ethylene oxide using KOH catalyst.

product distribution. As mentioned previously, the acidity of a secondary alcohol is much less than that of the corresponding primary alcohol or monoethyleneoxy adduct. This means that the equilibrium concentration for the secondary alkoxide is much lower than for primary alkoxides. Therefore, the relative concentration of $RO^-/RO(CH_2CH_2O^-)$ will be much lower for secondary alcohols than for primary alcohols. Consequently, the product mixture obtained from using secondary alcohols has considerably more unreacted alcohol than when using primary alcohols for a given order of polyoxyethylation. This can be seen clearly in a comparison of product distributions obtained from the oxyethylation of 1-dodecanol compared with 2-dodecanol using basic catalysts (Fig. 6). Application of the Weibull–Nycander equation to these systems results in predicted distributions in good agreement with experimental results.

Commercially, oxyethylation of secondary alcohols must be conducted using a two-step method to overcome this basic mechanistic problem. This will be discussed further in the section on acid-catalyzed oxyethylation.

(c) Distributions Obtained Using Group II Bases. Product distributions for polyoxyethylene alcohols obtained using soluble group II metal bases have been shown to be narrower than those obtained using group I bases such as KOH. The first report that divalent metal cations were capable of producing different (and narrower) product distributions was by Yang et al. [73] using barium oxide as catalyst. Within a few years there were additional reports based on barium [74–78] and new ones based on the other group II bases, strontium [77,79], calcium [77,78,80,81], and magnesium [78,81–83]. These additional basic divalent metal catalysts were also shown to be capable of producing

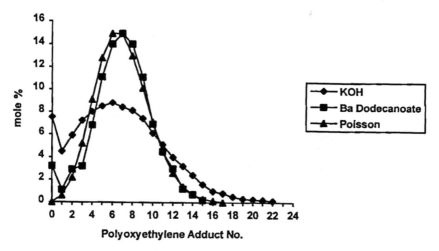

FIG. 7 Product distributions obtained from the reaction of dodecanol with 7 moles of ethylene oxide using KOH and barium dodecanoate compared with a Poisson distribution.

narrower product distributions compared with the group I metal bases. Because the solubility of the group II metal hydroxides is extremely low in most detergent range alcohols, solubilization techniques, such as the addition of phenols, ethylene glycol, or low molecular weight polyether alcohols, had to be used in order to produce active catalysts. Although the activity of these soluble group II bases varies (Ba ~ Sr > Ca >> Mg), the product distributions obtained using these catalyst systems at equivalent ethylene oxide/alcohol molar ratios are essentially identical. This suggests that the same mechanism is operating for each system. A comparison of oligomer distributions obtained using KOH and a barium dodecanoate catalyst system with a typical Poisson distribution for the same degree of oligomerization is shown in Fig. 7 [84].

Although less mechanistic and kinetic studies have been conducted using the group II basic catalysts, two controlling factors have been proposed in order to explain the narrower distributions obtained using soluble group II metal bases. One explanation emphasizes the effect of a change in the relative equilibria associated with the proton exchange reaction [Eq. (3), discussed previously], whereas the other addresses factors causing a decrease in the relative reaction rates of the growing polymer chain. Either of these factors would, a priori, result in a narrower product distribution compared with the factors already controlling the relatively broad distributions obtained using homogeneous group I basic catalysts.

One study suggests that the narrow distribution is not due to a kinetic origin but is instead related to the different stabilities of the polyoxyethylene anion ion pair compared with the alcoholate ion pair [84]. In this approach different values of the proton exchange equilibrium [Eq. (3) above] for the first several adducts have been assumed. For KOH, the same equilibrium constant is assumed for all polyoxyethylene adducts. For divalent metal catalysts, the equilibrium constant for the first four adducts is assumed to decrease from 5.0 to 2.05. Justification for this explanation is given by assuming a tight ion pair for the monovalent ion pairs. In contrast, it is suggested that divalent cations can interact with more ether oxygens in growing chains, thus significantly affecting the stability and the equilibrium concentration of the growing chains.

The other study has shown in kinetic experiments that only monovalent catalyst metal ions are subject to the Weibull–Tornquist complexation effect described previously, whereas divalent metal ions are not subject to the effect [67]. Because the divalent cations are not subjected to the broadening effect of the Weibull–Tornquist sequestering complexation, they are capable of producing narrower product distributions. Distribution coefficients derived from products obtained using KOH and a calcium-based catalyst system using a Natta–Mantica relationship clearly show a decreasing reactivity of the growing polyoxyethylene anions for the calcium system compared with the KOH system. It is suggested that the two polyoxyethylene chains associated with the calcium cation prevent either chain from entering into an intramolecular coordination, which is necessary for a Weibull–Tornquist broadening effect. The result is a narrower product distribution.

Recognition that the group II metal alkoxide catalysts were capable of producing narrower product distributions when using alcohol as an initiator compared with group I metal catalysts led to the development of additional heterogeneous narrow-range catalyst systems. Before the advent of this development in catalyst research, the only method of producing product distributions narrower than those obtained from KOH was to use acidic catalyst systems, which are normally undesirable due to the significant amount of side products that accompany their use. The narrow-range catalyst systems will be discussed in more detail in a later section.

In general, the Weibull–Nycander and modifications of this distribution equation have predicted the product mixtures obtained using basic catalysts quite well. Deviations from these predictions can be explained in most cases by steric hindrance associated with the initiator. For example, slightly different distributions are obtained during the ethoxylation of nonylphenol vs. dinonylphenol. Because the latter substrate has more steric constraint in the vicinity of the phenolic anion, its conversion to the monoadduct is slower than that of nonylphenol [85]. These differences are also known for the ethoxylation of primary alcohols wherein there exists branching at the alfa carbon [42]. In such cases, the product mixtures of branched polyoxyethylene alcohols will always contained more unreacted alcohol than polyoxyethylene alcohols derived from straight chain or normal alcohols at equivalent ethylene oxide/alcohol molar ratios.

2. Acid-Catalyzed Ring Opening of Ethylene Oxide

(a) Mechanism of Reaction. As stated previously, polyoxyethylene surfactants are not produced commercially using acidic catalysts due to the formation of significant concentrations of undesirable side products. In essentially all cases reported, polyoxyethylene glycol, dioxane, and 2-methyl-1,3-dioxolane are produced during polyoxyethylation using acidic catalysts [86–88]. In many cases the reaction products are either colored or lead to colored products in subsequent conversions, such as during sulfation. This is due to the high concentrations of carbonyls and acetals formed during reaction.

The mechanism of ethylene oxide ring opening under acidic conditions differs significantly from the base-catalyzed mechanism. The main difference is that the catalyst species is primarily associated with ethylene oxide using acidic catalysts. As discussed previously for base-catalyzed reactions, the basic catalyst species is primarily associated with the initiator. This difference has profound effects on the product composition, notably the almost identical product distributions obtained in the polyoxyethylation of secondary alcohols compared with primary alcohols. As noted earlier, using basic catalysts secondary alcohols are more slow to react with ethylene oxide compared with primary alcohols, and the product distributions of secondary alcohol ethoxylates are much broader (with more residual alcohol) than primary alcohol ethoxylates at equivalent orders of polyoxyethylation.

The mechanism of reaction using acidic catalysts is dependent on the strength of the acid catalyst used. The fastest reaction step using any acidic catalyst is the reaction of the ethylene oxide with the acid. This is true when using either a Bronsted or a Lewis acid:

$$\underset{H_2C - CH_2}{\overset{\displaystyle O}{\diagup\diagdown}} + Y^{-x} \longrightarrow \underset{H_2C - CH_2}{\overset{\displaystyle O}{\overset{\displaystyle Y^{+x}}{\diagup\diagdown}}}$$

where $Y = H$ or metal and $x \geq 1$

However, at this point the mechanism differs depending on the strength of the acid. For strong acids (e.g., CF_3SO_3H) the rate-determining step is the ring opening of the conjugate acid of ethylene oxide, followed by reaction with the nucleophile:

$$\underset{\substack{H_2C \overline{} CH_2}}{\overset{\overset{\displaystyle H}{\overset{|}{\underset{\diagup\diagdown}{O+}}}}{}} \quad \xrightarrow{\text{slow}} \quad HOCH_2CH_2^+$$

$$RXH \; + \; HOCH_2CH_2^+ \; \xrightarrow{\text{fast}} \; RXCH_2CH_2OH \; + \; H^+$$

With this mechanism the reaction rate is independent of the structure or concentration of the initiator. Such reaction mechanisms are known as S_N1 reactions [89], wherein the bond breaking (or ring opening) of ethylene oxide is essentially complete and in the transition state the bond formation with the nucleophile is negligible. Sterically hindered initiators, e.g., secondary alcohols, will react as rapidly as less sterically encumbered ones, e.g., primary alcohols, and the product distribution will be essentially the same.

For weak acids the ring opening of ethylene oxide is either partial or nonexistent. In this case the involvement of the initiator in the transition state is considerable, leading to a bimolecular (S_N2) reaction mechanism [30]:

$$\underset{\substack{H_2C \overline{} CH_2}}{\overset{\overset{\displaystyle H}{\overset{|}{\underset{\diagup\diagdown}{O+}}}}{}} + RXH \longrightarrow \left[\begin{array}{c} \overset{\displaystyle H\,\delta+}{\underset{\diagup}{O}} \\ H_2C \overline{} CH_2\,\delta+ \\ | \\ XH\,\delta+ \\ | \\ R \end{array} \right] \rightarrow HOCH_2CH_2XR + H^+$$

In this case there is discrimination between initiators based on their relative steric hindrance. For example, secondary alcohols will react less rapidly than primary alcohols and the product distributions will be somewhat, but not substantially, different.

Except for the strongest and weakest acids, most catalyst systems operate somewhere between a pure S_N1 and S_N2 mechanism. The factors that influence the type of ring opening and the mechanistic implications have been reviewed [90]. These include temperature, solvents, and the steric constraints and electronic features of the initiator.

There are several plausible mechanisms for the formation of the major side products that accompany acid-catalyzed ethylene oxide ring opening. The polyoxyethylene chain is highly subjective to acid-catalyzed cleavage, potentially leading to the production of dioxane, 2-methyl-1,3-dioxolane, or polyoxyethylene glycols with broad molecular weight ranges. However, these side products need not result from chain degradation only. They could also be formed from secondary reactions of ethylene oxide. Hydration of ethylene oxide to form ethylene glycol and isomerization of ethylene oxide to acetaldehyde could lead to the formation of polyoxyethylene glycol, dioxane, and 2-methyl-1,3-diox-olane as shown below.

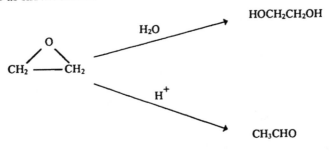

(b) Distribution Equations for Acid-Catalyzed Polyoxyethylation. The mechanism of reaction discussed above for the acid-catalyzed reaction of ethylene oxide with alcohols (or other initiators) suggests that ethylene oxide should add to all activated hydrogen-containing species (RXH or $RX(CH_2CH_2O)_xH$) at essentially the same rate. Therefore, the ratio of chain initiation to chain propagation should be unity ($k_p/k_i = 1$). In this special case the Poisson distribution discussed previously should fit well with the experimental data. This is the simplest distribution equation to use and requires only one parameter for the calculation, i.e., the mole ratio of ethylene oxide to active hydrogen-containing compound.

$$[RX(EO)_nH] = [RXH]_0\, e^{-u}(u^n/n!)$$

In most cases the Poisson distribution equation fits the experimental results well, especially for the case of primary alcohols. This is illustrated in Fig. 8 [105]. However, the fact that the Poisson equation fits most acid-catalyzed reaction products well, product distributions obtained using secondary and tertiary alcohols are slightly less peaked than the Poisson equation would predict. This suggests that k_p/k_i is slightly greater than 1. In these cases modification of the Weibull–Nycander equation, while more complicated, can be made to fite the data much more precisely. In these cases the parameter r is used, which is the ratio of the rate of propagation to the rate of initiation.

$$r = k_p/k_i$$

In practice r is an adjustable parameter that is varied to fit the experimental data. However, the value of r that is determined for a reaction gives some mechanistic insight into that particular reaction. For a Poisson distribution $r = 1$. As previously stated, this indicates that the rate of chain initiation and propagation are equivalent. However, r values slightly greater than 1 have been determined for the acid-catalyzed reaction of secondary and tertiary alcohols, thus indicating that these alcohols react with ethylene oxide slightly more slowly than their adducts with ethylene oxide. This can be explained by the increased steric hindrance of these alcohols and the mechanism being somewhere between a S_N1 and S_N2 reaction. A number of r values have been determined from a study of acid-catalyzed reaction of ethylene oxide with alcohols [40] with r values ranging from 1.0 for CH_3OH to 2.6 for t-butanol. It should be stated that while

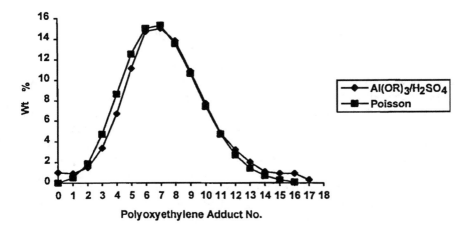

FIG. 8 A comparison of a product distribution obtained from the reaction of dodecanol with 7 moles of ethylene oxide using an acidic catalyst with a Poisson distribution [105].

some of these r values are higher than 1, each product mixture is still relatively narrow range in its oligomeric distribution. Using basic catalysts, the r value of t-butanol is 64–65, indicating a very broad range of polyoxyethylene adducts and considerable unreacted alcohol in the product mixture at low ethylene oxide/alcohol molar ratios used [68].

Since polyoxyethylation of secondary and tertiary alcohols using acidic catalysts tends to produce a narrower range of oligomeric adducts, these catalysts have been applied to the commercial preparation of secondary alcohol ethoxylates. However, the side reactions that accompany the use of acid catalysts to date prevent these materials from being produced commercially in a single step using acid catalysts. However, secondary alcohol ethoxylates are produced commercially by a two-step process initially using an acid catalyst followed by a basic catalyst. In the first step ~2 moles of ethylene oxide is added to a secondary alcohol using an acidic catalyst (usually $SnCl_4$) to produce a low molecular weight "seed ethoxylate." This intermediate is washed with water to remove side products and catalyst residues, residual secondary alcohol is removed via distillation, and the dried product is further reacted with ethylene oxide using a basic (KOH) catalyst. The product obtained has a narrow oligomer distribution, only slightly broader than a Poisson distribution (e.g., $r = 1.2$), and is used in the industrial surfactants market as the Tergitol series produced by Union Carbide. At this time there have been no reports of a catalyst system capable of selectively producing a narrow oligomeric distribution of secondary or tertiary alcohols in one step. This includes the new narrow-range catalyst systems.

B. Distributions Obtained Using Narrow-Range Catalysts

As mentioned previously, the utility of the polyoxyethylene alcohols in commercial applications depends not only on the R, X, and x value of the final product [$RX(CH_2CH_2O)_xH$], but also on the chain length distribution. Interest has been shown in obtaining products with a particularly narrow range of the polyoxyethylene adducts [1], with low concentrations of free alcohol and higher molecular weight adducts. For example, some commercial applications require very low levels of free alcohol in the product to prevent problems such as volatile emissions (pluming) in spray tower applications when preparing powdered laundry product. This need has been filled by using "topped" or "stripped" commercial products that have been obtained by removal of the light ends via distillative procedures [91]. In recent years the capability to produce polyoxyethylene alcohols with extremely low levels of free alcohol and lower molecular weight adducts has also been achieved catalytically.

This section will address progress that has been made over the last couple of decades in the field of new narrow-range catalyst development. Although numerous patents have issued describing the preparation and use of these catalysts [93–112], relatively few kinetic studies have been published discussing the narrow-range catalyst systems [67,84,92]. For convenience, the narrow-range catalysts will be divided into sections according to structural similarities, e.g., acid-activated metal alkoxides, metal phosphates, and activated metal oxides. From experimentally obtained product distributions, some insight can be gained concerning the reaction mechanism and thus requirements for adjustment in the Weibull–Nycander and other relationships that must exist to properly fit the data.

1. Acid Activation of Metal Alkoxide Catalyst Systems

One of the earliest reports on a nonacidic catalyst system that was capable of producing adduct distributions narrower than conventional KOH-based systems was by Yang et al. [73]. This described how barium salts, mainly the hydroxide or carboxylates, produced narrower than expected product distributions (see Fig. 7). It was later proposed [84] that group II catalysts were capable of producing narrower distributions due to the complexation of the cation by the polyether oxygens in the chain, resulting in a change in the equilibrium between deprotonated initiator and polyoxyethylene anion. This, in effect, meant that $K_{eq} > 1$, which by the previous discussion described the effect of producing a narrower distribution.

After the observation that barium was capable of producing narrower distributions, it followed rapidly that the other group II bases were found capable of doing this, albeit with different catalyst activities claimed. Systems based on strontium [77,79,93,94], calcium [77,78,80,81,94], and magnesium [78,81–83,94] were developed and reported. One interesting aspect was the observation that each of these systems produced essentially the same product distributions using a given initiator and at comparable ethylene oxide/alcohol molar ratios.

Rapidly after this development, several research groups found that the chain length distribution could be made even narrower if the group II metal alkoxide were "activated" by partially neutralizing the group II metal alkoxide [95–101]. The reaction mixture was not completely neutralized because less than stoichiometric amounts of acid (usually sulfuric or phosphoric) was used to activate the catalyst. Strong mineral acids were claimed to produce the most active catalyst systems. In some cases these catalyst systems were observed to be essentially homogeneous, as catalyst solutions in the alcohol initiator were optically clear. In many cases, however, slurries were obtained during catalyst preparation, indicating that the catalytically active species was indeed heterogeneous.

Although the structure of the catalyst system was not well characterized, some authors suggested that the catalytically active species might appear as follows:

$$Ca(OR)_2 + H_3PO_4 \longrightarrow Ca(OR)^+ H_2PO_4^- + ROH$$

Undoubtedly, the actual chemical equilibrium in solution is much more complicated than that shown above. Nevertheless, some partially neutralized species
must be the active catalyst because control experiments using inorganic calcium
sulfate or calcium phosphate were shown to be catalytically inactive [95,96].
Complete neutralization of the calcium alkoxide also led to inactive catalyst
systems.

Shortly after these developments, activation of other metal alkoxides using
organic and inorganic acids was reported. Activation of aluminum alkoxide
with both sulfuric acid [102–104] and phosphoric acid [105] was reported to
produce very narrow product distributions. These were mildly acidic catalyst
systems and produce either Poisson or narrow-than-Poisson distributions. Therefore, the aluminum-based systems could be accurately described using the
Poisson equation.

Further optimization of the acid-activated metal alkoxide systems led to the
first examples of polyoxyethylene adduct distributions that were significantly
narrower than Poisson distributions. Careful activation of calcium alkoxide–
based systems with phosphoric acid were the first examples of this type of
system [95,96]. Adduct distributions of products prepared using 7 moles of
ethylene oxide per mole of alcohol initiator were obtained that contained as
much as 25 wt% of the product containing one adduct (e.g., $RO(CH_2CH_2O)_7H$)
[106]. As shown in Fig. 9 the 7-mole adduct of a product prepared using a catalyst that has a Poisson distribution has only 16 wt% of the product.

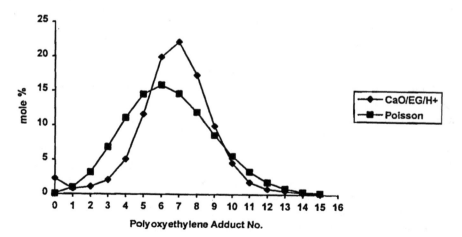

FIG. 9 Product distributions obtained from the reaction of dodecanol with 6.5 moles of
ethylene oxide using an activated CaO catalyst system compared with a Poisson
distribution.

2. Heterogeneous Metal Phosphate Catalyst Systems

Activation of the basic group II alkoxides with phosphoric acid led to heterogeneous catalyst systems that were capable of producing extremely peaked product distributions. Although the structure of these catalyst systems was not well understood, it was clear that the active catalysts were not simple group II inorganic phosphate salts. Control reactions using calcium phosphate or magnesium phosphate as catalyst showed these simple phosphate salts to be inactive. This was somewhat unfortunate because the simple salts would be much cheaper and simpler to use than the somewhat complicated recipes [95,96] required to obtain active, peaking catalysts.

However, activation of basic lanthanum or rare earth alkoxide catalyst systems [29] led to the development of the first solid, inorganic metal phosphate catalyst system that was not only active but also capable of achieving the same extremely narrow product distribution as that for the calcium-based systems [107,108]. Partial neutralization of lanthanum alkoxide with phosphoric acid led to an increase in peak maximum concentration of the main adduct from 9 wt% for the basic lanthanum alkoxide system to 16 wt% for the partially neutralized system using ethylene oxide/alcohol molar ratios of 7.0. However, preparation of inorganic lanthanum phosphate by one of a number of routes [107] led to a solid, white powder catalyst capable of achieving a peak maximum of the main adduct of ~25 wt% at ethylene oxide/alcohol molar ratios of 7.0. A comparison of product distributions using various lanthanum catalyst systems is illustrated in Fig. 10. At least for the rare earth metal phosphates, the active catalyst species is indeed the simple inorganic phosphate.

It should be pointed out that the unactivated basic lanthanum alkoxide catalyst system affords product mixtures with the same oligomer distribution as basic group I metal alkoxide catalysts [29]. This is illustrated in Fig. 11. Thus, this trivalent cation acts more like a monovalent cation than a divalent (group II) cation in polyoxyethylation reactions. This could be viewed as unexpected since divalent cations, such as barium or calcium, were shown previously to produce narrower oligomer distributions than monovalent cations. It is possible that complexion of La^{3+} by the polyoxyethylene adducts more closely resembles the behavior with K^+ than divalent cationic species, thus leading to broad oligomer product distributions. Nevertheless, activation of lanthanum alkoxide with phosphoric acid dramatically changes the catalyst species to one that produces very narrow oligomer distributions.

Application of the same techniques developed for the rare earth–based systems led to inorganic barium phosphate narrow-range catalyst systems [109,110]. While the barium-based systems produced similar distributions to the activated calcium phosphate and rare earth or lanthanum phosphate–based systems, catalyst preparation was somewhat complicated and the catalyst

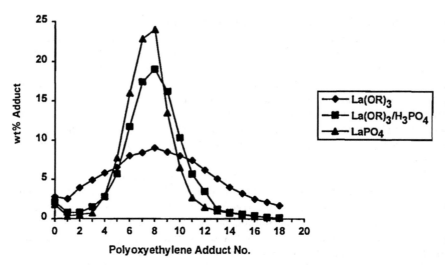

FIG. 10 A comparison of product distributions obtained by activating lanthanum compounds with phosphoric acid [29,107].

activity of the barium systems was found to be less than that of the calcium and lanthanum systems.

3. Heterogeneous Mixed Metal Oxide Catalyst Systems

Recently, heterogeneous mixed metal oxide systems [111–118] have been claimed to produce narrow product distributions. Several of these systems

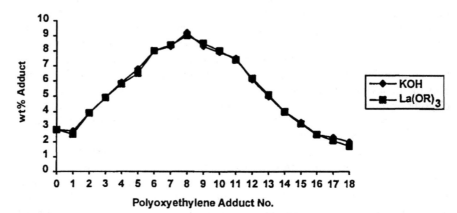

FIG. 11 A comparison of product distributions obtained using KOH and La(OR)$_3$ as catalysts for the reaction of 7 moles of ethylene oxide per mole of dodecanol.

involve the activation of normally inactive group II metal phosphates (see above) by solubilization or partial solubilization of the phosphates with carboxylic acids. These systems are relatively new and no mechanistic or kinetic work has been published.

The mechanism of reaction using the narrow-range catalyst systems described above is not well understood. This is presumably due to the fact that they are for the most part heterogeneous catalyst systems, which are normally much more difficult to understand than homogeneous reaction systems. Nevertheless, basic principles applied to bifunctional acid–base heterogeneous catalyst systems could be applied to these catalyst systems. It is possible that the mechanism involves activation of the active hydrogen containing compound by a basic site on the catalyst, whereas the ethylene oxide could be activated by a neighboring acidic site. The structure in Fig. 12 has been proposed as a possible intermediate with mixed metal oxide catalyst for the reaction of ethylene oxide with alcohols or methyl esters [119]. The overall reaction could be concerted with deprotonation of the alcohol assisted by the oxide occurring simultaneously with the partial ring cleavage activated by the metal cation. This type of mechanism is often proposed in bifunctional metal oxide catalyst systems. The same mechanism could be operating in the metal phosphate catalyst systems. As shown in Fig. 13, the lanthanum cation could activate the ethylene oxide in concert with the phosphate anion deprotonating the alcohol species.

Although the above reaction mechanism might describe adequately each actual ring opening step, the overall mechanism must account for one important fact. At low orders of polyoxyethylation (e.g., at ethylene oxide/alcohol molar ratios <2), the product distributions obtained using heterogeneous narrow-range ethoxylate catalysts are very similar to those of conventional based-catalyzed products. However, at ethylene oxide/alcohol molar ratios >2, the distributions are quite narrow. This can be seen in Fig. 14 for oligomer distributions obtained using lanthanum phosphate catalysts compared with conventional catalysts. This strongly suggests that the relative reaction rate of ethylene oxide addition to growing chains decreases with an increase in the length of the polyoxyethylene anion chain. This is in contrast to conventional homogeneous base-catalyzed reactions wherein the polyoxyethylene anion chains react at

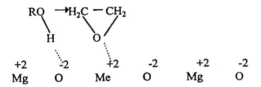

FIG. 12 Proposed bifunctional acid–base activation of reactant using mixed metal oxide catalyst; Me = metal, but not Mg.

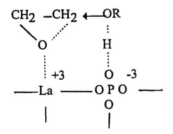

FIG. 13 Proposed activation of reactants using metal phosphate catalyst.

equivalent rates. This could be explained by differences in the rates of adsorption-desorption on the solid catalyst surface, wherein the higher molecular weight chains are less able to compete for a catalyst site and slower to desorb. Thus the overall reaction mechanism might be explained by a base-catalyzed reaction wherein adduct adsorption-desorption from the catalyst surface controls the adduct composition. This would mean that the individual reaction rates would approximate the following ordering:

$$k_1 < k_2 < k_3 = k_4 > k_5 > k_6 \text{ etc.}$$

Therefore, the individual reaction rates of the initiator and low molecular weight adducts would increase to a maximum and then decrease with increasing

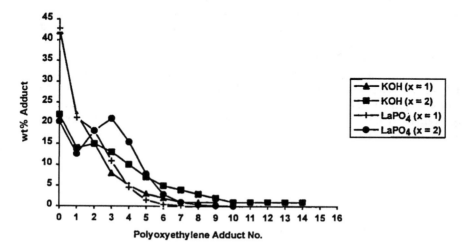

FIG. 14 A comparison of product distributions obtained using KOH and LaPO$_4$ as catalyst for the reaction of ethylene oxide with a 50:50 mixture of dodecanol and tridecanol alcohols (x = number of moles of EO/ROH) [28,107].

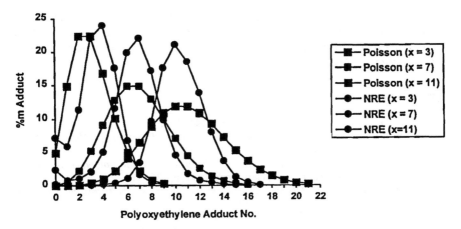

FIG. 15 A comparison of Poisson distributions (■) with NRE catalyst distributions (●) of the reaction of ethylene oxide with a 50:50 mixture of dodecanol and tetradecanol at EO/ROH molar ratios varying from 3 to 11 [1].

chain length. A steady decrease in the relative rates would be necessary to explain why peaking does not significantly decrease with an increase in the degree of polyoxyethylation. For example, the adduct peak maximum increases to ~25 wt% of the product mixture using narrow-range catalysts at ethylene oxide/alcohol molar ratios of 2–3 and is maintained at that value even as the average molecular weight of the product increases (to approx. $x = 11$). In contrast, the peak maximum of products prepared using catalysts that follow a Poisson distribution, where all reaction rates are identical, decreases steadily with increasing polyoxyethylation. This is illustrated in Fig. 15 comparing product distributions obtained using narrow-range and Poisson-type catalysts over the ethylene oxide/alcohol molar ratio range of 3–11.

There is some experimental evidence to support this possible mechanism. For each heterogeneous catalyst system described, the actual overall rate of ethylene oxide consumption decreases with reaction time. This is the opposite of the result using homogeneous catalyst systems. This is consistent with the suggestion that the higher molecular weight adducts react with ethylene oxide slower than the lower molecular weight species.

Distribution equations describing the product mixtures obtained using these heterogeneous catalyst systems must therefore contain parameters that account for different (and decreasing) individual reaction rates for the individual steps. Therefore, a simple value for r used in the Weibull–Nycander equation will not accurately describe these systems. More complicated equations such as Natta–Mantica must be used. At this time there are no definitive reports that describe these heterogeneous narrow-range catalyst systems with the same degree of thoroughness as that achieved by Weibull to define the mechanism of reaction of ethylene oxide with alcohols using conventional basic group I homogeneous catalyst systems [71].

REFERENCES

1. K. W. Dillan, J. Am. Oil Chemists Soc. *62*:1144 (1985).
2. S. Winstein and R. B. Henderson, *Heterocyclic Compounds*, Vol 1 (R. C. Elderfield, ed.), Wiley, New York, 1950, p. 1.
3. F. E. Bailey, and J. V. Koleske, *Alkylene Oxides and Their Polymers*, Surfactant Science Series, Marcel Dekker, New York, 1991, pp. 39–51
4. J. March, in *Advanced Organic Chemistry: Reactions, Mechanisms and Structure*, McGraw-Hill, New York, 1968, p. 290.
5. H. Vogel, Angewandte Chemie *72*:4–25 (1960).
6. R. A. Nelson and R. S. Jessup, J. Res. Natl. Bur. Std. *48*:206 (1952).
7. L. E. St. Pierre, in *Polyethers, Part 1, Polyalkylene Oxides* (N. G. Gaylord, ed.), Wiley-Interscience, New York, 1963, Chap. 3.
8. *Ethylene Oxide User's Guide*, Shell Chemical Company, September 1995.
9. G. L. Cunningham, A. W. Boyd, R. J. Myers, W. D. Gwinn, and W. I. LeVan, J. Chem. Phys. *19*:676 (1951).
10. T. E. Turner and J. A. Howe, J. Chem. Phys. *24*:924 (1956).
11. C. Hirose, Bull. Chem. Soc. Jpn *47*(6):1311–1318 (1974).
12. Occupational Safety and Health Administration, 25 CFR Ch. XVII 1910.1047.
13. H. Vaino, D. McGregor, and E. Heseltine, Meeting of the IARC Working Group on Some Industrial Chemicals, Scand. J. Work Environ. Health *20*:227–229.
14. W. M. Generoso, C. T. Cain, L. A. Hughes, G. A. Sega, P. W. Braden, D. G. Gosslee, and M. D. Shelby, Ethylene oxide dose and dose-rate effects in the mice dominant lethal test, Environ. Mutagen. *8*:1–7.
15. W. M. Snellings, R. R. Maranpot, J. P. Zelenak, and C. P. Laffoon, Teratology study in Fischer 344 rats exposed to ethylene oxide by inhalation, Toxicol. Appl. Pharmacol. *64*:476–481.
16. L. Goldberg, *Hazard Assessment of Ethylene Oxide*, CRC Press, Boca Raton, FL, 1986.
17. Occupational Safety and Health Administration, 29 CFR 1910.1047.
18. D. Conrad, Bundesgesundheitsblatt *9*:139–141 (1963).
19. Y. Hashigushi, Tokyo Kogyo Shikensho, *60*(3):85–91 (1965).
20. J. Osugi, M. Okusima, and M. Hamanoue, Koatsu Gasu *8*(4):201–206 (1971).
21. L. G. Hess and V. V. Tilton, Ind. Eng. Chem. *42*(6):1251–1258 (1950).
22. B. Pesetsky, Chem. Eng. Prog. Loss. Prev. *13*:132–141 (1980).
23. N. Shachat and H. L. Greenwald, in *Nonionic Surfactants* (M. J. Schick, ed.), Marcel Dekker, New York, 1966, pp. 28–31, 37.
24. W. B. Satkowski and C. G. Hsu, Chem. Eng. *49*:1875 (1957).
25. J. Falbe, *Surfactants in Consumer Products*, Springer-Verlag, Berlin, 1987, p. 88.
26. R. O. Coclough, G. Gee, W. C. E. Higginson, J. B. Jackson, and M. Litt, J. Polym. Sci. *34*:171–179 (1959).

27. P. Sallay, L. Farkas, J. Morgos, and I. Rusznak, Period. Polytech. Chem. Eng. *32*(1–3):169–171 (1988).
28. P. Sallay et al., J. Am. Oil Chemists Soc. *62*(4):824–827 (1985).
29. C. L. Edwards, U.S. Patent 5,059,719 to Shell Oil Co. (1991).
30. J. March, in *Advanced Organic Chemistry: Reactions, Mechanisms and Structure*, McGraw-Hill, New York, 1986, p. 251.
31. Ibid. pp. 288–289.
32. Ibid. p. 281.
33. F. Patat, E. Cremer, and O. Bobleter, J. Polym. Sci. *12*:489 (1954).
34. C. L. Edwards, U.S. Patent 4,575,569 to Shell Oil Co. (1986).
35. A. N. Wrigley, F. D. Smith, and A. J. Stirton, J. Am. Oil Chemists Soc. *34*:39 (1957).
36. E. Santacesaria, M. Di Serio, R. Garaffa, and G. Addino, Ind. Eng. Chem. Res. *31*:2413–2418 (1992).
37. E. Santacesaria, M. Di Serio, L. Lisi, and D. Gelosa, Ind. Eng. Chem. Res. *29*:719 (1990).
38. N. Shachat and H. L. Greenwald, in *Nonionic Surfactants* (M. J. Schick, ed.), Marcel Dekker, New York, 1966, p. 19.
39. J. Hine and M. Hine, J. Am. Chem. Soc. *74*:5266 (1952).
40. G. J. Stockburger and J. D. Brandner, J. Am. Oil Chemists Soc. *40*:590 (1963).
41. G. Gee, W. C. E. Higginson, P. Levesley, and K. J. Taylor, J. Chem. Soc. 1338 (1959).
42. P. R. Geissler and Adrain E. Johnson, Jr., J. Am. Oil Chemists Soc. *67*:541–546 (1990).
43. C. A. Carter, U.S. Patent 2,870,220 (1959).
44. J. March, in *Advanced Organic Chemistry: Reactions, Mechanisms and Structure*, McGraw-Hill, New York, 1968, pp. 220–221.
45. R. B. Bening and C. L. Willis, U.S. Patent 5,391,663 to Shell Oil Co. (1995).
46. Y. Oshiro and S. Komari, Kogyo Kagaku Zasshi *65*:1830 (1962).
47. K. Nagase and K. Sakaguchi, Kogyo Kagaku Zasshi *64*:1031 (1961).
48. N. N. Lebedev and M. M. Smirnova, Kinetika i Kataliz *2*:519 (1961).
49. N. Bortnick, L. S. Luskin, M. D. Hurwitz, W. E. Craig, L. J. Exner, and J. Mirza, J. Am. Chem. Soc. *78*:4039 (1956).
50. A. T. Ballun, J. N. Schumacher, G. E. Kapella, and J. V. Karabinos, J. Am. Oil Chemists Soc. *31*:20 (1954).
51. A. J. O'Lenick, Jr., and J. K. Parkinson, J. Soc. Cosmet. Chem. *44*(6):319–28 (1993).
52. A. J. Lowe and B. Weibull, J. Polym. Sci. *12*:493 (1954).
53. P. J. Flory, J. Am. Chem. Soc. *62*:1561 (1940).
54. P. J. Flory, *Principles of Polymer Chemistry*, Cornell University Press, Ithaca, NY, 1953, Chap. 10.
55. B. Wesslen, E. Andersson, and K. Holmberg, J. Am. Oil Chemists Soc. *66*(8):1107–1112 (1989).
56. J. Morgos, P. Sallay, L. Farkas, and I. Rusznak, J. Am. Oil Chemists Soc. *60*(11):1905–1907 (1983).
57. G. L. Goeke and F. J. Karol, U.S. Patent 4,193,892 to Union Carbide Corporation (1980).

58. W. T. Reichle, U.S. Patent 4,667,013 to Union Carbide Corporation (1987).
59. L. Gold, J. Chem. Phys. *20*:1651 (1952).
60. G. Natta and M. Simonetta, Rend. 1st Lombardo Sci. Lettere A. *78*:307 (1945).
61. H. J. R. Maget, J. Polym. Sci. *57*:773 (1962).
62. B. Weibull and B. Nycander, Acta Chem. Scand. *8*:847 (1954).
63. L. Gold, J. Chem. Phys. *28*:91–99 (1958).
64. G. Natta and E. Mantica, J. Am. Chem. Soc. *74*:3152 (1952).
65. H. K. Krummel and R. M. Wise, U.S. Patent 4,098,818 to Procter & Gamble Co. (1978).
66. G. Tischbirek, *Proc. of 3rd Int. Congr. Surface Activity*, Cologne, Germany, I, 126 (1960).
67. H. Hreczuch, G. Bekierz, K. Poland, and J. Szymanowski, Tenside Surf. Det. *32*:55–60 (1995).
68. Y. Ishii, S. Sekiguchi, and A. Hayakawa, Kogyo Kagaku Zasshi *65*:1041 (1962).
69. B. Weibull and J. Tornquist, Chem., Phys. Chem. Anwendungstech. Grenzflaechenaktiven Stoffe, 6th Ber. Int. Kongr., 1973.
70. P. Sallay, J. Morgos, L. Farkas, I. Rusznak, G. Veress, and B. Bartha, Tenside Det. *17*:298–300 (1980).
71. B. Weibull, Acta Chem. Scand. *49*(3):207–216 (1995).
72. J. Morgos, P. Sallay, L. Farkas, Laszlo and I. Rusznak, J. Am. Oil Chemists Soc. *63*(9):1209–1210 (1986).
73. K. Yang, G. L. Nield, and P. H. Washecheck, U.S. Patent 4,210,764 to Conoco Corp. (1980).
74. K. Yang, U.S. Patent 4,239,917 to Conoco Corp. (1980).
75. J. H. McCain, European Patent 115083 to Union Carbide Corp. (1984).
76. J. H. McCain and L. F. Theilling, European Patent 26544 to Union Carbide Corp. (1981).
77. D. J. Foster and J. H. McCain, European Patent 26546 and 26547 to Union Carbide Corp. (1981).
78. A. Behler and U. Ploog, Fett Wiss. Techno. *92*(3):109–114 (1990).
79. K. Yang, G. L. Nield, and P. H. Washecheck, U.S. Patent 4,223,164 to Conoco Corp. (1980).
80. C. L. Edwards, U.S. Patent 4,396,779 to Shell Oil Co. (1983).
81. K. Yang, G. L. Nield, and P. H. Washecheck, European Patent 85167 to Conoco Corp. (1981).
82. C. L. Edwards, U.S. Patent 4,375,564 to Shell Oil Co. (1983).
83. C. L. Edwards, U.S. Patent 4,465,877 to Shell Oil Co. (1984).
84. E. Santacesaria, M. Di Serio, R. Garaffa, and G. Addino, Ind. Eng. Chem. Res. *31*:2419–2421 (1992).
85. R. C. Mansfield, J. E. Locke, and K. A. Booman, J. Am. Oil Chemists Soc. *38*:289 (1961).
86. H. F. Drew and J. R. Schaeffer, Ind. Eng. Chem. *50*:1253 (1958).
87. K. Nagase and K. Sakaguchi, Kogyo Kagaku Zasshi *64*:1199 (1961).
88. D. J. Worsfold and A. M. Eastham, J. Am. Chem. Soc. *79*:897–900 (1957).
89. J. March, in *Advanced Organic Chemistry: Reactions, Mechanisms and Structure*, McGraw-Hill, New York, 1968, pp. 256–262.
90. R. E. Parker and N. S. Issacs, Chem. Rev. *59*:737 (1959).

91. G. Smith, W. M. Sawyer, and R. C. Morris, U.S. Patent 3,682,849 to Shell Oil Co. (1972).
92. J. Zhu and Z. Jin, Riyong Huaxue Gongye 6:281–286 (1987).
93. K. Yang, G. L. Nield, and P. H. Washecheck, U.S. Patent 4,306,093 to Conoco Corp. (1981).
94. K. Yang, G. L. Nield, and P. H. Washecheck, U.S. Patent 4,302,613 to Conoco Corp. (1981).
95. J. H. McCain and L. F. Theiling, Jr., U.S. Patent 4,453,022 to Union Carbide Corp. (1984).
96. J. H. McCain and L. F. Theiling, Jr., U.S. Patent 4,453,023 to Union Carbide Corp. (1984).
97. L. Cheng, Z. Wu and X. Sun, Riyong Huaxue Gongye 4:185–191 (1991).
98. J. H. McCain, S. W. King, R. J. Knogf, C. A. Smith, and C. F. Hauser, European Patent 361619 A2 to Union Carbide Corp. (1989).
99. E. C. Y. Nieh, European Patent 347064 A1 to Texaco Development Corp. (1989).
100. A. Behler, H. Raths, and U. Ploog, German Patent DE 4,225,136 A1 to Henkel (1992).
101. E. L. Brace, M. L. Shannon, and D. L. Wharry, U.S. Patent 4,835,321 to Vista Chem. Co. (1989).
102. C. L. Edwards, U.S. Patent 4,721,816 to Shell Oil Co. (1988).
103. C. L. Edwards, U.S. Patent 4,825,009 to Shell Oil Co. (1989).
104. C. L. Edwards, U.S. Patent 4,931,205 to Shell Oil Co. (1989).
105. C. L. Edwards, U.S. Patent 4,721,817 to Shell Oil Co. (1988).
106. R. J. Knopf and L. F. Theiling, Jr., U.S. Patent 4,886,917 to Union Carbide Corp. (1989).
107. C. L. Edwards, U.S. Patent 5,057,627 to Shell Oil Co. (1991).
108. C. L. Edwards and R. A. Kemp, U.S. Patent 5,057,628 (1991).
109. R. Wijngaarden, K. Latjes and J. Van Schaik, U.S. Patent 5,175,374 to Shell Oil Co. (1992).
110. B. J. Breukel and E. Gerard, European Patent 665206 A1 to Shell International Res Mij BV (1995).
111. H. Nakaya, I. Adachi, N. Aoki, and H. Kanao, Japanese Patent 03185095 A2 to Lion Corp. (1989).
112. T. Tsukui, Y. Yanagi, and H. Takeguchi, Japanese Patent 63280036 A2 to Mitsubishi Petrochemical Co. (1987).
113. German Patent Application DE 4325237 A1 to BASF AG (1995).
114. German Patent Application DE 4010606 to Henkel (1990).
115. German Patent Application DE 4138631 to Henkel (1991).
116. T. Imanaka, J. Kono, H. Nagumo, H. Tamura, and T. Tanaka, WO9517248-A1 to Kao Corp. (1995).
117. D.H. Champion and G.P. Speranza, U.S. Patent 5,110,991 to Texaco Chemical Co. (1991).
118. T. S. Sandoval and P. A. Schwab, U.S. Patent 5,220,077 to Vista Chemical Co. (1993).
119. I. Hama, presented to Am. Oil Chemists Soc. spring meeting, Anaheim, CA (1993).

2

Polyoxyethylene Alkylphenols

ROBERT M. WEINHEIMER and PIERRE T. VARINEAU
Research and Development, Union Carbide Corporation,
South Charleston, West Virginia

I. INTRODUCTION

Polyoxyethylene alkylphenols have played a major role in many markets for over 50 years. Early documentation of the chemistry of polyoxyethylene alkylphenols appeared in 1937 [1]. Academic studies of polyoxyethylene octylphenols appeared in the 1940s [2,3]. The markets for polyoxyethylene alkylphenols have consistently grown and today represent a substantial portion of the nonionic surfactant market [4]. Nonylphenol and octylphenol are the most commonly used hydrophobes with dodecylphenol less commonly used.

Polyoxyethylation of alkylphenols allows the manufacture of a line of products with a wide range of solubility and performance characteristics. Polyoxyethylene alkylphenols, like other classes of nonionic surfactant, find use in a wide variety of cleaning products. They are also found in a number of specialty applications, such as emulsion polymer manufacture, where other nonionics are not particularly well suited.

Consistent with interest in environmental performance, a significant body of environmental performance data has been generated in recent years. Environmental monitoring studies have demonstrated that the rates of degradation of polyoxyethylene alkylphenols are sufficiently high to avoid accumulation in the environment. In addition, studies have demonstrated high removal rates of polyoxyethylene alkylphenol from wastewater.

II. ALKYLPHENOLS

Polyoxyethylene alkylphenols are made from two major raw materials, an alkylphenol and ethylene oxide. As ethylene oxide is a raw material common to many classes of nonionic surfactants, this chapter will focus on the alkylphenol raw materials. Readers are referred to an excellent review by Dever et al. for a discussion of ethylene oxide [5].

Although many alkylphenols are of commercial interest in many industries, only two are used extensively as surfactant hydrophobes. Nonylphenol, by far the largest volume surfactant-related alkylphenol, had a worldwide demand of about 200,000 tons in the early 1990s. Approximately 25,000 tons per year of octylphenol is produced. A third alkylphenol, dodecylphenol, is of comparatively lower volume; the volume is difficult to determine as much of this material is used captively [6].

A. Nonylphenol

Polyoxyethylene nonylphenols are produced from para-substituted nonylphenol. Although *ortho*-nonylphenol can be produced by proper choice of alkylation catalyst, the para isomer is preferred as it is more readily polyoxyethylated [7].

The nonyl group is a highly branched propylene trimer. Due to rearrangements arising from proton transfers, this trimer is a mixture of isomers characterized in Table 1. Branching in the alkyl chain is further enhanced by proton

TABLE 1 Olefin Types in Propylene Trimer [104]

Structure	Wt%
$RCH=CH_2$	1
$RCH=CHR$	14
$CH_2=CR_2$	8
$RCH=CR_2$	35
$R_2C=CR_2$	42

rearrangement during alkylation. Although linear nonyl groups are often mentioned in relation to these products, none are used commercially at this time [8]. However, even in reportedly linear alkylphenols, double bond shifts can result in the formation of a variety of isomers [9].

While alkylation of phenol can occur in the absence of catalyst, a judicious catalyst choice produces the desired isomer distribution and more complete olefin conversion. A review of recent literature shows a considerable level of effort devoted to catalyst systems for alkylphenol production. Boron trifluoride has been used as an effective catalyst and was the catalyst of choice in the past [10]. Acid-activated clays have also been mentioned [11]. Montmorillonite clay acidified by pretreatment with mineral acids has a high selectivity for *p*-alkylphenol [12]. Kiedik et al. describe the production of nonylphenol with the use of porous or gelling cationites [13]. Knifton describes the use of heteropolyacids for adding mixed nonenes to phenol. The catalyst is stable at elevated temperatures allowing adiabatic reaction [14]. Zeolites as catalyst for the alkylation of phenols is becoming an area of active interest. A recent Japanese patent [15] notes that zeolites reduce the need for reactor anticorrosion protection compared to more acidic catalysts. Zeolites are effective for the manufacture of alkylphenols from mixed octenes, mixed nonenes, or mixed dodecenes [16]. Kapustin et al. describe the use of a catalyst consisting of porous sulfonated copolymers of styrene and divinylbenzene claiming that yields and purity are increased [17]. It is thought that catalysts of this type are used in current commercial practice.

In the past, batch processes were used to produce nonylphenol. These typically were performed in stirred tank reactors with catalyst suspended in the reaction mixture. Such processes are becoming obsolete in favor of continuous processes.

The following describes a more modern process [18]:

A mixture of phenol and propene trimer (molar ratio 1.7:1) is preheated to 70°C and pumped to the top of the first reactor at an hourly rate of about 30 mass units per 1 mass unit of catalyst. The product leaving the first reactor with a temperature of 120°C is cooled to 100°C and fed to the second reactor. The temperature of the reaction product leaving the second reactor is about

125°C. Product compositions after the first and second step are (in wt%): propene trimer 28.9 and 3.8; phenol 46.1 and 27.5; nonylphenol 22.4 and 65.8; and dinonylphenol 2.6 and 2.9; respectively. The product of the second reaction step is purified by distillation. Recovered phenol and propene trimer, as well as dinonylphenol, are recycled.

As mentioned above, sulfonated polystyrene-polydivinylbenzene ion exchange resins are commonly used in continuous alkylation processes as they are highly selective for monoalkylation and can be readily arranged in fixed beds. They are reported to have long service lives, important for use in fixed bed reactors. The upper operating temperature allowable with cation exchangers is about 140°C [19]. The alkylation is quite exothermic resulting in the need for heat removal. In a continuous process, excess phenol is commonly used to reduce the temperature rise in the reactor. Excess phenol also tends to drive the alkylation to the monoalkylate, the preferred species for polyoxyethylation. Use of excess phenol also allows operation at or near adiabatic conditions. As noted in the process description, distillation is frequently required to purify the product. This is accomplished in a three-column distillation train. Excess phenol is removed in the first column. The second column removes light materials such as unreacted alkene. Heavy byproducts such as dialkylphenols are removed in the third column in addition to recovering monoalkylphenol overhead. Because excessive reboiler temperatures can lead to dealkylation, the distillation is conducted under vacuum [20].

Finished nonylphenol is stored in heated tanks made of stainless steel or phenolic lined carbon steel. These tanks are inerted with nitrogen to limit discoloration caused by product oxidation [21]. Bulk shipments are made in rail cars or tank trucks, likewise made of stainless steel or phenolic lined carbon steel. As nonylphenol is a liquid at room temperature, small shipments of nonylphenol are transported in drums or tote tanks.

B. Octylphenol

Octylphenol used in surfactant production shares many of the characteristics of nonylphenol. Of greatest importance is that it is predominantly both para-substituted and highly branched. One outstanding difference is that the alkyl group is a relatively pure material, diisobutylene. There are two isomers of diisobutylene, but alkylation with either isomer results in the same alkyl structure:

$$
\underbrace{
\begin{array}{c}
CH_3 \\
| \\
CH_3-C-CH=C-CH_3 \\
| \quad\quad | \\
CH_3 \quad CH_3
\end{array}
\;+\;
\begin{array}{c}
CH_3 \\
| \\
CH_3-C-CH_2-C=CH_2 \\
| \quad\quad\quad | \\
CH_3 \quad\quad CH_3
\end{array}
}_{\text{Diisobutylene Isomers}}
\;\xrightarrow{H^+}\;
\underbrace{
\begin{array}{c}
CH_3 \quad\quad CH_3 \\
| \quad\quad\quad\; + \\
CH_3-C-CH_2-C-CH_3 \\
| \quad\quad\quad | \\
CH_3 \quad\quad CH_3
\end{array}
}_{\text{Diisobutylene Carbocation}}
\tag{1}
$$

$$CH_3-\underset{\underset{CH_3}{|}}{\overset{\overset{CH_3}{|}}{C}}-CH_2-\overset{+}{\underset{\underset{CH_3}{|}}{C}}-CH_3 \quad + \quad \underset{\text{Phenol}}{\underset{\text{Diisobutylene Carbocation}}{\bigcirc\!\!-OH}} \quad \longrightarrow \quad CH_3-\underset{\underset{CH_3}{|}}{\overset{\overset{CH_3}{|}}{C}}-CH_2-\underset{\underset{CH_3}{|}}{\overset{\overset{CH_3}{|}}{C}}-\bigcirc\!\!-OH \quad (2)$$

Diisobutylene Carbocation   Phenol   *p-tert*-Octylphenol

The general process description for the manufacture of nonylphenol applies to octylphenol as well. Detailed process conditions such as reactant and catalyst ratios and operating temperatures may be slightly different. The purification steps are very similar to those for nonylphenol. The materials compatibility and product quality issues associated with nonylphenol also apply to octylphenol. Because octylphenol has a high freezing point, it must be shipped in insulated trucks or rail cars. These vehicles must be equipped to allow melting of the product should it freeze in transit. The great difference in freezing points of octylphenol and nonylphenol provides a quick method for surfactant manufacturers to confirm the identity of inbound shipments of alkylphenol.

C. Dodecylphenol

Dodecylphenol is made from propylene tetramer and phenol. This tetramer is an even more complicated mixture of isomers than the propylene trimer used in nonylphenol manufacture. The manufacturing process is similar to that described for octylphenol and nonylphenol.

D. Properties

In spite of the purification processes employed on alkylphenols, there is commonly some measurable level of ortho isomer and dialkylate in the finished product. Typically, alkylphenols contain at least 90% para isomer, less than 10% ortho isomer, and a trace of dialkylphenol. Higher purity grades are available [22]. Typical properties are shown in Table 2. Note that octylphenol, due to the relative purity of the octyl group, is a solid at room temperature. Due to the heterogeneity of their alkyl groups, nonylphenol and dodecylphenol are liquids.

TABLE 2 Physical Properties of Alkylphenols [105]

Property	4-*tert*-Octylphenol	4-Nonylphenol	4-Dodecyl-phenol
Typical assay (%)	90–98	90–95	89–95
Molecular weight	206	220	262
Physical form at 25°C	Solid	Liquid	Liquid
Boiling point (°C)	290	310	334
Freezing point (°C)	81	<20	<20
Flash point (°C)	132	146	>100
Molten color (APHA)	200	100	500

The environmental and toxicologic characteristic of alkylphenols are discussed in the environmental issues section later in this chapter.

III. POLYOXYETHYLENE ALKYLPHENOLS

Polyoxyethylene alkyphenols are important surfactants as evidenced by the large number of producers of these products listed in Table 3. Several companies, including Union Carbide, Huntsman, and Rhône-Poulenc, have extensive product lines whereas several others supply only a few products.

A. Manufacture

Polyoxyethylene alkylphenols are produced in a semibatch process at temperatures ranging from 100°C to 180°C at a reactor pressure between 5 and 6 bar gauge with alkaline catalyst concentrations of 0.1% to 0.03% by weight. Sodium hydroxide and potassium hydroxide are most commonly used to catalyze the polyoxyethylation. Acid catalysts are not often used due to byproduct formation [23].

The semibatch process is commonly used for production of polyoxyethylated nonionic surfactants. The reaction vessel is typically agitated with means for heating and cooling, either internally by coil or jacket or externally in a circulation loop. The reactors is inerted with nitrogen to remove oxygen and then charged with alkylphenol and catalyst. The catalyst can be added as solid beads, flakes, or aqueous solution. Water is generated during the catalysis step:

$$R\!-\!\!\left\langle\bigcirc\right\rangle\!\!-\!OH \;+\; n\,MOH \qquad\longrightarrow$$

$$\tag{3}$$

$$R\!-\!\!\left\langle\bigcirc\right\rangle\!\!-\!O^-\,M^+ \;+\; (1\text{-}n)\quad R\!-\!\!\left\langle\bigcirc\right\rangle\!\!-\!OH \;+\; n\,H_2O$$

$$M = \text{Sodium or Potassium}$$

where n is the mole ratio of base to alkylphenol and is typically 0.005–0.02. The catalyzed alkylphenol is heated to reaction temperature and purged with nitrogen to reduce the water level generally to less than 0.1% by weight. Water removal is important if polyethylene glycol formation is to be minimized. After drying, ethylene oxide feed to the reactor is started. The ethylene oxide may be added via a sparger in the bottom of the reactor, it may be mixed with the bulk liquid in a external pumped circulating loop; or it may be added directly to the vapor space of the reactor. It is important to disperse the ethylene oxide into the bulk liquid quickly and thoroughly, and different reaction systems accomplish this in various special sparger designs, reactor agitators, reactor circulation, spray circulation systems, or eductor mixers. Producers must avoid a vapor space explosion, and this is done by eliminating ignition sources or by using

TABLE 3 Polyoxyethylene Alkylphenol Producers [106,107]

Trademark	Supplier	Location
Ablunol NP	Taiwan Surfactant Corp.	Taiwan
Agrimul 70-A	Sidobre-Sinnova S.A.	France
Agrisol PX	Akcros Chemicals	England
Akypo OP	CHEM-Y GmbH	Germany
Akyporox NP, OP	CHEM-Y GmbH	Germany
Alkasurf NP, OP	Rhône-Poulenc, North American Chem.	USA
Antarox LF-222	Rhône-Poulenc, North American Chem.	USA
Arkopal N Grades	Hoechst AG/Zentrale Werbung	Germany
Armul 900 (series)	Witco Corp.	USA
Atlas EMJ	Atlas Refinery	USA
Atlas G-4809, 5002	ICI Surfactants	England
Atlox 4898, 4911	ICI Surfactants	England
Atlox 775	ICI Surfactants	USA
Atsolyn T	Atsaun Chemical Corp.	India
Basopon LN	BASF Corp.	USA
Berol (series)	Akzo Nobel Surface Chemistry	USA
Berol (series)	Berol Nobel AB	Sweden
Caloxylate N-9	Pilot Chemical Co.	USA
Capcure Emulsifier 65	Henkel Corp./Functional Prod. Grp.	USA
Carsonon N	Lonza Inc.	USA
Cedepal CA, CO	Stepan Canada Inc.	Canada
Cenegen NWA	Crompton & Knowles	USA
Cenekol FT Supra	Crompton & Knowles	USA
Chemax NP, OP, DNP	Chemax Inc.	USA
Chimipon NA	Auschem S.p.A	Italy
Cirrasol AEN-XZ	ICI Surfactants	England
Corexit CL	Exxon Chemical Company	USA
Dehydrophen PNP, POP	Pulcra, S.A.	Spain
Delonic NPE, OPE	DeForest Enterprises, Inc.	USA
DeSonic S, N, D	Witco Corp.	USA
Dowfax 9N	Dow Europe SA	Switzerland
Eccoscour RC	Eastern Color & Chemical Co.	USA
Eccoterge EO-100	Eastern Color & Chemical Co.	USA
Eleminol HA-100, 161	Sanyo Chemical Industries, Ltd.	Japan
Emulan NP, OP, PO	BASF Aktiengesellschaft	Germany
Emulsifier 632/90%	Ethox Chemicals	USA
Emulsit	Dai-ichi Kogyo Seiyaku Co., Ltd.	Japan
Emulsogen A, N-600	Hoechst Celanese Corp.	USA
Emulson AG (series)	Auschem S.p.A	Italy
Ethylan	Akcros Chemicals	England
Etophen Series	Zschimmer & Schwarz	Germany
Geronol AG-821	Rhône-Poulenc, North American Chem.	USA
Gradonic N-95	Graden Chemical	USA
Hetoxide NP	Heterene, Inc.	USA

TABLE 3 (Continued)

Trademark	Supplier	Location
Hostapal N, CVS	Hoechst Celanese Corp.	USA
Hyonic NP, OP	Henkel Corp./Functional Prod. Grp.	USA
Iberscour W Conc.	A. Harrison & Co.	USA
Iberwet W-100	A. Harrison & Co.	USA
Iconol NP, OP	BASF Corp.	USA
Igepal CA, CO	Rhône-Poulenc, North American Chem.	USA
Igepal NP, O, OD, RC	Rhône-Poulenc Chimie (France)	France
Igepal SS-837	Rhône-Poulenc, North American Chem.	USA
Lipocol NP-9 USP	Lipo Chemicals	USA
Liponox NC	Lion Corp.	Japan
Lutensol AP	BASF Aktiengesellchaft	Germany
Macol NP OP, DNP	PPG Industries	USA
Makon 4–6, 7–11, 12–50	Stepan Europe	France
Makon NI 10/Nl 20/Nl 30	Stepan Europe	France
Makon, Makon OP-9	Stepan Company	USA
Marlophen 80, 800	Hüls AG	Germany
Marlophen DNP, N, P	Hüls AG	Germany
Marlophen NP, P1	Hüls America	USA
Marlowet 4900, 5641	Hüls AG	Germany
Marlowet BIK, ISM, TM	Hüls AG	Germany
Minemal 320, 325, 330	Nippon Nyukazai Co., Ltd.	Japan
Mulsifan RT37, RT 231	Zschimmer & Schwarz	Germany
Neutronyx 656	Stepan Company	USA
Newcol 500, 600, 700, 800	Nippon Nyukazai Co., Ltd	Japan
Nikkol NP, OP	Nikko Chemicals Co., Ltd.	Japan
Nissan Dispanol N-100	NOF Corporation	Japan
Nissan Nonion HS, NS	NOF Corporation	Japan
Noigen EA	Dai-ichi Kogyo Seiyaku Co., Ltd.	Japan
Nonal	Toho Chemical Industry Co., Ltd.	Japan
Nonarox	Seppic	France
Nonicol 100, 190	Atsaun Chemical Corp.	India
Nonionic E	Calgene Chemical, Inc.	USA
Nonipol BX, D-160	Sanyo Chemical Industries, Ltd.	Japan
Norfox NP, OP	Norman, Fox & Co.	USA
NP-55-60, 80, 85, 90	Hefti Ltd.	Switzerland
Nutrol 600 (series)	Boehme Filatex Canada Inc.	Canada
Octapol	Sanyo Chemical Industries, Ltd.	Japan
Polystep OP, F	Stepan Company	USA
Quimipol ENF	CPB-Companhia Petroquimica Do	Portugal
Remcopal (series)	Ceca SA	France
Renex 600	ICI Surfactants	England
Rioklen	Auschem S.p.A	Italy
Sabofen AF	Sabo S.p.A.	Italy
Sermul EN (series)	Servo Delden, B.V.	Holland
Simulsol NP 575	Seppic	France

TABLE 3 (Continued)

Trademark	Supplier	Location
Sinopol (series)	Sino-Japan Chemical Co.	Taiwan
Solegal W Conc.	Hoechst AG/Zentrale Werbung	Germany
Soprophor	Rhône-Poulenc, North American Chem.	USA
Soprophor 37, 40/D	Rhône-Poulenc Geronazzo S.P.A.	Italy
Soprophor BSU, CY/8, S	Rhône-Poulenc, North American Chem.	USA
Sterox	Monsanto Co.	USA
Surflo S	Exxon Chemical Company	USA
Surfonic N, OP	Huntsman Corporation	USA
Syn Fac	Milliken Chemical	USA
Synperonic N, NP, NPE1800, OP	ICI Surfactants	England
Synthrapol N	ICI Surfactants	USA
Syntopon (series)	Witco SA	France
T-DET N, O, DD	Harcros Organics	USA
Tergitol NP (series)	Union Carbide Corporation	USA
Teric (series)	ICI Australia Operations Pty, Ltd.	Australia
Tex-Wet 1155	Intex Chemical	USA
TN Cleaner S	Tokai Seiyu Ind. Co., Ltd.	Japan
Triton N, X (series)	Union Carbide Corporation	USA
Trycol 6000 (series)	Henkel Corp./Emery Organic Prod. Grp.	USA
Value 3608, 3700	Marubishi Oil Chemical Co., Ltd.	Japan
Wettol EM 2	BASF Aktiengesellschaft	Germany
Witbreak DRB	Witco Corp.	USA

nitrogen dilution to limit the ethylene oxide concentration to a point below its flammable limit at the reactor operating temperature and pressure. Many producers choose a design and operating conditions to eliminate both concerns. The ethylene oxide feed rate is dependent on the heat removal capability and the reactor design pressure and temperature, plus any constraints imposed by limiting the ethylene oxide vapor space concentration to below the flammable limit.

When the alkylphenol has been polyoxyethylated to the desired extent, the reaction mixture is held at reaction temperature until the residual ethylene oxide concentration in the liquid product has been reduced to an acceptable level. Specification tests can then be conducted to assure product quality. The product is then neutralized and posttreated as described below [24].

High molecular weight products may require a multistep manufacturing process. In most cases, a minimum level of alkylphenol must be charged to the reactor in order to properly cover the agitator and heat transfer surfaces. This minimum level of alkylphenol will grow in volume as it is polyoxyethylated; its volume may equal that of the reactor before the required degree of polyoxyethylation is achieved. In this case, a portion of the reaction mixture may be removed from the reactor to leave room for further product growth, and the removed portion may be stored and used for future production. Alternatively, a lower molecular weight polyoxyethylene alkylphenol may be used as the starting alcohol in Eq. (3).

Upon completion of the polyoxyethylation, the reaction mixture must be neutralized. Commonly used acids include acetic, phosphoric, propionic, and lactic. If necessary, color may be reduced by bleaching with hydrogen peroxide. Finally, the product may be filtered to remove any insoluble salts formed during neutralization.

B. Product Composition

Due to the high acidity of alkylphenols, the first mole of ethylene oxide is added with great selectivity. The relative acidity of the alkylphenol compared to the polyoxyethylene chain results in a nearly complete conversion of alkylphenol to a polyoxyethylene (1) before polyoxyethylene (2) alkylphenol forms in appreciable amounts. The second and subsequent moles of ethylene oxide add as described by a Poisson distribution.

The general reaction scheme

$$R\text{—}\langle\text{—}\rangle\text{—OH} + n\ CH_2\text{—}CH_2 \longrightarrow R\text{—}\langle\text{—}\rangle\text{—O-(CH}_2CH_2O)_nH \quad (4)$$

can be described in two steps if n is >1:

$$R\text{—}\langle\text{—}\rangle\text{—OH} + CH_2\text{—}CH_2 \longrightarrow R\text{—}\langle\text{—}\rangle\text{—O-CH}_2CH_2OH \quad (5)$$

$$R\langle\text{—}\rangle\text{-O-CH}_2CH_2OH + m\ CH_2\text{—}CH_2 \longrightarrow R\langle\text{—}\rangle\text{-O-CH}_2CH_2O\text{-(CH}_2CH_2O)_mH \quad (6)$$

The average degree of ethoxylation represented in Eq. (4), then, is composed of the individual oligomers

$$R\text{—}\langle\text{—}\rangle\text{—O-CH}_2CH_2O\text{-(CH}_2CH_2O)_xH \quad (7)$$

where x is 0, 1, 2, 3, . . . , and the term $(x+1)$-mer indicates the total moles of ethylene oxide in the molecule.

The mol% of an individual $(x+1)$-mer is calculated by

$$\text{mol\% } (x+1)\text{-mer} = \left(\frac{e^{-m}m^x}{x!}\right)(100) \quad (8)$$

and the wt% by

$$\text{wt\% } (x+1)\text{-mer} = \left(\frac{e^{-m}m^x}{x!}\right)\left(\frac{M_{x+1}}{M_n}\right)(100) \tag{9}$$

where $e = 2.71828$
 $m = n + 1$
 $x = 1, 2, 3, \ldots$
 $M_{x+1} = $ molecular weight of $(x+1)$-mer
 $M_n = $ molecular weight of product

Values of molecular weight distributions calculated from these equations are in Table 4 for polyoxyethylene nonylphenols and in Table 5 for polyoxyethylene octylphenols. The correlation between molecular weight distribution calculated from the Poisson distribution and that determined analytically shows good agreement as seem in Table 6. *ortho*-Alkylphenol and dialkylphenol are always present in the raw materials; therefore these calculated values are close to but not exactly equal to the actual composition of the product.

While polyoxyethylene alkylphenols of different molecular weights can be blended together to yield a product of a known molecular weight, the molecular weight distribution of the blend may be quite different from that of a product made by directly polyoxyethylating the hydrophobe. Table 7 shows the results of blending a polyoxyethylene (5) nonylphenol with a polyoxyethylene (15) nonylphenol in proportions leading to a product with the molecular weight of a polyoxyethylene (10) nonylphenol. Clearly, the resulting bimodal molecular weight distribution of the blend is quite different from that of polyoxyethylene nonylphenol (10).

Polyoxyethylene glycols (PEGs) are commonly formed during surfactant manufacture due to side reactions in the polyoxyethylation process. In most cleaning applications, byproduct PEG acts as little more than a diluent. However, if the polyoxyethylene alkylphenol is to be derivatized by, for example, sulfation, then the resulting PEG disulfate is not at all surface active and is undesirable. The presence of PEG in the surfactant also reduces the solubility of polyoxyethylene alkylphenols in nonpolar solvents. PEG can also have an effect on polyoxyethylene alkylphenol product appearance as levels of PEG >0.5% can lead to the formation of a hazy bottom layer on extended storage.

Byproduct PEG in a polyoxyethylene alkylphenol can originate with residual water in the reaction mixture; this water is polyoxyethylated along with the intended hydrophobe. Small amounts of water can enter the process as a trace component of the alkylphenol. A second source of water is the residue from reactor and line cleaning. Finally, water is formed during the catalysis step in the polyoxyethylene process. Water from all of these sources is removed in a

TABLE 4 Poisson Distribution (Wt%) Polyoxethylene Nonylphenol

x-mer	MW	Mole ratio (n)										
		2	3	4	5	6	7	8	9	10	12	15
1	264	31.53	10.15	3.32	1.10	0.37	0.12	0.04	0.01			
2	308	36.79	23.68	11.61	5.13	2.14	0.87	0.34	0.13	0.05	0.01	
3	352	21.02	27.07	19.91	11.72	6.12	2.97	1.37	0.61	0.27	0.05	
4	396	7.88	20.30	22.40	17.58	11.48	6.69	3.61	1.84	0.90	0.20	0.02
5	440	2.19	11.28	18.67	19.54	15.95	11.15	7.02	4.09	2.25	0.60	0.07
6	484	0.48	4.96	12.32	17.19	17.55	14.72	10.81	7.20	4.45	1.45	0.20
7	528	0.09	1.81	6.72	12.50	15.95	16.06	13.75	10.47	7.29	2.90	0.52
8	572	0.01	0.56	3.12	7.74	12.34	14.92	14.90	12.96	10.15	4.94	1.13
9	616		0.15	1.26	4.17	8.31	12.05	14.04	13.96	12.30	7.31	2.13
10	660		0.04	0.45	1.99	4.95	8.61	11.70	13.29	13.18	9.58	3.55
11	704		0.01	0.14	0.85	2.64	5.51	8.74	11.34	12.65	11.24	5.30
12	748			0.04	0.33	1.27	3.19	5.91	8.77	11.00	11.94	7.17
13	792			0.01	0.12	0.56	1.69	3.65	6.19	8.73	11.59	8.86
14	836				0.04	0.23	0.82	2.07	4.02	6.38	10.35	10.07
15	880				0.01	0.09	0.37	1.09	2.42	4.32	8.56	10.60
16	924				0.00	0.03	0.16	0.53	1.35	2.72	6.59	10.39
17	968					0.01	0.06	0.25	0.71	1.60	4.75	9.52
18	1012						0.02	0.11	0.35	0.89	3.21	8.20
19	1056						0.01	0.04	0.16	0.46	2.05	6.65
20	1100							0.02	0.07	0.23	1.24	5.11
21	1144							0.01	0.03	0.11	0.71	3.72
22	1188								0.01	0.05	0.38	2.57
23	1232									0.02	0.20	1.70
24	1276									0.01	0.10	1.07
25	1320										0.05	0.65
26	1364										0.02	0.37
27	1408										0.01	0.21
28	1452											0.11
29	1496											0.06
30	1540											0.03
31	1584											0.01
32	1628											0.01

Calculated from Eq. (9)

x-mer	MW	Mole ratio (n)		
		20	30	40
6	484	0.01		
7	528	0.02		
8	572	0.05		
9	616	0.13		
10	660	0.30		
11	704	0.61		
12	748	1.11		
13	792	1.86	0.01	
14	836	2.88	0.02	
15	880	4.11	0.05	
16	924	5.46	0.10	
17	968	6.80	0.19	
18	1012	7.94	0.34	
19	1056	8.75	0.57	
20	1100	9.11	0.91	0.01
21	1144	9.00	1.37	0.02
22	1188	8.46	1.97	0.04
23	1232	7.58	2.69	0.06
24	1276	6.48	3.52	0.11
25	1320	5.31	4.40	0.19
26	1364	4.17	5.27	0.31
27	1408	3.14	6.07	0.47
28	1452	2.28	6.72	0.71
29	1496	1.60	7.18	1.01
30	1540	1.08	7.39	1.41
31	1584	0.70	7.34	1.88
32	1628	0.44	7.06	2.43
33	1672	0.27	6.57	3.04
34	1716	0.16	5.93	3.69
35	1760	0.09	5.19	4.33
36	1804	0.05	4.41	4.95

x-mer	MW	Mole ratio (n)		
		20	30	40
38	44	0.01	2.92	5.94
39	88	0.01	2.28	6.23
40	132		1.73	6.37
41	176		1.28	6.35
42	220		0.93	6.17
43	264		0.65	5.86
44	308		0.45	5.42
45	352		0.30	4.90
46	396		0.20	4.33
47	440		0.13	3.75
48	484		0.08	3.17
49	528		0.05	2.62
50	572		0.03	2.13
51	616		0.02	1.69
52	660		0.01	1.31
53	704		0.01	1.00
54	748			0.75
55	792			0.55
56	836			0.40
57	880			0.28
58	924			0.20
59	968			0.13
60	1012			0.09
61	1056			0.06
62	1100			0.04
63	1144			0.02
64	1118			0.02
65	1232			0.01
66	1276			0.01
66	3124			0.01
67	3168			0.00

TABLE 5 Poisson Distribution (Wt%) Polyoxyethylene Octylphenol

x-mer	MW	Mole ratio (n)										
		2	3	4	5	6	7	8	9	10	12	15
1	250	31.28	10.01	3.26	1.07	0.36	0.12	0.04	0.01			
2	294	36.79	23.54	11.49	5.05	2.11	0.85	0.34	0.13	0.05	0.01	
3	338	21.15	27.07	19.82	11.62	6.06	2.93	1.35	0.60	0.26	0.05	
4	382	7.97	20.40	22.40	17.52	11.41	6.63	3.57	1.82	0.89	0.19	0.02
5	426	2.22	11.37	18.74	19.54	15.90	11.09	6.96	4.05	2.22	0.59	0.07
6	470	0.49	5.02	12.41	17.24	17.55	14.69	10.76	7.15	4.42	1.43	0.20
7	514	0.09	1.83	6.78	12.57	15.99	16.06	13.72	10.43	7.25	2.88	0.52
8	558	0.01	0.57	3.16	7.80	12.40	14.95	14.90	12.94	10.12	4.91	1.12
9	602		0.15	1.28	4.21	8.36	12.09	14.07	13.96	12.28	7.28	2.12
10	646		0.04	0.46	2.01	4.99	8.65	11.74	13.32	13.18	9.55	3.53
11	690		0.01	0.15	0.86	2.66	5.55	8.78	11.38	12.67	11.22	5.28
12	734			0.04	0.33	1.29	3.22	5.94	8.80	11.02	11.94	7.15
13	778			0.01	0.12	0.57	1.71	3.67	6.22	8.76	11.60	8.84
14	822				0.04	0.23	0.83	2.09	4.04	6.41	10.37	10.06
15	866				0.01	0.09	0.38	1.10	2.43	4.34	8.58	10.60
16	910					0.03	0.16	0.54	1.36	2.74	6.62	10.40
17	954					0.01	0.06	0.25	0.72	1.61	4.77	9.54
18	998						0.02	0.11	0.35	0.89	3.23	8.22
19	1042						0.01	0.04	0.16	0.47	2.06	6.67
20	1086							0.02	0.07	0.23	1.24	5.12
21	1130							0.01	0.03	0.11	0.71	3.73
22	1174								0.01	0.05	0.39	2.58
23	1218									0.02	0.20	1.71
24	1262									0.01	0.10	1.08
25	1306										0.05	0.65
26	1350										0.02	0.38
27	1394										0.01	0.21
28	1438											0.11
29	1482											0.06
30	1526											0.03
31	1570											0.01
32	1614											0.01

Calculated from Eq. (9)

x-mer	MW	Mole ratio (n) 20	30	40
6	470	0.01		
7	514	0.02		
8	558	0.05		
9	602	0.13		
10	646	0.30		
11	690	0.60		
12	734	1.10		
13	778	1.85	0.01	
14	822	2.86	0.02	
15	866	4.09	0.05	
16	910	5.45	0.10	
17	954	6.78	0.19	
18	998	7.93	0.34	
19	1042	8.74	0.57	
20	1086	9.11	0.91	0.01
21	1130	9.01	1.37	0.02
22	1174	8.47	1.97	0.03
23	1218	7.59	2.69	0.06
24	1262	6.49	3.51	0.11
25	1306	5.32	4.39	0.19
26	1350	4.18	5.27	0.31
27	1394	3.15	6.07	0.47
28	1438	2.29	6.72	0.71
29	1482	1.60	7.17	1.01
30	1526	1.08	7.39	1.40
31	1570	0.70	7.35	1.88
32	1614	0.44	7.07	2.43
33	1658	0.27	6.58	3.04
34	1702	0.16	5.93	3.68
35	1746	0.09	5.19	4.34
36	1790	0.05	4.41	4.95

x-mer	MW	Mole ratio (n) 20	30	40
38	1878	0.01	2.92	5.93
39	1922	0.01	2.28	6.23
40	1966		1.74	6.37
41	2010		1.29	6.35
42	2054		0.93	6.18
43	2098		0.66	5.86
44	2142		0.45	5.42
45	2186		0.30	4.91
46	2230		0.20	4.34
47	2274		0.13	3.75
48	2318		0.08	3.17
49	2362		0.05	2.63
50	2406		0.03	2.13
51	2450		0.02	1.69
52	2494		0.01	1.32
53	2538		0.01	1.00
54	2582			0.75
55	2626			0.55
56	2670			0.40
57	2714			0.28
58	2758			0.20
59	2802			0.13
60	2846			0.09
61	2890			0.06
62	2934			0.04
63	2978			0.02
64	3022			0.02
65	3066			0.01
66	3110			0.01
66	3110			0.01
67	3154			0.00

TABLE 6 Comparison of Poisson Distribution and Measured Molecular Weight Distribution for Triton X-100

Oligomer number	Percent by weight [108]	
	Poisson distribution[a]	Measured[b]
1	0.09	0.00
2	0.44	0.9
3	1.4	2.5
4	3.3	4.6
5	6.0	6.9
6	9.2	9.5
7	11.9	11.7
8	13.5	12.7
9	13.5	12.3
10	12.0	11.2
11	9.7	9.1
12	7.2	6.8
13	4.9	4.8
14	3.1	3.1
15	1.8	2.0
16	0.99	1.2
17	0.51	0.7
18	0.25	0.0
19	0.11	0.0
20	0.05	0.0
21	0.02	0.0
22	0.01	0.0

[a] Calculated from Eq. (9).
[b] By gas chromatography.

drying step prior to the introduction of ethylene oxide. Enyeart [25] presents a detailed discussion of the PEG formation process. For a monofunctional hydrophobe such as an alkylphenol, and assuming that residual water in the reaction system is the only source of PEG production, the wt% of PEG in the product, wt% PEG, can be calculated:

$$\text{wt\% PEG} = \frac{H \cdot MW_B}{MW_T} \left[1 + \frac{44(100)n}{H \cdot MW_B + 900} \right] \tag{10}$$

where　H = wt% water based on alkylphenol
　　　　MW_B = molecular weight of alkylphenol
　　　　n = degree of polyoxyethylation
　　　　$MW_T = MW_B + 44(n)$, the theoretical molecular weight of the surfactant

TABLE 7 Effect of Blending Polyoxyethylene Nonylphenols on Molecular Weight Distribution

		Poisson distributions			
		Mole ratio (*n*)			Blend
Mole ratio	Mol. wt.	5	15	10	5/15
1	264	1.10	0.00	0.00	0.37
2	308	5.13	0.00	0.05	1.71
3	352	11.72	0.00	0.27	3.90
4	396	17.58	0.02	0.90	5.87
5	440	19.54	0.07	2.25	6.55
6	484	17.19	0.20	4.45	5.86
7	528	12.50	0.52	7.29	4.51
8	572	7.74	1.13	10.15	3.33
9	616	4.17	2.13	12.30	2.81
10	660	1.99	3.55	13.18	3.03
11	704	0.85	5.30	12.65	3.82
12	748	0.33	7.17	11.00	4.89
13	792	0.12	8.86	8.73	5.95
14	836	0.04	10.07	6.38	6.73
15	880	0.01	10.60	4.32	7.07
16	924		10.39	2.72	6.93
17	968		9.52	1.60	6.35
18	1012		8.20	0.89	5.47
19	1056		6.65	0.46	4.44
20	1100		5.11	0.23	3.41
21	1144		3.72	0.11	2.48
22	1188		2.57	0.05	1.72
23	1232		1.70	0.02	1.13
24	1276		1.07	0.01	0.71
25	1320		0.65		0.43
26	1364		0.37		0.25
27	1408		0.21		0.14
28	1452		0.11		0.07
29	1496		0.06		0.04
30	1540		0.03		0.02
31	1584		0.01		0.01
32	1628		0.01		0.00
Mol. wt.		440	881	660	660

Blend of 33.3% polyoxyethylene (5) nonylphenol and 66.7% polyoxyethylene (15) nonylphenol.

TABLE 8 Theoretical Polyoxyethylene Glycol Content in Polyoxyethylene Nonylphenol[a]

Mole ratio	Initial water content, percent based on nonylphenol					
	0.01	0.02	0.05	0.10	0.20	0.50
2	0.08	0.15	0.38	0.75	1.47	3.47
3	0.10	0.19	0.48	0.96	1.87	4.40
4	0.11	0.23	0.56	1.12	2.18	5.12
5	0.13	0.25	0.63	1.24	2.43	5.70
6	0.14	0.27	0.68	1.35	2.63	6.17
7	0.15	0.29	0.73	1.43	2.80	6.56
8	0.15	0.31	0.76	1.51	2.95	6.89
9	0.16	0.32	0.79	1.57	3.07	7.18
10	0.17	0.33	0.82	1.62	3.17	7.43
15	0.19	0.37	0.92	1.81	3.55	8.29
20	0.20	0.39	0.98	1.93	3.77	8.81
30	0.21	0.42	1.04	2.06	4.02	9.41
40	0.22	0.43	1.08	2.13	4.17	9.74
70	0.23	0.46	1.13	2.23	4.36	10.20

[a] Calculated from Eq. (10).

TABLE 9 Theoretical Polyoxyethylene Glycol Content in Polyoxyethylene Octylphenol[a]

Mole ratio	Initial water content, percent based on octylphenol					
	0.01	0.02	0.05	0.10	0.20	0.50
2	0.08	0.15	0.37	0.74	1.45	3.42
3	0.10	0.19	0.47	0.93	1.83	4.32
4	0.11	0.22	0.55	1.08	2.12	5.00
5	0.12	0.25	0.61	1.20	2.36	5.55
6	0.13	0.26	0.66	1.30	2.55	5.99
7	0.14	0.28	0.70	1.38	2.70	6.35
8	0.15	0.29	0.73	1.45	2.84	6.66
9	0.15	0.31	0.76	1.51	2.95	6.93
10	0.16	0.32	0.79	1.56	3.05	7.15
15	0.18	0.35	0.87	1.73	3.38	7.95
20	0.19	0.37	0.93	1.83	3.58	8.42
30	0.20	0.40	0.99	1.95	3.81	8.95
40	0.21	0.41	1.02	2.01	3.94	9.25
70	0.21	0.43	1.06	2.10	4.12	9.66

[a] Calculated from Eq. (10).

The PEG content of polyoxyethylene nonylphenol and octylphenol given a variety of initial water levels is shown in Tables 8 and 9. Clearly, the removal of water from the catalyzed alkylphenol prior to polyoxyethylation is important if PEG formation is to be minimized. In commercial practice, the water content of the catalyzed alkylphenol is reduced to no more than 0.05% prior to ethoxylation.

In addition to PEG, small amounts of acetaldehyde, formaldehyde, and 1,4-dioxane are formed. These materials are present at ppm levels [26]. Mechanisms for formation of these byproducts are unclear; however, it is well known that ethylene oxide can isomerize to form acetaldehyde [27] and that acidic catalysis of the polyoxyethylation reaction is linked to 1,4-dioxane formation [28].

IV. PROPERTIES

A. Physical Properties

The physical state of a polyoxyethylene alklyphenol at ambient conditions depends greatly on its degree of polyoxyethylation. As seen in Table 10, the pour point of polyoxyethylene nonylphenols climbs with increasing degree of polyoxyethylation. Products with more than about 15 moles of polyoxyethylation are solids at room temperature. It is common practice to make the higher mole products available for sale as concentrated aqueous solutions. Adding water to a typical level of 30% in the final product leads to a lower freezing liquid that is easier to handle.

The pour point of polyoxyethylene octylphenols (Table 11) passes through a minimum as the degree of ethoxylation increases. Recall that octylphenol has a freezing point of about 80°C. For low levels of polyoxyethylation, the polyoxyethylene octylphenol properties reflect the octylphenol properties and specifically the high pour point. As the degree of polyoxyethylation increases,

TABLE 10 Typical Physical Properties of Polyoxyethylene Nonylphenol [109]

Tergitol	Mole ratio	Pour point (°C)	Kinematic viscosity (mPa·s)			Specific gravity (20/20°C)
			20°C	40°C	60°C	
NP-4	4	−40	445	105	8	1.031
NP-6	6	−12	373	99	9	1.035
NP-7	7	−6	338	97	10	1.055
NP-8	8	−3	325	102	11	1.056
NP-9	9	0	318	105	12	1.057
NP-10	10.5	7	327	105	13	1.062
NP-13	13	16	410	125	15	1.071
NP-40	40	48	Solid	Solid	48	Solid

TABLE 11 Typical Physical Properties of Polyoxyethylene Octylphenol [110]

Triton	Mole ratio	Pour point (°C)	Viscosity (mPa·s)			Specific gravity (25/25°C)
			25°C	60°C	100°C	
X-15	1	−9	790	45	8	0.985
X-35	3	−23	370	38	8	1.023
X-45	5	−26	290	36	9	1.040
X-114	7.5	−9	260	40	11	1.054
X-100	9.5	7	240	43	13	1.065
X-102	12	16	330	50	15	1.071

the polyoxyethylene chain plays a greater role in the pour point. As with poly-oxyethylene nonylphenols, the higher oligomers are solids at room temperature. These products are also available as aqueous blends.

The liquid polyoxyethylates of both nonylphenol and octylphenol are color-less to pale yellow and are clear to slightly hazy. Haze can result from the salt formed while neutralizing the reaction mixture after concluding the poly-oxyethylation step. Sodium phosphate and potassium phosphate, for example, result from catalyzing the polyoxyethylation with sodium hydroxide or potas-sium hydroxide, then neutralizing with phosphoric acid. These phosphate salts are not soluble in nonionic surfactants. Present at levels of about 0.05% by weight, they can impart some haze to the product and with time may settle. This settling has little if any effect on product performance. As noted earlier, haze can also result from high levels of PEG.

Another appearance issue is noted with polyoxyethylene alkylphenols and other classes of nonionic surfactants. Recall that these products are polydis-perse. From Table 4 we can see, for example, that the polyoxyethylene (9) nonylphenol contains 5% of the polyoxyethylene (15) and higher oligomers. At temperatures below ambient but above the pour point some of these higher oligomers can solidify, then settle. The result can be a hazy layer on the bottom of the container. Because there may be some segregation of product in a drum, it is prudent to assure that the contents are well mixed if the drum has been left to stand for more than a few days in colder temperatures. Should a drum freeze, either partially or completely, the entire contents of the drum should be melted, then mixed well prior to use. Failure to mix the entire contents of the drum might lead to the use of a nonrepresentative sample with uncertain results.

Heated circulated storage tanks are recommended for bulk storage of poly-oxyethylene alkylphenols. As these products are subject to color development caused by oxidation, temperatures should not exceed 50°C and the tank head space should be inerted with nitrogen. Concentrated aqueous solutions of polyoxyethylene alkylphenols are corrosive; products sold in this form must be stored in lined steel, glass fiber–reinforced plastic, or stainless steel tanks.

The viscosity of most polyoxyethylene alkylphenols is about 200–300 mPa·s at room temperature. Higher ethoxylates at temperatures close to their pour points may be more viscous. Lower ethoxylates may also be higher in viscosity. This is a reflection of the high viscosity of nonylphenol or the physical state of octylphenol. The viscosity of these products decreases with increasing temperature (see Tables 10 and 11).

The density or specific gravity increases with increasing degree of polyoxyethylation. The rate of change is greatest for lower ethoxylates. These measurements can be used to follow the polyoxyethylation reaction.

B. Solution Properties

Polyoxyethylene alkylphenols exhibit a number of interesting properties related to their solution phase behavior. Most obvious among these properties to users of these products are the cloud point and aqueous solution viscosity characteristics.

The cloud point is a very useful and important property of polyoxyethylene alkylphenols. The cloud point is the temperature above which an aqueous solution of water-soluble nonionic surfactant becomes turbid. It is widely agreed [29–38] that the micellar molecular weight of a polyoxyethylated nonionic surfactant increases with temperature due to reduced surfactant solubility and increased hydrophobicity caused by dehydration of the polyoxyethylene chain. This increase in micellar molecular weight becomes very pronounced as the solution temperature approaches the cloud point. As temperature increases above the cloud point, the solution may separate into two phases—one surfactant-rich, the other water-rich.

Polyoxyethylene alkylphenols with 8–15 moles of polyoxyethylation exhibit this cloud point phenomena (see Tables 12 and 13). The cloud point depends

TABLE 12 Cloud Points of Polyoxyethylene Nonylphenols [111]

Tergitol	Mole ratio	Cloud point (°C)
NP-4	4	Insoluble
NP-6	6	Insoluble
NP-7	7	20
NP-8	8	43
NP-9	9	54
NP-10	10.5	63
NP-13	13	83
NP-40	40	>100

TABLE 13 Cloud Points of Polyoxyethylene Octylphenols [112]

Triton	Mole ratio	Cloud point (°C)
X-15	1	Insoluble
X-35	3	Insoluble
X-45	5	Insoluble
X-114	7.5	22
X-100	9.5	65
X-102	12	88
X-165	16	>100
X-305	30	>100
X-405	40	>100

strongly on the degree of polyoxyethylation and is an excellent method of assuring product quality during manufacture. Indeed, cloud point is the principal specification against which water-soluble polyoxyethylene alkylphenols are sold.

The cloud point is somewhat dependent on the concentration of the test solution. Sadaghiania and Khan [39] show that the cloud point of Triton X-100 drops from 69.5°C at a very low surfactant concentration to 66°C at about 0.2% by weight. The cloud point is then constant at 66°C through a concentration of 2% by weight. Above 2% by weight, the cloud point of Triton X-100 gradually rises to 90°C at a concentration of 60% by weight. For use as a production specification, it is important to choose a test concentration whereby cloud point is not a strong function of concentration. Polyoxyethylene alkylphenols with more than about 15 moles of polyoxyethylation are too water-soluble to exhibit cloud points; their solutions are clear to the boiling point. They are usually specified by molecular weight as calculated from hydroxyl number. Polyoxyethylene alkylphenols with fewer than 8 moles of polyoxyethylation are not readily water-soluble. Preferably, they can be specified by molecular weight. Another common but less reliable method is to measure their cloud point in a polar solvent such as an alcohol or a glycol ether. If specified by cloud point, the product is dissolved in a fixed volume of an alcohol such as isopropanol. Water is then titrated at fixed temperature into the alcoholic surfactant solution until the alcohol is diluted to the point that it no longer solubilizes the surfactant. The volume of water required to bring on insolubility is then reported as the cloud point [40].

The aqueous cloud point of polyoxyethylene alkylphenol solutions will be reduced in the presence of many salts. Figure 1 shows the effects of electrolytes on the cloud point of Tergitol NP-9. Sodium chloride is sometimes used to make the cloud points of higher ethoxylates more distinct. One must be careful to properly prepare the cloud point solution since the salt concentration can strongly affect the cloud point. Schott et al. [41] show that acids raise the cloud

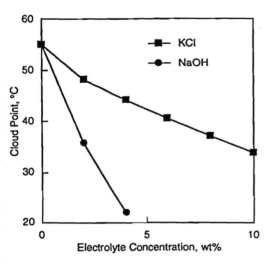

FIG. 1 Effect of electrolyte on the cloud point of 1 wt% Tergitol NP-9.

point of Triton X-100. Han et al. [42] show that many urea derivatives also increase the cloud point of this surfactant.

The aqueous solution viscosity of concentrated polyoxyethylene alkylphenol solutions can be very interesting. As surfactant concentration increases, the aqueous viscosity also increases. At sufficiently high concentrations the mixture may gel due to the formation of a complex between the surfactant and water. At even higher surfactant concentrations the mixture is once again a viscous liquid. Table 14 and 15 illustrate this behavior. Figure 2 is a phase diagram for Triton X-114 [43] showing a liquid crystalline region from about 50% by weight to

TABLE 14 Aqueous Solution Viscosities (mPa·s) of Polyoxyethylene Nonylphenol at 25°C [113]

Mole ratio	Active concentration by weight in water								
	10%	20%	30%	40%	50%	60%	70%	80%	90%
7	135	137	204	240	3047	38800	95600	338000	330
8	29	224	615	966	1300	9020	43000	364	305
9	9	76	527	1315	1690	4560	14300	26500	289
10	3	26	312	Gel	Gel	2055	19600	602	307
13	2	6	53	Gel	Gel	Gel	1035	467	303
15	2	8	35	439	Gel	Gel	Gel	450	315
40	4	28	182	—	—	1775	1071	585	Solid
70	10	53	223	718	1172	1790	1258	Solid	Solid

TABLE 15 Aqueous Solution Viscosities (mPa·s) of Polyoxyethylene Octylphenol [114]

Triton	Mole ratio	Temp. (°C)	Active concentration by weight in water				
			10%	30%	50%	70%	90%
X-114	7.5	25	100	280	500	27000	320
X-100	9.5	25	2	80	Gel	530	280
X-102	12	25	3	20	Gel	Gel	310
X-100	9.5	50	3	40	110	100	80
X-102	12	50	3	9	Gel	110	80

about 80% by weight at 25°C. From Table 15 it can be seen that the viscosity of Triton X-114 is very high in this region. LaMesa et al. [44] studied the phase behavior of polyoxyethylene (10) nonylphenol and found, at room temperature, an apparent hexagonal mesophase from 37% to 55% by weight. They also found an apparent lamellar mesophase between 63% and 72% by weight. From Table 14, the hexagonal mesophase coincides with gelation of the surfactant solution

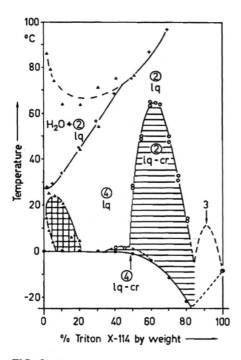

FIG. 2 Phase diagram of Triton X-114.

TABLE 16 Solubility of Polyoxyethylene Nonylphenols in Various Solvents [115]

Tergitol	Mole ratio	Butyl Carbitol		Butyl Cellosolve		Ethanol		Methylene chloride		Mineral spirits		Propylene glycol		Toluene	
		10%	50%	10%	50%	10%	50%	10%	50%	10%	50%	10%	50%	10%	50%
NP-4	4	M	M	M	M	M	M	M	M	M	M	M	M	M	M
NP-6	6	M	M	M	M	M	M	M	M	TPL	TPL	M	M	M	M
NP-7	7	M	M	M	M	M	M	M	M	TPL	TPL	M	M	M	M
NP-8	8	M	M	M	SH	M	M	M	M	TPL	TPL	M	M	M	M
NP-9	9	M	SH	M	M	M	M	M	M	TPL	TPL	M	M	M	M
NP-10	10	M	M	M	M	M	M	M	M	TPL	TPL	M	M	M	M
NP-13	13	M	M	M	M	M	M	M	M	TPL	TPL	M	M	M	M
NP-40	40	TPL	S	SH	S	M	M	M	HZ	TPSL	S	SH	TPSL	M	TPSL

All concentrations % v/v, all measurements at 25°C.

HZ, hazy; M, miscible; S, solid; SH, slightly hazy, settling; TPL, two-phase liquid; TPSL, two-phase liquid/solid.

TABLE 17 Solubility of Polyoxyethylene Octylphenols in Various Solvents [116]

Triton	Mole ratio	Polar organic solvents (alcohols, glycols, ethers, ketones, etc.)	Aromatic hydrocarbons	Aliphatic hydrocarbons
X-15	1	Miscible	Miscible	Soluble
X-35	3	Miscible	Miscible	Soluble[a]
X-45	5	Miscible	Miscible	Soluble[b]
X-114	7.5	Miscible	Miscible	Insoluble
X-100	9.5	Miscible	Miscible	Insoluble
X-102	12	Miscible	Insoluble	Insoluble
X-165	16	Miscible	Insoluble	Insoluble
X-305	30	Miscible	Insoluble	Insoluble
X-405	40	Miscible	Insoluble	Insoluble

[a] Hazy at 1% in isooctane.
[b] Can be solubilized by the addition of a suitable coupling agent such as ethylene glycol, butyl alcohol, hexylene glycol, or oleic acid.

whereas the lamellar mesophase coincides with a region of high viscosity. The area between these two mesophases is described as a solution phase, the viscosity of which is much lower. A similar result was measured by Beyer in the Triton X-100/D_2O system. The hexagonal phase exists below about 30°C whereas the lamellar phase exists only below about 5°C [45].

On a practical level, this phase behavior can affect surfactant dissolution rate. As a droplet of surfactant begins to hydrate and dissolve, a layer of water and surfactant forms around the droplet. As the local concentration enters this gel range, diffusion of surfactant and water is greatly inhibited, thus slowing dissolution. Gelation can be minimized by raising the temperature of dissolution: note from Table 15 that Triton X-100 does not appear to gel at 50°C. This gelation phenomenon is not unique to polyoxyethylene alkylphenols; it is also seen with alcohol ethoxylates.

Polyoxyethylene alkylphenols are generally soluble in polar organic solvents and aromatic hydrocarbons. They are generally not soluble in aliphatic hydrocarbons. Exceptions are polyoxyethylene alkylphenols with fewer than 5 moles of ethoxylation that are soluble up to several percent in aliphatic hydrocarbons. Tables 16 and 17 summarize the solubility of various polyoxyethylene alkylphenols in a number of different solvents.

C. Surface Properties

Surface tension reduction and micelle formation are among the key features of all surfactants including polyoxyethylene alkylphenols. Polyoxyethylene alkylphenols reduce surface tension very effectively as shown by typical values for equilibrium surface tensions of 0.1% solutions in Tables 18 and 19. For poly-

TABLE 18 Surface Tension of Polyoxyethylene Nonylphenol Solutions [117]

Tergitol	Mole ratio	Surface tension (mN/m), 0.1% active at 25°C
NP-7	7	30
NP-8	8	30
NP-9	9	30
NP-10	10.5	31
NP-13	13	34
NP-40	40	45

oxyethylene nonylphenols and octylphenols, surface tension is at its lowest when the product is barely soluble. Surface tension rises steadily as the degree of polyoxyethylation increases.

In recent years, researchers such as Hua and Rosen [46,47] have described the maximum bubble pressure technique for measuring dynamic surface tension. The dynamic surface tension of the polyoxyethylene alkylphenol solutions in Table 20 are not as low as the equilibrium values in Tables 18 and 19. Unlike equilibrium surface tension, where the surface tension is measured after an equilibrium period measured in minutes, dynamic surface tension measurement allows surface tension determination at a surface age of a fraction of a second. Joos et al. [48] developed a model that shows that results such as these can be explained by a surfactant diffusion-controlled mechanism. It appears that equilibrium surface tensions may not be adequate predictors of surfactant performance in cases where rapid wetting is required. For example, Green and Green [49] report that dynamic surface tension as measured by the maximum bubble pressure technique correlated well with the effectiveness of herbicide solutions applied to weeds. Performance did not correlate with equilibrium surface tension.

TABLE 19 Surface Tension of Polyoxyethylene Octylphenol Solutions [118]

Triton	Mole ratio	Surface tension (mN/m), 0.1% active at 25°C
X-45	5	28
X-114	7.5	29
X-100	9.5	30
X-405	12	32
X-165 70% active[a]	16	35
X-305 70% active[a]	30	37
X-405 70% active[a]	40	37
X-705 70% active[a]	70	39

[a] Measured at 0.1% active concentration, corrected for water content in product as sold.

TABLE 20 Dynamic Surface Tension of Polyoxyethylene Alkylphenols [119]

Bubble frequency	Dynamic surface tension (mN/m), 0.1%, 25°C	
	Triton N-101[a]	Triton X-100[b]
Equilibrium	30	30
2 Hz	36	35
8 Hz	39	37

[a] Polyoxyethylene (9.5) nonylphenol.
[b] Polyoxyethylene (9.5) octylphenol.

TABLE 21 Surface Properties of Polyoxyethylene Nonylphenols at 25°C

Mole ratio	Critical micelle concentration [120], (μmol)	Area per molecule [121], (nm^2)	Aggregation number [122]
5	36	4.8	—
6	58	5.5	—
9	70	6.0	—
10	—	—	100
11	82	6.7	—
15	93	7.4	52
30	169	—	19

The micellar behavior of polyoxyethylene alkylphenols is summarized in Tables 21 and 22. Critical micelle concentration (CMC) increases with the degree of polyoxyethylation. Polyoxyethylene nonylphenols have lower CMCs than the polyoxyethylene octylphenols with the same degree of polyoxyethylation. That the difference in alkyl group size can lead to such a large difference in CMC has been recognized for some time [50]. From the Gibbs equation, the area occupied by the molecule at the air–water interface can be calculated. The aggregation number of polyoxyethylene nonylphenol micelles drops with

TABLE 22 Surface Properties of Polyoxyethylene Octylphenols at 25°C

Mole ratio	Critical micelle concentration [123], (μmol)	Area per molecule [124], (nm^2)
7	184	5.8
8	247	6.4
9	290	6.6
10	320	8.0
16	430	—
40	810	—

TABLE 23 Ross–Miles Foam Heights of Polyoxyethylene Nonylphenols [125]

	Foam height (mm)	
Mole ratio	Initial	5 min
6	15	10
9.5	80	60
10.5	110	80
15	130	110
20	120	110
30	120	105
40	115	105
50	100	85
100	75	50
Sodium lauryl sulfate [127]	160	140

0.1% Solutions, 25°C.

TABLE 24 Ross–Miles Foam Heights of Polyoxyethylene Octylphenols [126]

	Foam height (mm)	
Mole ratio	Initial	5 min
4.5	20	10
7.5	60	40
9.5	110	100
12	120	100
Sodium lauryl sulfate [127]	160	140

0.1% Solutions, 25°C.

increasing degree of polyoxyethylation. Many researchers have studied the aggregation number of Triton X-100, polyoxyethylene (9.5) octylphenol: in water at or near 25°C values range from 100 (measured by Dwiggins et al. [51] by ultracentrifugation) to 111 (measured by Mankowich [52] by light scattering) to 140 (reported by Kushner and Hubbard [53], again by light scattering).

TABLE 25 Draves Wetting Times for Polyoxyethylene Nonylphenols [128] at 25°C

		Wetting time (s)		
Tergitol	Mole ratio	0.05%	0.10%	0.20%
NP-8	8	27	8	3
NP-9	9	21	8	3
NP-10	10.5	23	9	2
NP-13	13	26	16	8

TABLE 26 Draves Wetting Times for Polyoxyethylene Octylphenols [129] at 25°C

Triton	Mole ratio	Active Ingredient (wt%) for wetting time of		
		10 s	25 s	50 s
X-35	3	0.36	0.18	0.11
X-45	4.5	0.10	0.055	0.034
X-114	7.5	0.096	0.050	0.031
X-100	9.5	0.092	0.048	0.028
X-102	12	0.123	0.064	0.045
X-165	16	0.780	0.330	0.170

The foaming characteristics of polyoxyethylene alkylphenols are given in Tables 23 and 24. Initial Ross–Miles foam heights of polyoxyethylene nonylphenols pass through a maximum at a mole ratio of about 10, at all mole ratios foam height drops slightly in 5 min. Polyoxyethylene (6) nonylphenol foams very little as it is barely soluble in water at 25°C. By comparison, sodium lauryl sulfate, an anionic, has significantly higher initial and 5-min foam heights. Similar trends are noted for polyoxyethylene octylphenols.

The wetting performance of polyoxyethylene alkylphenols is given in Tables 25 and 26. In Table 25, the times required to achieved wetting in the Draves test are listed for several products at three concentrations. Wetting occurs most rapidly with products that are just water-soluble; performance drops rapidly with increasing degree of polyoxyethylation. Table 26 reports the concentration of surfactant required to achieve a given wetting time; lower wetting concentrations are considered to be better. Again, optimum performance occurs with products that are just soluble in water under test conditions.

V. APPLICATIONS

The uses of polyoxyethylene alkylphenols cover a wide a range of industries from cleaning systems to paints and coatings to agricultural chemicals. A summary of product uses is given in Table 27. Specific examples of interesting uses are described below.

A. Detergents

Cleaning systems can include many things, from cleaning of clothing to soil remediation. Dry cleaning detergents can use polyoxyethylene alkylphenols, often in combination with other surfactants [54]. Liu et al. [55] found polyoxyethylene nonylphenols and octylphenols with mole ratios of oxyethylene of 9–12 to the most effective surfactants in their study of the removal of polycyclic aromatic hydrocarbons from soil. Osberghaus and Kresse describe powdered dry cleaning systems for carpets using polyoxyethylene alkylphenols [56].

TABLE 27 Applications of Polyoxyethylene Alkylphenols [130]

Agriculture	*Paint*
Emulsifiers and wetters	Emulsion polymerization of latexes
Herbicidal adjuvants	Latex stabilization
Fruit washes	Pigment wetting and dispersion
Cleaners	*Paper*
Household detergents	Pulping
Dry cleaning detergents	Absorbent papers
Industrial hand cleaners	De-inking wastewater
Detergents/sanitizers	Wet-felt washing
Solvent degreasers	Adhesives
Metal cleaners	
Hard surface cleaners	*Textiles*
Commercial laundry detergents	Greige goods scouring
	Bleaching
Dust wetting	Carbonizing of wool
Coal mines	Scouring agent
Ceramic industries	Desizing agent
Foundries	Wetting and rewetting agents
	Polyethylene softener emulsifiers
Leather	Mineral oil and solvent emulsifiers
Hide soaking	Fiber lubricant emulsifiers
Degreasing	Dye-leveling agents
Fat liquor stabilization	Resin bath additives
Tanning and dyeing	

Aqueous laundry detergents can also use polyoxyethylene alkylphenols. Schwarts [57] describes the use of a mixture of polyoxyethylene dodecylphenol and nonylphenol in a detergent for the machine washing of delicate cotton articles. Merrill and Wood [58] describe the use of polyoxyethylene alkylphenols in combination with an anionic surfactant for use in low-foaming powdered laundry detergents. Harrison and Weller [59] discuss the use of polyoxyethylene alkylphenols in liquid laundry detergents containing anionic and cationic surfactants. These cleaners not only are good detergents but also have very effective microbiocidal properties due to the inclusion of the cationic surfactant.

Polyoxyethylene alkylphenols can also clean hard surfaces. Scardera et al. [60] invented an emulsion cleaner to clean surfaces contaminated by chemical warfare agents. Polyoxyethylene (6) alkylphenol is suggested to serve as an emulsifier for solvents that are used in this emulsion cleaner. Fuggini and Streit [61] invented a hard surface cleaner that effectively removes dirt and grease, leaving a protective barrier to repel soil. The preferred composition includes 2% of polyoxyethylene (9) nonylphenol. Magyar [62] includes polyoxyethylene nonylphenol and octylphenol among the possible ingredients in an automotive cleaner well suited to cleaning whitewall tires, vinyl tops, and fabric interiors. Another automotive cleaner is described by Smith [63]; this one can contain polyoxyethylene nonylphenol with 9–10 moles of polyoxyethylation and is safe for use on clear-coated alloy wheels and painted surfaces. Polyoxyethylene alkylphenols are also useful in industrial metal cleaners. Sturwold [64] describes their use in cleaners to remove grease, oil, dirt, scale, and metal fines from metals.

It has generally been the experience of the authors that polyoxyethylene nonylphenols are very effective for the cleaning of textiles. On the other hand, polyoxyethylene octylphenols are more widely used for cleaning hard surfaces.

B. Metalworking

Metalworking fluids cool and lubricate in cutting, grinding, and metal forming operations. Modern fluids include oils containing surfactants that act as emulsifiers when added to water by the end-user. Although the predominant emulsifiers are sodium petroleum sulfonates, polyoxyethylene nonylphenols are also used in hard water systems or to provide special emulsion characteristics [65].

C. Paints and Coatings

Polyoxyethylene alkylphenols are also found in paints and coatings. They are frequently used as stabilizers for latex particles; they are almost always used in conjunction with anionic surfactants. Polyoxyethylene alkylphenols used for stabilization are the higher mole ratio products, such as polyoxyethylene (15) through polyoxyethylene (70), since their physical size contributes greatly to steric stabilization. Polyoxyethylene alkylphenols are also used as wetting agents in finished paints. Typical products are polyoxyethylene (7) through polyoxyethylene (10). It is generally accepted that polyoxyethylene alkylphenols are much more effective in this application than polyoxyethylene alcohols. In the experience of the authors, polyoxyethylene octylphenols are more generally used than polyoxyethylene nonylphenols in the manufacture of latexes used in paints and coatings, whereas polyoxyethylene nonylphenols are more commonly used in the formulation of finished paints and coatings.

D. Agriculture

Surfactants can serve several purposes in agricultural formulations. They may serve as emulsifiers or dispersants when a concentrate of water-insoluble active materials is diluted for use. The wetting ability of surfactants is used to enhance the leaf wetting characteristics of spray adjuvants. In a discussion of surfactant use in herbicide dispersions, Sonntag [66] shows that polyoxyethylene nonylphenols and octylphenols are very effective at reducing the surface tension and contact angle of herbicide solutions on plant leaves.

VI. ENVIRONMENTAL ISSUES

A comprehensive review of the environmental and toxicologic aspects of alkylphenols and polyoxyethylene alkylphenols is given in Sylvia Talmage's *Environmental and Human Safety of Major Surfactants* [67]. An extensive text on biodegradation of surfactants was written by Swisher [68].

A. Biodegradation

A vast number of laboratory-based biodegradation studies have been performed on polyoxyethylene alkylphenols with the degree of biodegradation ranging from zero [69] to 100% [70], depending on the test methods used and the duration of the experiments. A few of the known biodegradation test results for alkylphenols and the polyoxyethylene alkylphenols are summarized in Table 28.

It is generally accepted that the rates of biodegradation of polyoxyethylene nonylphenols and polyoxyethylene octylphenols are slower than those of nonionic surfactants produced with linear hydrophobes, such as the polyoxyethylene alcohols [71]. However, studies of environmental levels of alkylphenols as well as the degree of their removal from wastewater suggest that the biodegradation rates are sufficient enough to prevent accumulation in the aquatic environment [72]. Extensive research is underway internationally, within government (U.S. Environmental Protection Agency and the European Union), academia, and industry [through the Chemical Manufacturer's Association (CMA)] with the aim of understanding the potential risk of alkylphenols and polyoxyethylene alkylphenols to the environment [73].

Biodegradation of the polyoxyethylene alkylphenols can occur via either an aerobic or an anaerobic pathway. The aerobic pathway generates lower mole-weight ethoxylates, carboxylic acid derivatives of both the ethoxylate and hydrophobe chain [74,75], unspecified intermediate biodegradation products, and ultimate biodegradation products (water and carbon dioxide). There is no evidence that alkylphenol is generated during the aerobic biodegradation process. The anaerobic pathway generates lower mole-weight ethoxylates, alkylphenol, unspecified intermediate metabolites, and ultimately, carbon dioxide and methane. It is presumed that anaerobic digestion of the carboxylic acid intermediates may also produce alkylphenol in the route to ultimate biodegradation. The major source of alkylphenol in the environment is through anaerobic digestion of the parent surfactant and the subsequent biodegradation intermediates.

B. Risk Assessment

In its simplest form, risk assessment is carried out by comparing the levels of a compound in the environment (i.e., exposure levels) with measured toxicity levels (i.e., "effect" levels). The definition of "risk" is somewhat contentious,

TABLE 28 Summary of Select Biodegradation Studies

Compound	Test	Endpoint (analysis)	%Biodegradation	Ref.
NPE-9	28-day closed bottle	28-day BOD as % ThOD	45(28 day only)	131
	28-day Screening (OECD)	% DOC removal (7/14/21/28)*	61/66/70/72	131
NPE-13	28-day Screening (OECD)	% DOC removal (7/14/21/28)	66/56/70/72)	131
OPE-10	20-day BOD	BOD ad % ThOD 5/10/15/20	30/40/49/51	131
	28-day closed bottle (OECD) 5/15/28	% BOD 11/34/36	11/34/36	131
	28-day screening,	% DOC removal 7/14/21/28	58/62/80/90	131
OPE-8	Semi continuous activated sludge	%DOC removal	40% in 1 day	132
NPE-4	—	—	58% in 1 day	132
NPE-9	—	—	55% in day	132
APE-8	Continuous-flow activated sludge	UV spectroscopy	>76%	133
OPE-9	River water dieaway	Cobalt thiocyanate	68% in 11 days	134
NPE-9	—	—	91% in 16 days	134
NP	Sludge treated soil	HPLC/UV	80% in 3 weeks	135
NPE-1	—	—	—	135
NPE-2	—	—	—	135
NPE-9	Continuous-flow activated sludge	Cobalt thiocyanate	98–100% in 8 hours	136
	—	Foam	>95% in 8 hours	136
	—	Radiolabel^{3}H$_2$O production (EO chain label)	10–42% in 8 hours	136
	—	Radiolabel^{14}CO$_2$ production (hydrophobe label)	47–59% in 8 hours	136

*7/14/21/28 represent the days on which the system was sampled.
NPE-n, polyoxyethylene (n) nonylphenol; OPE-n, polyoxyethylene (n) octylphenol; APE-n, polyoxythylene (n) alkylphenol; NP, nonylphenol.

but if exposure levels are sufficiently lower than the effect levels, then the compound is generally not considered to pose an unacceptable environmental risk. Factors in an assessment of risk include use patterns, route of entry into the environment, fate of the material and its metabolites, biodegradation pathways (and rates), actual levels in the environment, routes of exposure to the organism, potential exposure levels, potential for uptake (and elimination) by the organism, and the number of species tested for "effects."

For risk assessment, the following quantities are commonly used:

NOEC: No Observable Effects Concentration, the level at which no observable effect is noted for a given species (for a specific endpoint, such as growth or survival)

LOEC: The Lowest Observable Effects Concentration, the lowest concentration at which effects are observed

MATC: Maximum Allowable Toxicant Concentration, the geometric mean of the NOEC and LOEC

Generally, a comparison of the MATC (used predominantly in the United States) or the NOEC (used predominantly in Europe) with measured or predicted environmental levels gives an estimate of the risk. If the environmental levels are sufficiently lower than the MATC or NOEC, then the risk to the environment is considered low [76].

C. Aquatic Exposure/Toxicity

1. Environmental Levels of Alkylphenols and Polyoxyethylene Alkylphenols

Table 29 summarizes levels of alkylphenols and polyoxyethylene alkylphenols in various aquatic environmental compartments. The lyophilicity of the alkylphenols and their corresponding 1- and 2-mole ethoxylates cause the concentrations of these species to be two to four orders of magnitude higher in the sediment than in the water column. Actual levels measured in the sediment depend on the organic carbon content and particle size (i.e., a higher surface area allows for a higher concentration of alkylphenol adsorption per kilogram of soil).

Sediment levels of nonylphenol in the United States have ranged from <2.9µg/kg to 2900 µg/kg (average = 162, with 28% of the samples having less than the minimum detection limit, or MDL, of 2.9µg/kg) [77]. The corresponding water column concentrations of nonylphenol ranged from <0.11 to 0.64 (average = 0.12 with 70% of the samples having less than the MDL of 0.11µg/L).

In Switzerland (Glatt River), sediment concentrations of nonylphenol have ranged from 190 to 13,100 µg/kg, with corresponding water column concentrations ranging from <0.3 to 7.9 µg/L (60% of the samples had <2 µg/L) [78]. In Venice, Italy, the values in the sediments ranged from 5 to 42 µg/kg (with corresponding water column levels of 0.64–4.30 µg/L) [79]. It is important to note that the highest sediment and water column values observed (Grand Calumet River, United States, and Glatt River, Switzerland) represent highly polluted areas with little or no history of secondary treatment; rivers receiving treated discharges showed substantially lower levels of nonylphenol in both the sediment and water column.

Bottom-feeding species are potentially exposed to nonylphenol in the sediment interstitial pore water. Nonylphenol concentrations in the interstitial pore water are estimated from sediment concentrations using the following relationship [80]:

TABLE 29 Surface Water Exposure Concentrations of NPEs

	Compound	Levels(μg/L)	Ref.
Italy, several streams	NPE	0.64–4.30	137
UK estuaries	NP	<0.08 ($n=9$)	138
		3.1 ($n=1$)	138
UK rivers	NP	<MDL ($n=11$)	138
		0.2–1.6 ($n=12$)	138
		9.0 ($n=1$)	138
UK River Aire	NP	<MDL ($n=1$)	138
		24–53 ($n=5$)	138
US, 30 rivers	NP	<0.11–0.64μg/L ($N=98$)	139
		(70% of samples <MDL of 0.11)	139
	NPE-1	<0.06–0.60	139
	NPE-2	<0.07–1.20	139
	NPE-3-NPE-17	<1.6–14.9	139
US, subset of 30 rivers	NPE-1-C	<0.07–2.02μg/L($N=8$)	140
	NPE-2-C	<0.07–5.22 μg/L ($N=8$)	140
	NPE-3-C	<0.07μg/L($N=8$)	140
	NPE-4-C	<0.07μg/L($N=8$)	140
US, New Jersey drinking water sample ($n=1$)	NP	Not reported	141
	NPE-1, NPE-2	0.111–0.113μg/L	141
	NPE-3-NPE-7	0.062–0.129μg/L	141
	NPE-2-C	0.164μg/L	141
	NPE-3-C-NPE-7-C	0.004–0.016	141
Switzerland, Glatt River	NP	<0.3–7.9 (<$N=10$)	142,143,144, 145,146,147
		60% of samples <2μg/L	"
	NPE-1	<0.3–20($n=110$)	"
	NPE-2	<0.3–21 ($N=110$)	"
	NPE-1-C	<1–29 ($N=48$)	"
	NPE-2-C	2.0–59 ($N=48$)	"

NP, nonylphenol; NPE-n, polyoxyethylene (n) nonylphenol; NPE-n-C, polyoxyethylene (n) nonylphenol carboxylate.

$$C_w = C_{sed}/(K_{oc} \cdot F_{oc}) \qquad (11)$$

where C_w = interstitial aqueous concentration of the analyte
   C_{sed} = sediment concentration of the analyte
   F_{oc} = organic fraction of the sediment
   K_{oc} = partition coefficient of nonylphenol between the sediment and water = 3825

The calculated levels of alkylphenol in interstitial pore waters are roughly equivalent to the levels measured in the river water above the sediment. For example, the calculated interstitial water concentration of nonylphenol was 1.2 µg/L using the sediment concentration of 2900 µg/L (Grand Calumet River, United States). The river water concentration of nonylphenol for the Grand Calumet was measured at 0.64 µg/L.

2. Aquatic Toxicity of Alkylphenols and Polyoxyethylene Alkylphenols

The aquatic toxicity of the polyoxyethylene alkylphenols and their breakdown products are inversely proportional to the hydrophobicity of the product. Generally, alkylphenols are at least an order of magnitude more toxic to aquatic species than the 10-mole (and higher) ethoxylates of the alkylphenols. There is a relatively large database of toxicity data for the alkylphenols and, in particular, for nonylphenol. A summary of a few key acute and chronic aquatic toxicity levels is given in Tables 30 and 31.

The lowest MATC for nonylphenol is approximately 5 µg/L (for the Mysid Shrimp; Table 31). In the United States, the highest level of nonylphenol (including calculated sediment pore water concentrations) was 1.2 µg/L. These data came from the "Thirty Rivers Study," which was designed so that the measured environmental levels would adequately reflect the river concentrations of nonylphenol (and the polyoxyethylene alkylphenols) in the entire United States [81].

Because the highest level of nonylphenol in U.S. water is significantly lower than the lowest observed MATC, the risk of nonylphenol to the U.S. environment is considered low.

For the United Kingdom, the highest levels of nonylphenol were 23 to 53 µg/L (River Aire). For all of the other rivers sampled, the levels of nonylphenol were from 0.02 to 1.6 µg/L. The River Aire, like the Grand Calumet River (United States), is a highly polluted area receiving domestic and industrial

TABLE 30 Acute Effects (Aquatic)

Surfactant	No. species	Test	Level, (µg/L)
Nonylphenol:			
Fish	9	96-h LC_{50}	130–3000
Invertebrates	9	24- to 96-h LC_{-50}	43–3000
Algae	4	Various	25–7500
Polyoxyethylene nonylphenol:			
Fish	10	24- to 96-h LC_{-50}	1300–18,000
Invertebrates	17	24- to 96-h LC_{-50}	100–50,000
Algae	13	Various	210–100,000

Compiled from Ref. 148.

TABLE 31 Chronic Effects (Aquatic)

Organism	Endpoint	Compound	NOEC, (μg/L)	LOEC, (μg/L)
Fish:				
Fathead minnow	Growth, 28-day	NP	23	
	Survival, 33-day	NP	7.4	14
Rainbow trout	Growth/survival (91-day)	NP	6.3	10.3
Invertebrates:				
Daphnia magna	21-day assay	NP	24	
Mysid shrimp	Growth, 28-day	NP	3.9	6.7
	Survival, 29-day	NP	6.7	9.1
Ceriodaphnia dubia	Reprod., 7-day	NP	89	202
	Survival, 7-day	NP	202	377
Ceriodaphnia dubia	Reprod., 7-day	NP1.5	285	886
	Survival, 7-day	NP1.5	285	886
Mytilus edulis	35-day fertilization and reproductive effects	NP	No effects at >200μg/L	
Chironomus tentans	14-day sublethal effects:			
	interstitial water exposure	NP	39	
	dosed water exposure	NP	76	
	interstitial water	NP	143	
	dosed sediment	NP	20 mg/kg	
Plants:				
Selanastrum sp. Green algae		NP	694	1480
Lemna minor Duckweed		NP	901	2080

Data from Refs. 149–151.

waste with little or no secondary treatment [82]. In Switzerland the highest levels of nonylphenol were 7.9 μg/L [83] (60% of the samples had <2 μg/L nonylphenol).

D. Mammalian Toxicity

In general, alkylphenols are of a low order of acute toxicity (oral and cutaneous LD_{50} values >2 g/kg body weight) and subchronic toxicity (no specific target organ toxicity has been shown at high doses) [84]. The toxicity of the polyoxyethylene alkylphenols decreases as the degree of polyoxyethylation increases. The 1- and 2-mole ethoxylates show a similar toxicity to the starting alkylphenol. Although alkylphenols and the lower mole polyoxyethylene alkylphenols are generally considered only minimally toxic to mammals, they can cause severe skin and eye irritation with possible permanent damage to the cornea.

Alkylphenols and the low-mole polyoxyethylene alkylphenols have been shown to exhibit very weak estrogenic activity in screening assays [85,86]. At the time of writing, there is extensive research and evaluation of a large class of environmental estrogens, including those derived from natural sources, such as soybeans [87]. Presently, there is no consensus in the scientific community on the relevance of chemicals with estrogenic activity in the environment, particularly those such as nonylphenol which show a very low estrogenic response in in vivo screening assays.

A rigorous assessment of exposure of alkylphenols and polyoxyethylene alkylphenols to mammals has not been performed. Exposure through drinking water that had measured levels of 0.2 μg/L nonylphenol in one study is not considered a significant source of alkylphenol from a toxicity standpoint [88]. Polyoxyethylene alkylphenols are commonly applied directly to agriculture; however, the uptake of alkylphenol into common crop plants was not detected [89]. Thus, the ingestion of alkylphenols/polyoxyethylene alkylphenols through agriculture is minimal. Polyoxyethylene alkylphenols are also used in the manufacture of plastic food containers; however, the potential for absorption into food has not been quantified. Finally, it has been postulated that mammalian exposure may occur through the consumption of fish, which may concentrate alkylphenol from the water. Table 32 summarizes the bioconcentration factors of nonylphenol for fish, invertebrates, and algae. Actual levels of alkylphenols and polyoxyethylene alkylphenols ingested via fish consumption have not been determined. However, a single mallard duck that was captured and sacrificed contained levels of nonylphenol similar to the fish in the nearby river [90].

TABLE 32 Bioconcentration Factors

Organism	BCF	Ref.
Invertebrates:		
Shrimp	90–110	152
Mussels	1.4–13	153
	3430	154
General invertebrates, predicted from $\log K_{ow} = 4.45$ and 1.75% lipid	530	155
Fish:		
Stickleback	1250	156
Fathead minnow	270, 350	157
Salmon	10–280	158
General fish, predicted from $\log K_{ow}$ and 5% lipid	1510	159
Microalga	200–10,000	160
Aquatic plants	200–500 est	—

BCF = (concentration of test chemical in an organism)/(freely dissolved water concentration).

E. Other Environmental Issues

It has been demonstrated that activated sludge treatment techniques remove 90–97% of polyoxyethylene alkylphenols from water [91]. However, the efficiency of alkylphenol and polyoxyethylene alkylphenol removal depends on the treatment conditions, and a wide range of treatment capabilities have been reported (Table 33) [92]. Trickling filter waste treatment facilities do not effectively remove alkylphenols and polyoxyethylene alkylphenols, presumably because the contact time between the wastewater and active bacteria is relatively low [93].

Polyoxyethylene alkylphenols are commonly applied directly to the environment as agricultural adjuvants. Sludge from water treatment plants (using the activated sludge process), with measured levels of nonylphenol ranging from 0.01 to 1 g/kg (dry sludge) is often applied directly to crops as a fertilizer [94–96]. Data on the fate of this nonylphenol are very limited, but one study has shown that it is reduced to 90% of its original value within 20 days [97]. Marcomini found that for nonylphenol in sludge applied to land, >80% of the biodegradation took place within the first month [98]. Kirchman and Tengsved

TABLE 33 Wastewater Treatment of Polyoxyethylene Alkylphenols

Location	%Removal	Ref.
Italy, 3 WWTPs	February, average removal of all APEs = 96.9% (range 93.0–99.5%)	161
—	June, average removal of all APEs = 98.6% (range 98.3–98.8%)	161
—	Overall effluent conc: 0.74–12.0 µg/L NPE	161
US, 12 plants (1990s)	Average removal = 97.5% of NP and NPE combined (range 92.5–99.8%)	162
—	Overall NP conc:<0.20–15.3 µg/L	162
—	Overall NPE-1–NPE-17 conc:<5–261 µg/L	162
Canada, 4 plants, 8 final effluents	NPE range 0.80–15.1 µg/L: OP range 0.12–0.170 µg/L	163
UK, 16 sewage treatment plants, 8 rivers	NP dissolved range <0.10–5.4 µg/L (average = 0.90 µg/L)	164
Switzerland, several sites in the Glatt Valley	NP 77% removal, NPEs 31–98% removal	165
German treatment plants	Hosel (trickling filter) APE 70% removal (March), 75% removal (Sept)	166

APE, polyoxyethylene alkylphenol; NP, nonylphenol; NPE-n, polyoxyethylene (n) nonylphenol; OP, octylphenol.

[99] showed that there was little or no uptake of nonylphenol by agricultural crops.

The aerobic degradation of polyoxyethylene alkylphenols can result in carboxylic acid intermediates which are not detected using the normal HPLC/UV analytical techniques for alkylphenol and polyoxyethylene alkylphenols. Ahel and Giger have monitored these compounds and have postulated that they represent a major portion of polyoxyethylene alkylphenols in the environment [100]. Levels (where $n = 0.1$) ranged from 2 to 116 µg/L in the Glatt River, in Switzerland. A preliminary examination of these same compounds in U.S. rivers found much lower levels ranging from <0.04 to 6 µg/L [101].

The acute toxicity of the ether carboxylates (LC_{50}) ($n = 0.1$) has been determined to be approximately six times lower than for nonylphenol [102]. However, these data are based on one species of fish (*Oryzias latipse*—killfish), and no chronic effects have been measured.

In the United States, the risk of alkylphenols and polyoxyethylene alkylphenols to the environment is low. Even so, the Chemical Manufacturer's Association continues to sponsor research to more fully understand the environmental fate and effects of alkylphenols and polyoxyethylene alkylphenols. In addition, the U.S. Environmental Protection Agency, the United Kingdom's Department of the Environment, the U.S. Interagency Testing Committee, the World Health Organization through the International Programme for Chemical Safety, and the European Union are presently conducting environmental risk assessments of alkylphenols and their polyoxyethylene derivatives. Recent information on environmental issues may be obtained from Ref. 103.

ACKNOWLEDGMENTS

The authors thank Connie E. Carroll, Kimberly A. Fisher, Mark J. McGuiness, Albert F. Joseph, Frank H. Kasnick, Jane C. Kimble, Martin J. Kollmeyer, Thomas A. Maliszewski, Leonard A. Neubert, Jace D. Patton, Charles A. Smith, Barbara A. Stefl, M. Jane Teta, John P. Van Miller, James B. White, James B. Williams, Cheryl M. Wizda (Union Carbide Corporation), Carter G. Naylor (Huntsman Corporation), and Charles A. Staples (Assessment Technologies, Inc.) for their help in the preparation of this chapter.

REFERENCES

1. H. A. Bruson and O. Stein, U.S. Patent 2,075,018, to Rohm and Haas Co. (1937).
2. A. A. Green and J. W. McBain, J. Phys. Colloid Chem. *51*:286 (1947).

3. E. Gonick and J. W. McBain, J. Am. Chem. Soc. *69*:334 (1947).
4. A. S. Davidsohn and B. Milwidsky, *Synthetic Detergents, 7th ed.*, Longman, Essex, England, 1987, pp. 181–182.
5. J. P. Dever, K. F. George, W. C. Hoffman, and H. Soo, in *Kirk-Othmer Encyclopedia of Chemical Technology, 4th Ed., Vol. 20* (M. Howe-Grant, ed.), Wiley, New York, 1994, pp. 915–959.
6. J. Lorenc, G. Lambeth, and W. Scheffer, in *Kirk-Othmer Encyclopedia of Chemical Technology, 4th Ed., Vol. 2* (M. Howe-Grant, ed.), Wiley, New York, 1992, pp. 131–133.
7. C. R. Enyeart, in *Nonionic Surfactants* (M. Schick, ed.), Marcel Dekker, New York, 1967, p. 82.
8. J. Lorenc, G. Lambeth, and W. Scheffer, in *Kirk-Othmer Encyclopedia of Chemical Technology, 4th Ed., Vol. 2* (M. Howe-Grant, ed.), Wiley, New York, 1992, p. 135.
9. V. Ravendino and P. Sasseti, J. Chromatography *153*:181 (1975).
10. H. Fujii, Japanese Patent 48,015,840 to Sugai Chemical Ind. Co. Ltd. (1973).
11. C. L. Jarreau, Sr., U.S. Patent 4,055,605 to Calumet Petrochemical (1977).
12. J. F. Knifton and Y. E. Sheu, U.S. Patent 5,171,896 to Texaco Chemical Co. (1992).
13. J. Klimiec, J. Kolt, and J. Marszycki, Chinese Patent 1,082,530-A to Inst Ciezkiej Syntezy Orga (1994).
14. J. F. Knifton, U.S. Patent 5,300,703 to Texaco Chemical Co. (1994).
15. K. Yamauchi, K. Hashimoto, Y. Kera, S. Uemura, and M. Kajiwara, Japanese Patent 05,329,374 to Nitsusei Kagaku Kogyosho KK (1993).
16. P. Dai, J. F. Knifton, and Y. E. Sheu, U.S. Patent 5,276,215 to Texaco Chemical Co. (1994).
17. P. P. Kapustin, P. S. Belov, A. P. Vorozheikin, Y. I. Ryazanov, V. F. Podtikhov, K. D. Korenev, and V. A. Zavorotnyi, Soviet Union Patent 1,544,762 to Nizhnekamskneftekhim Industrial Enterprises and Moscow Institute of the Petrochemical and Gas Industry (1990).
18. H. Alfs et al., German Patent 2,346,273, to Chemische Werke Huls AG (1973).
19. H. Fiege, in *Ullman's Encyclopedia of Industrial Chemistry, Vol A19, 5th Ed.* (B. Elvers, S. Hawkins, and G. Schulz, eds.), VCH, New York, 1991, p. 321.
20. J. Lorenc, G. Lambeth, and W. Scheffer, in *Kirk-Othmer Encyclopedia of Chemical Technology, 4th Ed., Vol. 2* (M. Howe-Grant, ed.), Wiley, New York, 1992, pp. 126–127.
21. J. Lorenc, G. Lambeth, and W. Scheffer, in *Kirk-Othmer Encyclopedia of Chemical Technology, 4th Ed., Vol. 2* (M. Howe-Grant, ed.), Wiley, New York, 1992, p. 128.
22. J. Lorenc, G. Lambeth, and W. Scheffer, in *Kirk-Othmer Encyclopedia of Chemical Technology, 4th Ed., Vol. 2* (M. Howe-Grant, ed.), Wiley, New York, 1992, pp. 135–137.
23. P. Fanelli, 83rd Am. Oil Chemists Soc. Annual Meeting & Expo, Toronto, 1992.
24. P. Fanelli, 83rd Am. Oil Chemists Soc. Annual Meeting & Expo, Toronto, 1992.
25. C. R. Enyeart, in *Nonionic Surfactants* (M. Schick, ed.), Marcel Dekker, New York, 1967, pp. 62–70.
26. Igepal CO-430 Material Safety Data Sheet, Rhone-Poulenc Surfactants and Specialties (1993).
27. C. R. Enyeart, in *Nonionic Surfactants* (M. Schick, ed.), Marcel Dekker, New York, 1697, p. 68.
28. P. Fanelli, 83rd Am. Oil Chemists Soc. Annual Meeting & Expo, Toronto, 1992.

29. K. Kuriyama, Kolloid Z. *180*:55 (1962).
30. K. Kuriyama, Kolloid Z. *181*:144 (1962).
31. K. Kuriyama, H. Inoue, and T. Nakagawa, Kolloid Z. *183*:68 (1962).
32. R. R. Balmbra, J. S. Clunie, J. M. Corkill, and J. F. Goodman, Trans. Faraday Soc. *58*:1661 (1962).
33. P. H. Elworthy and C. B. Macfarlane, J. Chem. Soc. *1963*:907 (1963).
34. R. R. Balmbra, J. S. Clunie, J. M. Corkill, and J. F. Goodman, Trans. Faraday Soc. *60*:979 (1964).
35. C. W. Dwiggins, Jr., and R. J. Bolen, J. Phys. Chem. *65*:1787 (1961).
36. M. J. Schick, J. Phys. Chem. *67*:1796 (1963).
37. M. E. Ginn, F. B. Kinney, and J. C. Harris, J. Am. Oil Chemists Soc. *37*:183 (1960).
38. J. M. Corkill, J. F. Goodman, and S. P. Harrold, Trans. Faraday Soc. *60*:202 (1964).
39. A. S. Sadaghiania and A. Khan, J. Colloid Interf. Sci. *144*:191 (1991).
40. J. Cross, in *Nonionic Surfactants Chemical Analysis* (J. Cross, ed.), Marcel Dekker, New York, 1987, p. 381.
41. H. Schott, A. E. Royce, and S. K. Han, J. Colloid Interf. Sci. *98*:196 (1984).
42. S. K. Han, S. M. Lee, and H. Schott, J. Colloid Interf. Sci. *126*:393 (1988).
43. R. Heusch and F. Kopp, Ber. Bunsenges. Phys. Chem. *24*:806 (1987).
44. C. LaMesa, B. Sesta, A. Bonincontro, and C. Cametti, J. Colloid Interf. Sci. *125*:634 (1988).
45. K. Beyer, J. Colloid Interf. Sci. *86*:73 (1982).
46. X. Y. Hua and M. J. Rosen, J. Colloid Interf. Sci. *124*:652 (1988).
47. X. Y. Hua and M. J. Rosen, J. Colloid Interf. Sci. *141*:180 (1991).
48. P. Joos, J. P. Fang, and G. Serrien, J. Colloid Interf. Sci. *151*:144 (1992).
49. J. H. Green and J.M. Green, Brighton Crop Prot. Conf.—Weeds, Vol. 1, pp. 323–330, 1991.
50. L. Hsia, H. N. Dunning, and P. B. Lorenz, J. Phys. Chem. *60*:657 (1956).
51. C. W. Dwiggins, Jr,. R. J. Bolen, and H. N. Dunning, J. Phys. Chem. *64*:1175 (1960).
52. A. M. Mankowich, J. Phys. Chem. *58*:1027 (1954).
53. L. M. Kushner and W. D. Hubbard, J. Phys. Chem. *58*:1163 (1954).
54. M. Wentz, in *Detergency Theory and Technology* (W. G. Cutler and E. Kissa, eds.), Marcel Dekker, New York, 1987, pp. 463–465.
55. Z. Liu, S. Laha, and R. G. Luthy, Water Sci. Technol. *23*:475 (1991).
56. R. Osberghaus and F. Kresse, German Patent 3,444,959, to Henkel KGaA (1986).
57. L. F. Schwartz, U.S. Patent 4,588,516 (1986).
58. C. L. Merrill and D. L. Wood, U.S. Patent 5,034,158, to Shell Oil Co. (1991).
59. K. A. Harrison and J. M. Weller, U.S. Patent 4,810,409 to Sterling Drug, Inc. (1989).
60. M. Scardera, J. Floyd, and F. S. Natoli, U.S. Patent 4,744,917, to Olin Corp. (1987).
61. C. L. Fuggini and A. L. Streit, U.S. Patent 4,948,531, to Sterling Drug, Inc. (1990).
62. A. M. Magyar, U.S. Patent 4,670,171, to Pennzoil Co. (1987).
63. R. E. Smith, U.S. Patent 4,895,675, to Pro-Max Performance (1990).
64. B. J. Sturwold, U.S. Patent 4,675,125, to Cincinnati Vulcan Co. (1987).
65. J. C. Childers, in *Metalworking Fluids* (J. P. Byers, ed.), Marcel Dekker, New York, 1994.
66. N. O. V. Sonntag, in *Surfactants in Chemical/Process Engineering* (D. T. Wasan, M. E. Ginn, and D. O. Shah, eds.), Marcel Dekker, New York, 1988.

67. S. S. Talmage, in *Environmental and Human Safety of Major Surfactants, Alcohol Ethoxylates and Alkylphenol Ethoxylates* (The Soap and Detergent Association), Lewis, Boca Raton, FL, 1994.
68. R. D. Swisher, in *Surfactant Biodegradation*, Marcel Dekker, New York, 1987.
69. L. Vaicum, M. Cicei, and L. Stefanescu, Studii Epurarea Apelor *17*:27 (1976).
70. H. H. Tabuk and R. L. Bunch, *Proceedings of the Purdue Industrial Waste Conference, Vol. 36*, (1981), p. 888.
71. L. Kravetz, J. P. Salanitro, P. B. Dorn, and K. F. Guin, J. Am. Oil Chemists. Soc. *69*:692 (1995).
72. C. G. Naylor, J. P. Mieure, W. J. Adams, J. A. Weeks, F. J. Castalki, L. D. Ogle, and R. R. Romano, J. Am. Oil Chemists. Soc. *69*:692 (1995).
73. Chemical Manufacturer's Association Panel of Alkyphenol Ethoxylates, 1300 Wilson Boulevard, Washington, DC 20037.
74. A Corti, S. Frassinetti, G. Vallini, S. D'Antone, C. Fichi, and R. Solaro, Environ. Pollution *90*:83 (1995).
75. M. Ahel, W. Giger, and C. Schaffner, Water Res. *28*:1143 (1994).
76. J. A. Weeks, presented at Society for Environmental Toxicology and Chemistry, Denver, CO, November 1994.
77. C. G. Naylor, J. P. Mieure, W. J. Adams, J. A. Weeks, F. J. Castalki, L. D. Ogle, and R. R. Romano, J. Am. Oil Chemists Soc. *69*:692 (1995).
78. M. Ahel and W. Giger, Water Res. *28*:1143 (1994).
79. A. Marcomini, B. Pavoni, A. Sfriso, and A. A. Orio, Marine. Chem. *29*:307 (1990).
80. C. G. Naylor, J. P. Mieure, W. J. Adams, J. A. Weeks, F. J. Castalki, L. D. Ogle, and R. R. Romano, J. Am. Oil Chemists Soc. *69*:692 (1995).
81. G. Naylor, J. P. Mieure, W. J. Adams, J. A. Weeks, F. J. Castalki, L. D. Ogle, and R. R. Romano, J. Am. Oil Chemists Soc. *69*:692 (1995).
82. M. A. Blackburn and M. J. Waldock, Water Res. *29*:1623 (1995).
83. M. Ahel, Biogeochemical Behavior of Alkyphenol Polyethoxylates in the Aquatic Environment, PhD thesis, University of Zagreb, (1987).
84. S. S. Talmage, in *Environmental and Human Safety of Major Surfactants, Alcohol Ethoxylates and Alkylphenol Ethoxylates* (The Soap and Detergent Association), Lewis, Boca Ratio, FL, 1994.
85. A. M. Soto, H. Justicia, J. W. Wray, and C. Sonnenschein, Environ. Health Persp. *92*:167 (1991).
86. R. White, S. Jobling, S. A. Hoare, J. P. Sumpter, and M. G. Parker, Endocrinology *135*:175 (1994).
87. W. H. Benson and A. C. Nimrod, in a report submitted to the Chemical Manufacturers Association Panel on Alkylphenol Ethoxylates, 1995.
88. L. B. Clark, R. T. Rosen, T. G. Hartman, J. G. Louis, I. H. Suffet, R. L. Lippincott, and J. D. Rosen, Int. J. Environ. Anal. Chem. *47*:167 (1992).
89. H. Kirchmann and A. Tengsved, Swed. J. Agric. Res. *21*:115 (1991).
90. M. Ahel, J. Mcevoy, and W. Giger, Environ. Pollut. *79*:243 (1993).
91. C. G. Naylor, Textile Chemist and Colorist *27*:29 (1995).
92. R. R. Birch and S. G. Hales, Muench. Beitr. Abwasser, Fish.-Flussbiol. *44*:398 (1990).

93. D. Brown, H. de Henau, J. T. Garrigan, P. Gerike, M. Holt, E. Kunkel, E Matthijs, J. Waters, and R. J. Watkinson, Tenside *24*:14 (1987).

94. A. Marcomini, P. D. Capel, W. Giger, and H. Haeni, Naturwissenschaften *75*:460 (1988).

95. P. Diercxsens, M. Wegmann, R. Daniel, H. Haeni, and J. Tarradellas, Gas, Wasser, Abwasser *66*:123 (1987).

96. A. Marcomini, P. D. Capel, T. Lichtensteier, P. H. Brunner, and W. Giger, J. Environ. Qual. *18*:523 (1989).

97. H. Kirchmann, H. Aastroem, and G. Joensaell, Swed. J. Agric. Res. *21*:107 (1991).

98. A. Marcomini, P. D. Capel, T. Lichtensteiger, P. H. Brunner, and W. Giger, J. Environ. Qual. *18*:523 (1989).

99. H. Kirchmann and A. Tengsved, Swed. J. Agric. Res. *21*:115 (1991).

100. M. Ahel and W. Giger, Water Res. *28*:1143 (1994).

101. J. A. Field and R. L. Reed Environ. Sci. Tech. *30*:3544 (1996).

102. K. Yoshimura, J. Am. Oil Chemists Soc. *63*:1590 (1986).

103. A comprehensive update on polyoxyethylene alkylphenol environmental issues, including efforts by the United States Environmental Protection Agency and members of the European Community, can be obtained from the Chemical Manufacturer's Association Panel of Alkylphenol Ethoxylates, 1300 Wilson Boulevard, Washington, DC 20037.

104. C. R. Enyeart, in *Nonionic Surfactants* (M. Schick, ed.), Marcel Decker, New York, 1967, p. 52.

105. J. Lorenc, G. Lambeth, and W. Scheffer, in *Kirk-Othmer Encyclopedia of Chemical Technology, 4th Ed., Vol. 2* (M. Howe-Grant, ed.), Wiley & Sons, New York, 1992, p 115.

106. *McCutcheon's Volume 1: Emulsifiers and Detergents, North America Edition*, MC Publishing, Glen Rock, NJ, 1995.

107. *McCutcheon's Volume 1: Emulsifiers and Detergents, International Edition*, MC Publishing, Glen Rock, NJ, 1994.

108. P. T. Varineau, Union Carbide Corp., previously unpublished data.

109. Union Carbide Corp. Technical Brochure F-48601A.

110. Union Carbide Corp. Technical Brochure F-80183A ICD.

111. Union Carbide Corp. Technical Brochure F-48601A.

112. Union Carbide Corp. Technical Brochure F-80183A ICD.

113. G. W. Carr, Union Carbide Corp., previously unpublished data.

114. Rohm and Haas Company, Triton Surfactants (1960).

115. G. W. Carr, Union Carbide Corp., previously unpublished data.

116. Union Carbide Corp. Technical Brochure F-80183A ICD.

117. Union Carbide Corp. Technical Brochure F-48601A.

118. Union Carbide Corp. Technical Brochure F-80183A ICD.

119. C. M. Wizda, Union Carbide Corp., previously unpublished data.

120. R. M. Weinheimer, Union Carbide Corp., previously unpublished data.

121. R. M. Weinheimer, Union Carbide Corp., previously unpublished data.

122. P. Becher, J. Colloid Interf. Sci. *16*:49 (1961).

123. E. H. Crook, D. B. Fordyce, and G. F. Trebbi, J. Phys. Chem. *67*:1987 (1963).

124. E. H. Crook, G. F. Trebbi, and D. B. Fordyce, J. Phys. Chem. *68*:3592 (1964).

125. G. M. Gantz, in *Nonionic Surfactants*, (M. Schick, ed.), Marcel Dekker, New York, 1967, p. 738.
126. Union Carbide Corp. Triton Tergitol Specialty Surfactant Reference Chart (1995).
127. G. M. Gantz, in *Nonionic Surfactants* (M. Schick, ed.), Marcel Dekker, New York, 1967, p. 738.
128. Union Carbide Corp. Technical Brochure F-48601A.
129. Union Carbide Corp. Technical Brochure F-80183A ICD.
130. Union Carbide Corp. Technical Brochure F-48601A.
131. R. L. Blessing, Union Carbide Corp., unpublished internal study, 1994.
132. V. W. Saeger, R. G. Kuehnel, C. Linck, and W. E. Gledhill, Report No. ES-80-SS-46, Monsanto Industrial Chemicals Company, St Louis, MO, 1980.
133. J. Pudo and E. Erndt, Verh. Int. Verein. Limnol. *21*:1083 (1981).
134. J. Ruiz Cruz and M. C. Dobarganes, Garcia Grasas Aceites 27:309 (1976).
135. A. Marcomini, P. D. Chapel, T. Lichtensteiger, P. H. Brunner, and W. Giger, J. Environ. Qual. *18*:523 (1989).
136. L. Kravetz, H. Chung, K. F. Guin, W. T. Shebs, and L. S. Smith, Household Pers. Prod. Ind. *19*:46 (1982).
137. C. Crescenzi, A. DiCorcia, R. Samperi, and A. Marcomini, Anal. Chem. *67*:1797 (1995).
138. M. A. Blackburn and M. J. Waldock, Water Res. *29*:1623 (1995).
139. C. G. Naylor, J. P. Mieure, W. J. Adams, J. A. Weeks, F. J. Castalki, L. D. Ogle, and R. R. Romano, J. Am. Oil Chemists Soc. *69*:692 (1995).
140. P. T. Varineau, Union Carbide Corp., unpublished results. This was a study conducted by Prof. J. Field, Oregon State University.
141. L. B. Clark, R. T. Rosen, T. G. Hartman, J. G. Louis, I. H. Suffet, R. L. Lippincott, and J. D. Rosen, Int. J. Environ. Anal. Chem. *47*:167 (1992).
142. M. Ahel and W. Giger, Anal. Chem. *57*:1577 (1985).
143. M. Agel and W. Giger, Anal. Chem. *57*:2584 (1985).
144. M. Ahel and W. Giger, Proc. Seminar on Nonyphenol Polyethoxylates and Nonylphenol. Swedish Environmental Protection Agency Report 3907 (1991).
145. M. Ahel and W. Giger. Water Res. *28*:1143 (1994).
146. M. Ahel, T. Conrad, and W. Giger, Environ. Sci. Technol. *21*:697 (1987).
147. M. Ahel, Biogeochemical Behavior of alkyphenol polyethoxylates in the aquatic environment, PhD. Thesis, University of Zagreb, Zagreb (1987).
148. S. S. Talmage, in *Environmental and Human Safety of Major Surfactans, Alcohol Ethoxylates and Alkylphenol Ethoxylates* (The Soap and Detergent Association), Lewis, Boca Raton, FL, 1994.
149. S. S. Talmage, in *Environmental and Human Safety of Major Surfactants, Alcohol Ethoxylates and Alkylphenol Ethoxylates* (The Soap and Detergent Association), Lewis, Boca Raton, FL, 1994.
150. L. T. Brooke, D. Thompson, M. Kahl, D. Cox, R. L. Spehar, and R. Erickson, Toxicity and Bioaccumulation of Nonylphenol with Freshwater Organisms, 15th Annual Meeting of the Society of Environmental Toxicity and Chemistry, Denver, Colorado, 1994.
151. Work commissioned by the Chemical Manufacturer's Association (ABC Laboratories), Report Numbers 40597, 41756, and 41766 (1993–1995).
152. R. Ekelund, A. Bergman, A. Granmo, and M. Berggren, Environ Pollut. *64*:107 (1990).

153. D. W. McLeese, B. B. Sergent, C. D. Metcalfe, V. Zitcko and L. E. Burridge, Bull. Environ. Contam. Toxicol. *24*:575 (1980).
154. R. Ekelund, A. Bergman, A. Granmo, and M. Berggren, Environ Pollut. *64*:107 (1990).
155. F. A. Gobas, Ecol. Model. *69*:1 (1993).
156. R. Ekelund, A. Bergman, A. Granmo, and M. Berggren, Environ Pollut. *64*:107 (1990).
157. C. G. Naylor, J. P. Mieure, I. Morici, and R. R. Romano, presented at Third CESIO International Surfactants Congress, London, 1992.
158. D. W. McLeese, V. Zitko, D. B. Sergeant, L. Burridge, and C. D. Metcalfe, Chemosphere *9*:79 (1980).
159. F. A. Gobas, Ecol. Model. *69*:1 (1993).
160. M. Ahel, J. McEvoy, and W. Giger, Environ. Pollut. *79*:243 (1993).
161. C. Crescenzi, A. DiCorcia, R. Samperi, and A. Marcomini, Anal. Chem. *67*:1797 (1995).
162. C. G. Naylor, Textile Chemist and Colorist *27*:29 (1995).
163. H. B. Lee and T. E. Peart, Anal. Chem. *67*:1976 (1995).
164. M. A. Blackburn and M. J. Waldock, Water Res. *29*:1623 (1995).
165. M. Ahel, W. Giger, and M. Koch. Water Res. *28*:1131 (1994).
166. D. Brown, H. de Henau, J. T. Garrigan, P. Gerike, M. Holt, E. Kunkel, E Matthijs, J. Waters, and R. J. Watkinson, Tenside *24*:14 (1987).
167. R. Heusch and F. Kopp, Ber. Bunsenges. Phys. Chem. *24*:806 (1987).

3

Polyoxyethylene Alcohols

CHARLES L. EDWARDS Westhollow Technology Center,
Shell Chemical Company, Houston, Texas

I. INTRODUCTION

The most important class of commercially available nonionic surfactants is the polyoxyethylene alcohols derived from reaction of an alcohol with ethylene oxide. These materials are produced in an excess of a billion pounds a year and are used in such diverse applications as household and institutional laundry, textile scouring, pulp and paper manufacturing, oil field surfactants, agricultural spray adjuvants, and environmental clean-up. The type of alcohol used as the initiator and the length of the polyoxyethylene chain define the applications of these versatile commercial products. A primary driving force for the use of these surfactants is their ready biodegradability and overall environmental acceptability. Current and anticipated environmental pressures ensure their continued replacement of other surfactant materials.

Conventional polyoxyethylene alcohols are produced commercially by reaction of an alcohol with ethylene oxide using a basic catalyst, usually potassium hydroxide. The overall reaction can be expressed as

$$ROH \quad + \quad n\,CH_2\text{–}CH_2 \xrightarrow{\text{basic catalyst}} RO(CH_2CH_2O)nH$$

In this equation R represents a hydrophobic moiety that can be either linear or branched, and the alcohol can be in principle primary, secondary, or tertiary. However, polyoxyethylene alcohols based on secondary or tertiary alcohols are prepared using a two-step oxyethylation procedure described in Chapter 1. Whereas there are limited applications for polyoxyethylene alcohols derived from secondary alcohols, applications for polyoxyethylene tertiary alcohols are rare.

The number of moles n of ethylene oxide reacting with 1 mole of hydrophobe need not be an integral and is the number average degree of polymerization of ethylene oxide in the product. It is the structure and molecular weight of the R group as well as the value of n in the molecule that determines the application of the surfactant. Further details of the mechanism of this reaction are given in Chapter 1.

This chapter discusses polyoxyethylene alcohols based on primary and secondary aliphatic alcohols. Polyoxyethylene alcohols based on other types of alcohols, such as aromatic alcohols (alkylphenols), are discussed in detail in Chapter 2. The synthesis of the hydrophobic intermediates will precede a discussion of the preparation, properties, and applications of the polyoxyethylene alcohols.

II. SYNTHESIS OF HYDROPHOBIC INTERMEDIATES

A. Primary Alcohols

1. Natural Sources

Linear detergent range primary alcohols were first commercially prepared from naturally occurring fats and oils. The most common sources were coconut oil, palm kernel oil, sperm oil, castor oil, and tallow. Coconut oil and tallow are the most commercially attractive sources because most of the alcohols produced from these natural products are within the effective detergent range (C_{12}–C_{18}). Coconut oil is particularly attractive as a raw material because it contains 65–70% of C_{12} and C_{14} acid derivatives as well as significant amounts of C_8, C_{10}, C_{16}, and C_{18} acid derivatives. These natural sources have only even-numbered carbons and are almost exclusively linear.

The natural fatty alcohols can be produced from coconut oil by one of several process variations based on the triglycerides obtained from the crude oil.

Early routes were based on direct hydrogenolysis of the triglycerides or hydrogenolysis of the fatty acids obtained after hydrolysis of the triglycerides [1–10]. However, these routes require high-pressure and high-temperature operating conditions and are less selective than current process options. A more selective, low-pressure hydrogenolysis process [11–13] has been demonstrated that involves the hydrogenolysis of the methyl esters of the fatty acids obtained by esterification of the fatty acids with methanol. The products are a mixture of fatty alcohols and glycerol, with the latter being sold as a valuable side product. Methanol is recycled in the overall process, which is shown.

Hydrolysis of Triglyceride

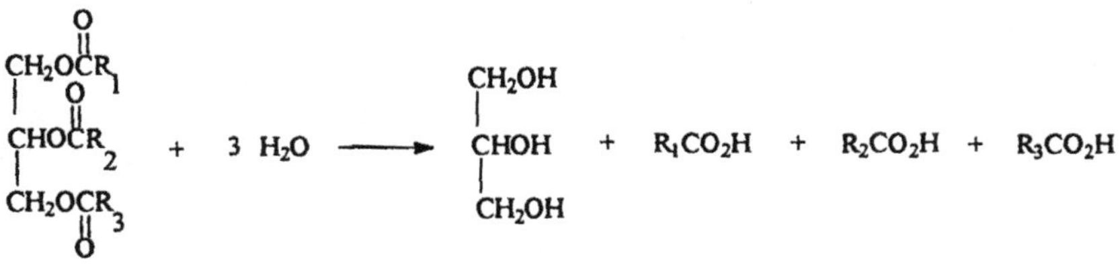

$$CH_2OCR_1 \quad\quad\quad\quad CH_2OH$$
$$CHOCR_2 \;+\; 3\,H_2O \longrightarrow CHOH \;+\; R_1CO_2H \;+\; R_2CO_2H \;+\; R_3CO_2H$$
$$CH_2OCR_3 \quad\quad\quad\quad CH_2OH$$

Esterification with Methanol

$$R'CO_2H \;+\; CH_3OH \longrightarrow R'CO_2CH_3 \;+\; H_2O$$

Hydrogenolysis

$$R'CO_2CH_3 \;+\; 2\,H_2 \longrightarrow R'CH_2OH \;+\; CH_3OH$$

The fatty alcohols produced from coconut oil are straight chain, saturated primary alcohols with even-numbered carbon atoms from C_8 to C_{18}. They are extremely pure, which has allowed their use in the personal care markets (e.g., cosmetics and shampoos). However, the cost and availability of linear alcohols based on natural sources fluctuates significantly, depending on the weather conditions of the geographic areas producing coconuts and palm kernels. For this and other strategic reasons, synthetic sources were developed for the production of aliphatic detergent range alcohols.

2. Synthetic Sources

Synthetic primary alcohols are based on petrochemical raw materials and are competitive in cost and performance with the natural fatty alcohols. The hydrocarbon chain of the synthetic alcohols can be either 100% linear or partially branched (20–25 wt% methyl branching at the 2 position) depending on the process used. The branching of the hydrophobe has a significant effect on the end use. The Ziegler alcohol process described in this section produces 100% linear alcohols with all even carbon numbers, making them competitive in markets previously dominated by natural fatty alcohols. The oxo and modified oxo (Shell hydroformylation) processes described below produce alcohols with various degrees of branching, making sulfates obtained from them particularly useful in cold water detergent applications.

All synthetic alcohols currently produced are based on one of several ethylene oligomerization process [14]. Originally, the olefin intermediates used in oxo-type processes were produced from wax cracking or chlorination-dehydrochlorination of paraffins, but these processes became less economically attractive with the advent of the ethylene oligomerization process. Either α- or internal olefin feedstocks can be used in the oxo or modified oxo processes, but the Ziegler alcohol process leads to the formation of 100% linear alcohols directly from ethylene without isolation of the detergent range olefin intermediate. This section describes processes that are currently used to produce synthetic primary alcohols.

(a) Source of Detergent Range Olefin Intermediates. Oxo- and modified oxo-based synthetic alcohol processes use detergent range olefins (C_{10}–C_{14}) as intermediates. These olefins are almost exclusively linear, but may be either α- or internal olefins. Because the detergent range olefins are such an important part of two of the three commercial processes used to produce synthetic alcohols, this section will address the current commercial processes to produce detergent olefin intermediates.

Chevron ethylene oligomerization process. Chevron produces high-purity linear α-olefins using a one-step Ziegler-type oligomerization process that produces a somewhat broad product mixture that follows the Schulz–Flory distribution [15]. The reaction involves an ethylene growth reaction using a soluble triethylaluminum (TEA) catalyst. The growth of the alkyl chain and displacement of the product olefin from the aluminum catalyst occur simultaneously during reaction. This is illustrated as follows:

<u>Chain Growth</u>

$$Al{\Large\lbrace}\begin{array}{l}CH_2CH_3\\-CH_2CH_3\\CH_2CH_3\end{array}\quad +\quad n\,CH_2{=}CH_2\quad\longrightarrow\quad Al{\Large\lbrace}\begin{array}{l}CH_2CH_2R_1\\-CH_2CH_2R_2\\CH_2CH_2R_3\end{array}$$

<u>Chain Displacement</u>

$$Al{\Large\lbrace}\begin{array}{l}CH_2CH_2R_1\\-CH_2CH_2R_2\\CH_2CH_2R_3\end{array}\quad +\quad CH_2{=}CH_2\quad\longrightarrow\quad Al{\Large\lbrace}\begin{array}{l}CH_2CH_3\\-CH_2CH_2R_2\\CH_2CH_2R_3\end{array}\quad +\quad R_1CH{=}CH_2$$

The chain displacement reaction is simplified to show only one replacement at a time. However, multiple replacements are occurring during reaction, which leads to the somewhat broad product distributions predicted by the Schulz–Flory equation.

$$K = \frac{\text{moles of } C_n^{2+} \text{ olefin}}{\text{Moles of } C_n \text{ olefin}}$$

This broad product slate requires that markets be obtained for the α-olefins produced outside the useful detergent range (i.e., $<C_{10}$ and $>C_{14}$) for maximum economic benefit. Balancing the product slate containing various α-olefins each with different economic value is always a marketing challenge.

Since the TEA catalyst is used in a relatively low concentration, it is not economical to attempt to recycle the catalyst after isolation of the olefins through distillation processes. Instead, it is treated with aqueous NaOH, which converts the material to sodium aluminate and is then sent for disposal. Further details of this process can be found in patents to Chevron and Gulf [16–18].

Ethyl's ethylene oligomerization-displacement process. The Ethyl process also uses a TEA-based ethylene oligomerization system. However, while the Chevron process is conducted under high-temperature conditions causing the growth and displacement reactions to occur simultaneously, the Ethyl process separates the two steps. This first step involves a low-temperature ethylene oligomerization reaction under conditions whereby no displacement occurs. The displacement reaction is conducted in a second high-temperature reaction. The Ethyl process uses larger quantities of TEA because the growth step is stoichiometric in aluminum.

First Stage

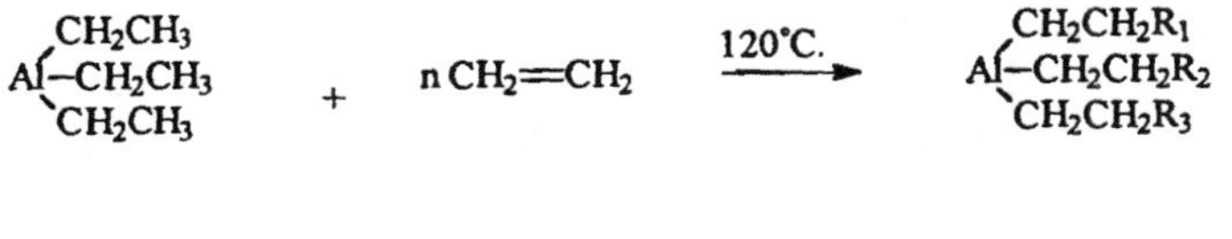

$$Al\begin{cases}-CH_2CH_3\\-CH_2CH_3\\-CH_2CH_3\end{cases} + \; n\,CH_2{=}CH_2 \quad \xrightarrow{\;120°C.\;} \quad Al\begin{cases}-CH_2CH_2R_1\\-CH_2CH_2R_2\\-CH_2CH_2R_3\end{cases}$$

Second Stage

$$Al\begin{cases}-CH_2CH_2R_1\\-CH_2CH_2R_2\\-CH_2CH_2R_3\end{cases} + \; 3\,CH_2{=}CH_2 \quad \xrightarrow{\;280°C.\;} \quad Al\begin{cases}-CH_2CH_3\\-CH_2CH_3\\-CH_2CH_3\end{cases} + \begin{matrix}R_1CH{=}CH_2\\R_2CH{=}CH_2\\R_3CH{=}CH_2\end{matrix}$$

This has two specific consequences. First, the TEA must be recovered and recycled in order for the process to be economical. In addition, the linear α-olefin product mixture is much narrower, following a Poisson distribution as shown below (see Chapter 1 for a complete discussion of the kinetics). This

TABLE 1 Typical Product Distributions
for the Ethyl Olefin Process [19]

Carbon number fraction	%
C_4	13
C_6–C_{10}	55
C_{12}–C_{18}	30
C_{20+}	2

means that the Ethyl process is somewhat more flexible in its ability to produce
the desired carbon number range for balancing market demands.

$$X_{(p)} = \frac{n^p e^{-n}}{p!}$$

where $X_{(p)}$ is the mole fraction of the chains containing added ethylene
        molecules
        p is the number of ethylene units added to a chain
        n is the average number of moles of ethylene reacted per equivalent
        of aluminum alkyl bonds

As with all TEA-based ethylene oligomerization reactions (Ziegler-type), the
linear α-olefin products resulting from the Ethyl process have an even number
of carbon atoms. A typical product distribution resulting from the Ethyl process
is shown in Table 1 [19]. The Ziegler-based chemistry has been extensively
reported [20–28].

Shell higher olefin process (SHOP). Shell's higher olefin process (SHOP) is
the most complex of the three commercial processes used to produce linear α-
olefins, but it is also the most flexible. The process is based on a proprietary
homogenous nickel-based catalyst system that produces the same geometric
(Schulz–Flory) distribution as the Chevron process. However, it also includes
chemistry to convert the less desirable lower and higher carbon number compounds to detergent range linear internal olefins by using a combination
isomerization-disproportionation (metathesis) scheme. The detergent range
internal olefins can be used in Shell's modified oxo process (described later) as
feed for their synthetic alcohol plants. This combination of the two processes
offers unique overall technology that is used to balance and maximize the prof-
itability of the overall product slate. A simplified block diagram of the SHOP
process is shown in Fig. 1 [29]. The ethylene oligomerization step is conducted
in a series of reactors using a proprietary nickel/phosphorus ligand catalyst
system in butanediol solvent at 90–100°C and at 1200–1500 psig ethylene pressure. The linear α-olefin product mixture is then fractionated to produce a C_6–
C_{20} range and a C_{20+} range of heavy ends. A majority of the linear α-olefins in

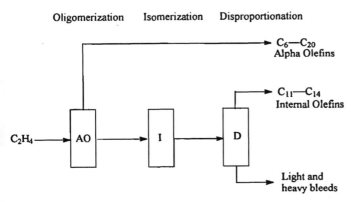

FIG. 1 Illustrative block diagram for SHOP.

the C_6–C_{20} range is further fractionated for sale in the high-value α-olefins markets, but the remainder along with the C_{20+} range is fed to the isomerization-disproportionation system converting the feedstock to internal olefins in the C_{11}–C_{14} range. These internal olefins are used for the production if polyoxyethylene alcohols using the modified oxo process described later in this section. Typical product distributions (K factors) for the ethylene oligomerization reaction are shown in Fig. 2.

The isomerization section contains a proprietary isomerization catalyst operating at 120°C, whereas the disproportionation section contains a proprietary catalyst system operating at 130°C. The isomerization and disproportionation reactions are as follows:

Isomerization Step

$$R-CH=CH_2 \quad \rightleftharpoons \quad Rx-CH=CH-Ry$$

(all possible isomers)

Disproportionation Step

$$
\begin{array}{c}
R_1CH=CHR_2 \\
+ \\
R_3CH=CHR_4
\end{array}
\quad \rightleftharpoons \quad
\begin{array}{c}
R_1CH \\
\| \\
R_3CH
\end{array}
+
\begin{array}{c}
R_2CH \\
\| \\
R_4CH
\end{array}
+
\begin{array}{c}
R_1CH \\
\| \\
R_4CH
\end{array}
+
\begin{array}{c}
R_2CH \\
\| \\
R_3CH
\end{array}
$$

Various aspects of the process along with recent process improvements have been extensively reported in the patent literature [30–41].

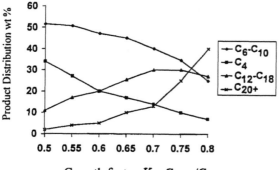

Growth factor, $K = C_{n+2}/C_n$, m/m

FIG. 2 Oligomer product distribution vs. K factor for SHOP [29].

The SHOP process differs from the Chevron and Ethyl processes in two major respects. First, it is based on a nickel/phosphorus ligand catalyst system instead of a TEA catalyst system. Second, it is a far more flexible process, as demonstrated by the fact that the $C_{11}-C_{14}$ internal olefins produced in the isomerization-disproportionation section are used almost exclusively in the Shell hydroformylation (modified oxo) process. Thus, the less commercially attractive internal olefins are upgraded when used to produce the synthetic alcohols.

The three commercial processes for producing linear α- and internal olefins described above can be used to produce olefins for sale or as intermediates for the production of primary synthetic alcohols. The next section will address processes for the production of synthetic primary alcohols.

(b) The Conventional Oxo (Hydroformylation) Process. The oxo or hydroformylation process is the addition of hydrogen and carbon monoxide to an olefin using a soluble cobalt metal catalyst at relatively high temperatures and pressures to produce an aldehyde containing one carbon more than the starting olefin. The aldehyde can be isolated for sale but is usually further hydrogenated in situ to produce the synthetic primary alcohol. The aldehydes and alcohols produced during hydroformylation are a mixture of linear and branched products. The reaction can be illustrated as follows:

$$RCH{=}CH_2 \ + \ H_2/CO \ \longrightarrow \ \underset{\underset{CHO}{|}}{RCH{-}CH_3} \ + \ RCH_2{-}CH_2{-}CHO$$

$$\underset{\underset{CHO}{|}}{RCH{-}CH_3} \ + \ RCH_2{-}CH_2{-}CHO \ \overset{H_2}{\longrightarrow} \ \underset{\underset{CH_2OH}{|}}{RCH{-}CH_3} \ + \ RCH_2{-}CH_2{-}CH_2OH$$

A typical conventional oxo- or hydroformylation process is conducted at 120–200°C and 2000–3000 psig using a soluble cobalt carbonyl catalyst. The synthesis gas (or H_2/CO) ratio is usually 1.0. The reaction rate is strongly dependent on the structure of the olefin [42]. The fastest reaction rates are achieved using lower molecular weight, linear α-olefins. Linear internal olefins react three to four times more slowly than linear α-olefins. Branching can reduce reaction rates by an order of magnitude. Several comprehensive reviews of the kinetics of hydroformylation have been reported [43–48].

Synthetic alcohols prepared using the oxo and modified oxo processes are significantly different from natural fatty alcohols and those produced from the Ziegler alcohol process. Synthetic alcohols derived from the modified oxo process have both odd- and even-numbered carbon atoms and contain some methyl branching (20–25%) in the 2 position. Natural fatty alcohols and Ziegler alcohols are 100% linear and are always even-numbered carbon alcohols. The structural differences lead to differences in physical properties and thus applications, providing some differentiation between synthetic alcohols obtained from the two types of processes.

(c) Modified Oxo (Shell Hydroformylation) Process. Discoveries made in the 1960s and 1970s by Shell led to ligand-modified cobalt catalyst systems with distinct advantages over the conventional oxo process. The conventional oxo process must be operated at low reaction temperatures with long reaction times and use linear α-olefin feedstock in order to minimize the production of the less biodegradable and thus less desirable branched alcohol isomers. Higher operating pressures are required using the conventional, unmodified cobalt catalyst system, because it is less stable than the ligand-modified cobalt catalyst system. Finally, the oxo catalyst is a poor hydrogenation catalyst, so that the hydrogenation of the intermediate aldehyde to alcohol must be accomplished in a second stage or step.

In contrast, the Shell's ligand-modified cobalt catalyst system can operate at higher reaction temperatures, lower overall operating pressure, and uses less expensive internal olefins as feedstock. While use of internal olefins in the conventional oxo process leads to significant increases in branching (up to 50%) in the alcohol, improved linearity (75–80%) is achieved using the ligand-modified process. Finally, the Shell catalyst produces alcohol in one step due to the ability of the catalyst to hydrogenate the intermediate aldehyde to alcohol. Each of these process improvements results in a significant economic benefit in capital and variable (catalyst) costs.

Shell's modified hydroformulation process (SHF) is based on a proprietary tertiary organophosphine-modified cobalt catalyst system that operates at 190°C and 1200 psig of H_2/CO in a 2:1 molar ratio. The overall chemistry is shown below and is described in detail in several publications and patents [49–54].

Cobalt Octacarbonyl   "Red Catalyst"

$- H_2 \updownarrow H_2$ $- H_2 \updownarrow H_2$

$$2 \ HCo(CO)_4 \quad \underset{CO}{\overset{R_3P}{\rightleftharpoons}} \quad 2 \ HCo(CO)_3R_3P$$

Active Oxo Catalyst   Active Ligand Modified Catalyst

(d) Ziegler Alcohol Process. The Ziegler alcohol process is based on the pioneering organoaluminum chemistry developed by Karl Ziegler and coworkers at the Max Planck Institute in Germany. As a result of this work, 100% linear primary alcohols are commercially produced by two companies (Vista and Amoco) using a three-step process. Ethylene is oligomerized in the presence of TEA, followed by air oxidation to form the aluminum alcoholate, followed by hydrolysis with dilute sulfuric acid to produce a narrow molecular weight range of alcohols and aluminum hydroxide.

The Vista (formally Conoco) process involves the oligomerization of ethylene under conditions (low temperature) that prevent chain displacement reactions, thus producing hydrophobic chains following a Poisson distribution as described previously. The overall reaction is as follows:

$$Al\begin{Bmatrix} CH_2CH_3 \\ -CH_2CH_3 \\ CH_2CH_3 \end{Bmatrix} + n \ CH_2=CH_2 \longrightarrow Al\begin{Bmatrix} (CH_2CH_2)xCH_2CH_3 \\ -(CH_2CH_2)yCH_2CH_3 \\ (CH_2CH_2)zCH_2CH_3 \end{Bmatrix}$$

$$Al\begin{Bmatrix} (CH_2CH_2)xCH_2CH_3 \\ -(CH_2CH_2)yCH_2CH_3 \\ (CH_2CH_2)zCH_2CH_3 \end{Bmatrix} + 3/2 \ O_2 \longrightarrow Al\begin{Bmatrix} O(CH_2CH_2)xCH_2CH_3 \\ -O(CH_2CH_2)yCH_2CH_3 \\ O(CH_2CH_2)zCH_2CH_3 \end{Bmatrix}$$

$$Al\begin{Bmatrix} O(CH_2CH_2)xCH_2CH_3 \\ -O(CH_2CH_2)yCH_2CH_3 \\ O(CH_2CH_2)zCH_2CH_3 \end{Bmatrix} + 3 \ H_2O \longrightarrow \begin{matrix} HO(CH_2CH_2)xCH_2CH_3 \\ + \\ HO(CH_2CH_2)yCH_2CH_3 \\ + \\ HO(CH_2CH_2)zCH_2CH_3 \\ + \\ Al(OH)_3 \end{matrix}$$

This process produces a mixture of alcohols with a significant amount of alcohols outside the detergent range. This presents a marketing challenge to Vista because overall economic success is dependent on the placement of each carbon fraction in a lucrative market. Modification of the Ziegler alcohol process to produce a narrower product distribution would be of significant economic benefit.

The Amoco (formally Ethyl or Albemarle) process produces a narrower-than-Poisson product distribution by combining the growth and displacement reactions in the overall process. In this process the growth reaction product, which has a Poisson product distribution, is reacted with a C_6–C_{10} olefin stream in a transalkylation scheme as follows:

Displacement or Transalkylation

$$R_1 \quad + \quad A \overset{R_3}{\underset{R_4}{\mid}}{-}R_2 \quad \rightleftharpoons \quad R_2 \quad + \quad A \overset{R_3}{\underset{R_4}{\mid}}{-}R_1$$

The product from this reaction contains heavy olefins and an aluminum alkyl stream (C_6–C_{10}) that can undergo additional growth reactions. By the proper combination of growth and transalkylation reactions, followed by fractionation and recyling of the appropriate olefin streams, the overall process can be managed to produce a majority of organoaluminum product in the C_{12}–C_{16} detergent range (85 wt% overall). The concept of combining various process steps and recyling streams in order to maximize the desired product slate is similar to the three-step SHOP process described previously to produce detergent range olefins. The overall process scheme is summarized in a simplified process diagram in Fig. 3 and described in patents [55,56].

The alcohols produced by the Ziegler alcohol process are 100% linear even-carbon-number alcohols. They are essentially the same as the natural fatty alcohols and can compete in most of the same markets. The alcohol product mixture follows the Poisson distribution (as with the Ethyl olefin process) and is subject to some shifting of the peak by changes in the process parameters. The alumina produced in the process is quite pure and is sold as a byproduct of the process. Therefore, the Zeigler alcohol process is not truly catalytic with respect to TEA, because it is consumed in the process. Finally, balancing of the alcohol product distribution is always a challenge, with the overall economics of the process being dependent on the value of the lower and higher molecular weight alcohols in their respective markets. The C_4–C_{10} alcohol fraction is used in the solvents and plasticizers markets, whereas the C_{16}+ fraction find various, but often lower value, markets.

B. Secondary Alcohols

Polyoxyethylene alcohols derived from secondary alcohols are used primarily in the industrial surfactants markets due to their increased liquidity and water

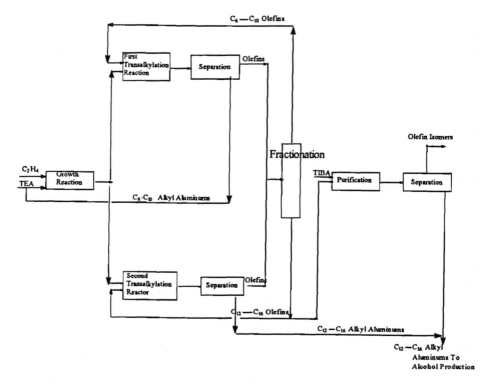

FIG. 3 Simplified diagram of the modified ethylene polymerization process [57].

solubility. However, they are reported to have somewhat lower biodegradability than their primary alcohol counterparts. There are two major processes reported for the preparation of secondary alcohols: paraffin oxidation using boric acid as mediator and olefin hydration using concentrated sulfuric acid. Union Carbide had commercially produced secondary alcohols by paraffin oxidation but now imports the secondary alcohol or its derivatives from Nippon Shokubai.

1. Paraffin Oxidation Using Boric Acid as Modifier

Until recently, Union Carbide produced secondary alcohols as intermediates for their polyoxyethylene alcohol series of Tergitol S nonionic and anionic surfactants. They do not market the secondary alcohols but instead use them captively to make the polyoxyethylene secondary alcohol surfactant. Union Carbide now imports the secondary alcohols or polyoxyethylene derivatives from Nippon Shokubai, who it is believed produces these secondary alcohols by paraffin oxidation in the presence of boric acid.

The oxidation of paraffins in the liquid phase without the use of modifiers or cocatalysts results in the formation of significant amounts of ketones, aldehydes, peroxides, esters, and acid side products. However, when the oxidation is conducted in the presence of boric acid, high yields of the secondary alcohol can be obtained. The explanation for the increase in selectivity is believed to be the trapping of the secondary alcohol by the boric acid to form borate esters, which are known to be quite resistant to further oxidation. However, water must be removed during the borate ester formation in order to achieve high selectivities. This process option has been investigated in detail and more information can be obtained from the patent literature [58–63].

$$H_3BO_3 \xrightarrow{\text{heat}} HBO_2 + H_2O$$

$$3\ ROH + 3\ HBO_2 \longrightarrow \underset{\substack{RO \diagup O \diagdown OR}}{\overset{\substack{OR \\ | \\ B \\ O \diagup \diagdown O \\ | \quad | \\ B \quad B}}{}} + 3\ H_2O$$

2. Hydration of Olefins

Secondary alcohols can also be obtained using a two-step process involving reaction of linear olefin with concentrated sulfuric acid to produce an alkylsulfuric acid (ASA) intermediate, which is subsequently hydrolyzed to the secondary alcohol in a second step.

$$R_1CH{=}CHR_2 + H_2SO_4 \longrightarrow \underset{\underset{SO_3H}{|}}{R_1CH{-}CH_2R_2}$$

Alkylsulfuric Acid (ASA)

(one of several isomers)

$$\underset{\underset{SO_3H}{|}}{R_1CH{-}CH_2R_2} + H_2O \longrightarrow \underset{\underset{OH}{|}}{R_1CH{-}CH_2R_2} + H_2SO_4$$

(one of several isomers)

The process is conducted in two steps in order to achieve high overall conversions to secondary alcohol. In the first step, concentrated sulfuric acid (96–100%) is added slowly with cooling to the olefin in a well-stirred reactor. The reaction goes to completion virtually instantaneously. The product mixture consists primarily of the ASA and some dialkyl sulfate. This product mixture is then hydrolyzed in a second step producing a two-phase medium consisting of primarily secondary alcohol and dilute sulfuric acid. The secondary alcohol is isolated by phase separation, treatment with dilute base, and distillation. The dilute sulfuric acid is reconstituted to concentrated sulfuric acid and recycled to the reaction scheme. This process has been studied periodically since the 1930s [64], and Shell used it in the 1940s to produce secondary alcohol sulfates ("Teepol") in Europe. Considerable work has been reported on olefin hydration [65,66].

Secondary alcohols can also be made by direct hydration of detergent range linear olefins using a strong homogeneous or heterogeneous (e.g., ion exchange) acid catalyst. This route has the disadvantage that the equilibrium of the reaction is not favorable for alcohol formation. Yields are restricted to ~30 mol% per pass over the catalyst, which means that significant recycling and separations can make this process uneconomical [67–70]. Although there are several other laboratory methods to prepare secondary alcohols, none are used commercially.

$$R_1CH{=}CHR_2 \quad + \quad H_2O \quad \underset{}{\overset{\text{acid catalyst}}{\rightleftharpoons}} \quad R_1CH{-}CH_2R_2$$
$$\underset{OH}{|}$$

(one of several isomers)

III. PREPARATION OF POLYOXYETHYLENE ALCOHOLS

Polyoxyethylene alcohols can be produced by a variety of laboratory routes, but they are produced commercially by reaction of the alcohol with ethylene oxide using a catalyst. The overall reaction can be expressed as follows:

$$ROH \quad + \quad n\ CH_2{-}CH_2 \quad \overset{\text{catalyst}}{\longrightarrow} \quad RO(CH_2CH_2O)nH$$

In this equation the number of moles n of ethylene oxide reacting with 1 mole of hydrophobe need not be an integral and is the number average degree of polymerization of ethylene oxide in the product. The choice of hydrophobe structure and molecular weight, the value of n, and the type of catalyst used determines the physical properties and thus applications of the polyoxyethylene alcohol produced. The choice of catalyst used in the process is particularly important, because it controls the distribution of the polyoxyethylene chain in the product. A more detailed discussion of the factors controlling the polyoxyethylene chain length distribution along with kinetics and mechanistic interpretations is found in Chapter 1. The following section briefly discusses the two main types of commercial products available to the surfactants market—conventional, and the new narrow-range or peaked polyoxyethylene alcohols.

A. Conventional Polyoxyethylene Alcohols

Conventional polyoxyethylene alcohols are produced using a basic catalyst, typically potassium hydroxide. The products obtained from this reaction consist

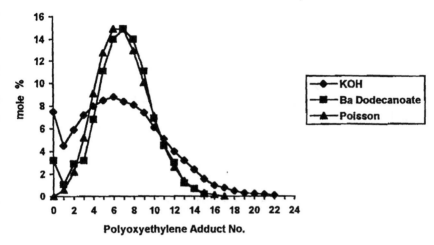

FIG. 4 Product distribution obtained from the reaction of dodecanol with 7 moles of ethylene oxide using either KOH or barium dodecanoate as catalyst and compared with a Poisson distribution [71].

of a mixture of polyoxyethylene alcohol oligomers. A typical product distribution is shown in Fig. 4 using potassium hydroxide as catalyst and compared with a product with a narrower chain length distribution obtained using activated catalysts described in Sec. B.

1. Catalyst

The vast majority of commercially available polyoxyethylene alcohols are produced using basic catalysts. Although many basic catalysts have been reported as useful for this reaction (71), the most commonly used catalyst is potassium hydroxide. Catalyst activity has been shown to be directly related to the size of the metallic cation within a specific family in the periodic table—the larger the ionic radius, the more active the catalyst (e.g., Li < Na < K < Rb < Cs for group I and Mg < Ca < Sr < Ba for group II metals. There are several reasons for using potassium hydroxide as the catalyst system: ease of handling, low cost, high activity, and high selectivity to desired products. The catalyst is usually added to the alcohol as either an aqueous potassium hydroxide solution (e.g., 50 wt%) or as a solid potassium flake. The catalyst mixture is activated by converting the potassium hydroxide to the potassium alkoxide of the alcohol to be reacted. In this manner, water is removed from the reaction mixture, thus minimizing the amount of polyethylene glycol produced as a side product.

$$ROH \; + \; KOH \; \longrightarrow \; RO^-K^+ \; + \; H_2O\uparrow$$

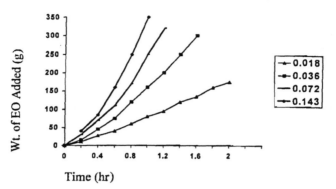

FIG. 5 Effect of KOH concentration on rate of polyoxyethylation at 135–140°C [72]. (Values are moles of KOH per mole of tridecyl alcohol used.)

As should be expected, an increase in catalyst concentration results in an increase in reaction rate. However, in a study wherein the concentration of potassium hydroxide catalyst was progressively doubled from 1.8 m% to 14.3 m% basis alcohol, the rate of reaction did not increase each time as expected [72]. The rate increase was greater when increasing the catalyst concentration at the lower catalyst concentrations than at the higher concentrations as shown in Fig. 5.

This can be explained by the fact that the ethylene oxide reacts with the charge-separated alkoxide ion instead of a covalently bound metal alkoxide. The degree of ionization is greater at lower catalyst concentrations than at the higher concentrations [73].

2. Temperature

Normal operating temperatures for the commercial production of polyoxyethylene alcohols is 140–180°C. An increase in reaction temperature leads to an expected increase in reaction rate using potassium hydroxide catalyst. However, the rate increase is not directly proportional to temperature [72,73] as can be seen in Fig. 6. The solubility of ethylene oxide in the catalyzed reaction mixture is lower at higher reaction temperatures at a given ethylene oxide partial pressure. Consequently, because the rate expression is approximately first order with respect to ethylene oxide concentration, temperature strongly affects the overall reaction rate due in part to its effect on ethylene oxide solubility.

3. Pressure

Ethylene oxide is an extremely hazardous chemical that is capable of explosion over a wide range of pressures. A detailed discussion of the physical properties and safe handling requirements of ethylene oxide is presented in Chapter 1. It is

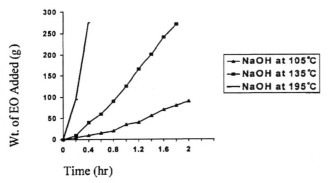

FIG. 6 Effect of temperature on rate of polyoxyethylation with NaOH as catalyst [72].

advisable to conduct polyoxyethylations using an inert gas such as nitrogen in order to ensure that the ethylene oxide in the vapor phase does not exceed the explosive limits. In commercial practice, the gas phase concentration of ethylene oxide is never allowed to exceed 40 mol%, the reported lower limit of the explosive region. In fact, it is advisable to conduct these reactions at gas phase ethylene oxide concentrations significantly below 40 mol%.

Most commercial polyoxyethylation reactors are batch reactors and operated at a total operating pressure of approximately 75 psia total reaction pressure (nitrogen + ethylene oxide). The partial pressure of ethylene oxide used at this total pressure is 15–25 psia. Studies have shown that the rate of reaction increases with a increase in ethylene oxide partial pressure [71].

4. Alcohol Structure

The structure of the alcohol significantly affects both reaction rate and the polyoxyethylene chain length distribution in polyoxyethylation. Linear primary alcohols react more rapidly than branched primary alcohols, with the position of the branch being of significantly importance. Primary alcohols, either linear or branched, react more rapidly than secondary or tertiary alcohols.

Due to the differences in reaction rates, the oligomer distribution of primary, secondary, and tertiary polyoxyethylene alcohols are significantly different. For example, a 7-mole polyoxyethylene adduct of the three types of alcohols would have dramatically different polyoxyethylene chain length distributions. Since the secondary and tertiary alcohols are significantly less reactive than their polyoxyethylene adducts, then the final products contain substantial amounts of residual secondary or tertiary alcohol with a very broad polyoxyethylene chain length distribution compared with primary alcohol products. This is illustrated in Fig. 7 for the polyoxyethylation of 1-dodecanol and 2-dodecanol using potassium hydroxide as catalyst and using an ethylene oxide/alcohol molar ratio of 7.

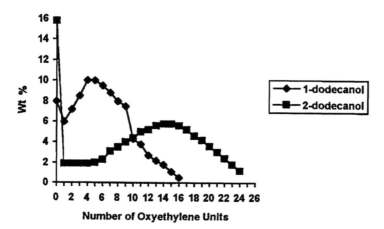

FIG. 7 Polyoxyethylation of 1-dodecanol and 2-dodecanol with 7 moles of ethylene oxide using KOH as catalyst [74].

As can be seen in the figure, polyoxyethylene secondary alcohols produced using potassium hydroxide as catalyst contain much higher concentrations of residual alcohol (16 wt%) compared with the primary alcohol materials (8 wt%). For this reason, polyoxyethylene secondary alcohols must be produced using a two-step (acidic, then basic catalyst) process as described in Chapter 1.

5. Commercial Production

Most commercial polyoxyethylation processes are conducted in batch operations. Traditional stirred tank reactors or the new Pressindustria SpA horizontal gas phase sparge-type reactors are used [75]. For maximum efficiency the alcohol/catalyst solution for the next batch reaction is prepared while the previous batch is in the process of polyoxyethylation. A typical simplified commercial reactor system is shown in Fig. 8.

In a typical commercial process, alcohol and 50 wt% aqueous potassium hydroxide are introduced to the catalyst reactor (approximately one-fifth the size of the batch reactor). The mixture is dried at 120–130°C under vacuum, usually while the previous batch is undergoing polyoxyethylation, until a residual water level of <200 ppm is attained. Then the catalyzed alcohol mixture is added to the reactor along with additional dried alcohol sufficient for the desired initial charge. The reaction temperature is increased to ~145°C at which time all external heating is terminated. Ethylene oxide and nitrogen are introduced to the reactor, which results in an exothermic reaction (ΔH=22 kcal/mole) that increases the temperature to the desired reaction temperature of ~160–180°C. Ethylene oxide is added at a rate sufficient to maintain reaction temper-

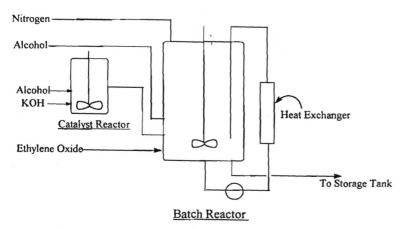

Batch Reactor

FIG. 8 Simplified diagram of a typical stirred tank polyoxyethylation reactor.

ature and the heat of reaction is removed by circulating a portion of the batch reactor contents to a heat exchanger. Safety procedures require that limits be placed on temperature and pressure. The gas phase ethylene oxide concentration is continually monitored by sensors, and if the reaction temperature decreases for some reason, ethylene oxide feed is immediately terminated to prevent a buildup of vapor phase ethylene oxide concentration.

After all of the required amount of ethylene oxide is added, the residual ethylene oxide is consumed by maintaining reaction temperature until there is no change in the reactor pressure. This usually requires approximately 30–60 min. Because many customers require that residual ethylene oxide be <5 ppm in the product, a nitrogen sparge of the product is optionally performed if product "aging" is insufficient to reduce ethylene oxide concentrations to the desired levels.

The reaction product is normally neutralized with acetic acid, but other homogeneous acids can be used on request. In some cases customers specify that there be no catalyst residues in the product, which can be accomplished by treating the product with sodium dihydrogen phosphate powder instead of acetic acid, and filtration of the product mixture. The potassium is removed by ion exchange with the inorganic material.

Modern commercial reactors are fully automated with computer-controlled operations. All current safety features are built into the control systems with automatic shutdowns set for over temperature and over pressure, as well as block valves to prevent catalyst mixtures from becoming backmixed with ethylene oxide supply.

B. Narrow-Range or Peaked Polyoxyethylene Alcohols

Conventional polyoxyethylene alcohols prepared using a basic catalyst such as potassium hydroxide are a mixture of polyoxyethylene oligomers with a broad chain length distribution. Several studies have shown that polyoxyethylene alcohols with a narrower polyoxyethylene chain length distribution than conventional are more effective in some detergency tests [76,77]. In these studies, the narrow range ethoxylates were made by laboratory techniques that were not suitable for commercial production. However, several synthetic alcohol manufacturers reported various catalyst systems that were capable of producing a much narrower polyoxyethylene chain length distribution than conventional in the 1980s and 1990s. These catalyst systems are listed and discussed in detail in Chapter 1. A comparison between the chain length distributions of conventional and narrow range polyoxyethylene alcohols is shown in Fig. 9.

Although there are not yet large-volume uses for narrow-range polyoxyethylene alcohols, several detergent manufacturers either have commercialized or are in the process of commercializing these materials. Narrow-range products have several inherent advantages over conventional materials. They usually have lower viscosity, a higher concentration of surface active oligomers in the effective detergent range, and lower free alcohol than their conventional counterparts. The latter is particularly important in applications wherein residual alcohol is undesirable (e.g., in spray drying applications).

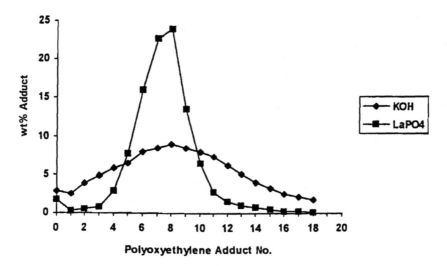

FIG. 9 A comparison of the oligomer distribution of conventional and narrow range polyoxyethylene alcohols [77].

Commercial narrow-range polyoxyethylene alcohol processes are very similar to the conventional process. Most, if not all, processes involve the use of a heterogeneous catalyst. A typical narrow-range polyoxyethylene alcohol process would begin by the preparation of a slurry of the heterogeneous catalyst in the alcohol to be polyoxyethylated. At this point the two process would be conducted in virtually the same manner as described previously. At the end of the ethylene oxide "aging" or soak-down phase, the catalyst could either be filtered for removal or left in the product, depending on customer requirements. The economics of the two processes would be similar if the heterogeneous catalyst could be left in the product. Catalyst removal could add 1–3 cents per pound to the cost of the process, depending on ease of removal and cost of disposal. Although 10–20 MM lb/year of narrow-range products is currently being sold, the future demand for these materials is uncertain.

IV. PROPERTIES OF POLYOXYETHYLENE ALCOHOLS

The physical properties of polyoxyethelene alcohols define their use in various applications. The most important physical properties of these materials are cloud point, pour point, foam profile, solubility, and wettability. The most common application of these materials is in emulsification and detergency. This section discusses various physical properties of surfactants and address how the structure of the polyoxyethylene alcohol affects these physical properties and applications.

A. Alcohol Structure and Surface Properties

The properties and applications of polyoxyethylene alcohols are dependent on the molecular weight and structure of the hydrophobe R,the number of moles of ethylene oxide (n) per mole of hydrophobe contained in the chain, and the chain length distribution (i.e., whether broad or narrow).

$$RO(CH_2CH_2O)_nH$$

The hydrophobe R is the oil soluble part of the surfactant. It can be either a linear or branched hydrocarbon chain obtained from natural or synthetic petrochemical sources. The hydrophobe is the basic "building block" of the surfactant molecule. The ethylene oxide chain is the hydrophile, or water-soluble portion, of the nonionic. In general, the larger the n value of the molecule, the more water-soluble is the nonionic.

1. Hydrophile–Lipophile Balance (HLB)

Polyoxyethylene alcohols are often classified according to their HLB or hydrophile–lipophile balance [79,80]. The HLB is an empirical measure of the emulsifying and solubilizing character of a nonionic surfactant. It is an

TABLE 2 HLB Ranges and Applications

HLB	EO content (wt%)	Application
4–6	20–30	Water in oil emulsion
7–15	35–75	Wetting agent
8–18	40–90	Oil in water emulsion
10–15	50–75	Detergent
10–18	50–90	Solubilizer

expression of the relationship of the size and strength of the polar (water-soluble) and nonpolar (oil-soluble) groups of a surfactant. The HLB number for a polyoxyethylene alcohol can be calculated from the following equation:

$$HLB = E/5$$

where E = the wt% of oxyethylene content in the nonionic surfactant. For example, the HLB of a $C_{12}H_{25}O(CH_2CH_2O)_nH$ surfactant where $n = 7$ is 62 ÷ 5, or 12.4. Oil-soluble surfactants have low HLBs and water-soluble surfactants have high HLBs. HLB numbers can be used as a rough guide for selection of a suitable nonionic surfactant in a specific application, e.g., water solubility, detergency, or emulsification of oils. Table 2 lists the suitable HLB ranges for various applications.

2. Cloud Point

The cloud point is a particularly important physical property of a surfactant because it determines the optimum condition for detergency and other applications. It is the temperature at which the nonionic surfactant oils out of solution or becomes insoluble due to the addition of heat. It is defined technically as that temperature at which a cloud or haze disappears when the warm, cloudy surfactant solution is slowly cooled (ASTM D2024 test method). Generally, nonionic surfactants exhibit optimum effectiveness when used at temperatures near their cloud points. As will be seen, the structure of the polyoxyethylene alcohol (i.e., both hydrophobe structure and EO content) determines the cloud point and application of the surfactant.

3. Pour Point

The pour point is a temperature 3°C above the temperature at which no movement of the product can be observed when cooled under prescribed conditions (ASTM D97 test method). This can be particularly important in the industrial applications area, especially in northern climates in the winter.

4. Solubility and Wettability

Solubility and wettability are particularly important properties because future projections are that surfactants are going to have to be more effective (soluble) at colder water wash temperatures. Wettability is important in industrial applications because wetting times need to be as short as possible in order to reduce cycle times.

5. Fundamental Physical Property and Detergency Studies

A number of fundamental detergency studies have been conducted to explore the effect of alcohol structure and molecular weight on surfactant properties. Early studies showed that detergency was strongly dependent on the length of the hydrophobe at a given polyethyleneoxy chain length [81,82,77]. A particularly interesting study was conducted comparing the effectiveness of a series of pure, single polyethylenoxy alcohols (e.g., $C_{12}O[CH_2CH_2O]_7H$) with conventional broad-range polyethylenoxy alcohols for oil soil removal. Specifically, a procedure using radiolabeled oily soils was used to compare the detergency of a series of pure, narrow-range, linear dodecanol ethoxylates containing from 3 to 8 moles of ethylene oxide (EO) and corresponding series of C_{12}–C_{13} alcohol ethoxylates with normal, broad polyoxyethylene distributions [82]. This study found that the pure, single EO alcohol ethoxylates significantly enhanced nonpolar soil removal from polyester/cotton fabrics with an optimum for detergency with the 5-mole EO adduct of dodecanol. However, there was no difference in detergency performance between the narrow-range and broad-range polyoxyethylene alcohols for polar soil (oleic acid) removal. These results suggest that the broad-range conventional polyoxyethylene alcohols that contain a significant fraction of the longer EO chain adducts, which are more water-soluble, are much less effective for oily soil removal. Thus, narrow-range polyoxyethylene alcohols contain more of the effective oligomers "concentrated" in the optimum detergency range compared with broad-range products for oily soil removal.

Recent studies [83,84] have been conducted addressing the effects of phase inversion temperature (PIT) and interfacial tension (IFT) on optimum detergency performance for oily soil removal using pure, homogeneous dodecyl ethoxylates (e.g., $C_{12}E_3$ and $C_{12}E_8$) and their blends. Phase inversion temperature is that temperature at which the transition from oil/water to water/oil phases occurs. More precisely, the PIT refers to the condition whereby equal volumes of oil and water are contained in the middle-phase microemulsion. At the PIT of a system ultralow interfacial tension occurs thus creating the conditions for maximum detergency. Detergency results were obtained in a study [83] that established the effect of temperature and the ethylene oxide content on the soil removal from polyester/cotton blends. Some of the results from this study are shown in Fig. 10 and 11. Additional studies [84,85] involving correlation of PIT to optimum detergency temperatures were conducted on polar and nonpolar soils using nonionic/anionic surfactant blends of pure polyoxyethylene alcohols (e.g., C_xE_y, where x is the number of carbons in hydrophobe and y is the number of EO moles per mole of hydrophobe).

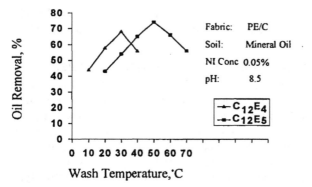

FIG. 10 Effect of temperature on oil soil removal from polyester/cotton using pure, homogeneous polyoxyethylene dodoecyl alcohols ($C_{12}E_4$ = 4 mole ethylene oxide adduct of dodecanol [83].

B. Commercially Available Products

There are many producers of polyoxyethylene alcohols and a large variety of products available under a large number of registered names. A representative but by no means comprehensive list of current manufacturers is shown in Table 3. A comprehensive listing can be found in McKutcheon's series of detergent manufacturers and sources [86]. The important physical and surface properties of commercially available products are the same as those discussed for single

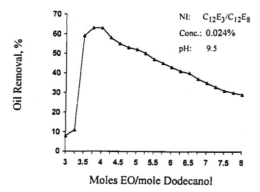

FIG. 11 Effect of ethylene oxide content on detergency using blends of $C_{12}E_3/C_{12}E_8$ [83].

TABLE 3 Major Polyoxyethylene Alcohol Manufacturers

Product	Alcohol type	Manufacturer
Alfonics	Normal, primary, Ziegler	Vista
Alkanol	Normal, primary, natural	DuPont
Basophor	Fatty alcohol	BASF
Bio-Soft EA	Fatty alcohol	Stepan
Brij	Normal, primary	ICI
Diazopon AN	Fatty alcohol	GAF
Dispersol A	Fatty alcohol	GAF
Emulphogene	Oxo, synthetic	Rhone-Poulenc
Genapol	Oxo, synthetic	Hoecsht-Celanese
Iconol	Branched decyl/tridecyl	BASF
Neodol	Oxo, synthetic	Shell
Novel	Narrow-range ethoxylates	Vista
Siponic	Fatty alcohol	Alcolac, Rhone-Poulenc
Surfonic	Oxo, synthetic	Huntsman
Tergitol	Secondary alcohol	Union Carbide

TABLE 4 Typical Physical Properties of Water-Insoluble Nonionics [80]

Product	Company	Hydrophobe	Avg. moles EO	EO, (wt%)	HLB	Pour point (°C)
Neodol 91-2.5	Shell	C_9,C_{10},C_{11}	2.7	42.3	8.5	−13
Neodol 1-3	Shell	C_{11}	3.0	43.3	8.7	−7
Neodol 1-5	Shell	C_{11}	5.0	56.1	11.2	6
Neodol 23-1	Shell	C_{12}/C_{13}	1.0	18.5	3.7	5
Neodol 23-3	Shell	C_{12}/C_{13}	2.9	39.6	7.9	1
Neodol 23-5	Shell	C_{12}/C_{13}	5.0	53.3	10.7	7
Neodol 25-3	Shell	$C_{12}/C_{13}/C_{14}/C_{15}$	2.8	37.3	7.5	3
Neodol 45-2.25	Shell	C_{14}/C_{15}	2.23	30.8	6.2	15
Alfonic 610-50R	Vista	$C_6/C_8/C_{10}$	3.1	50	10.0	−7
Alfonic 810-40	Vista	C_8/C_{10}	2.2	40	8.0	−5
Alfonic 1012-40	Vista	C_{10}/C_{12}	2.5	39	8.0	−5
Alfonic 12146C-30	Vista	C_{12}/C_{14}	1.9	30	6.0	7
Alfonic 12146C-40	Vista	C_{12}/C_{14}	2.0	40	8.0	4
Alfonic 1412-40	Vista	C_{14}/C_{12}	3.0	40	8.0	8
Alfonic 1216-22	Vista	$C_{12}/C_{14}/C_{16}$	1.3	22	4.4	4
Brij30	ICI	Lauryl	4	48.5	9.7	—
Genapol 26-L-1.6	Hoechst-Celanese	$C_{12}/C_{14}/C_{16}$	1.5	25	5.0	10
Genapol 26-L-3	Hoechst-Celanese	$C_{12}/C_{14}/C_{16}$	3.0	40	8.0	5
Genapol 26-L-5	Hoecsh-Celanese	$C_{12}/C_{14}/C_{16}$	5.0	53	10	6
Surfonic L12-3	Huntsman	C_{10}/C_{12}	3	45	9.0	−15
Surfonic L24-2	Huntsman	C_{12}/C_{14}	2	27	5.3	10
Surfonic L24-3	Huntsman	C_{12}/C_{14}	3	40	8.0	5

TABLE 5 Typical Physical Properties of Water-Soluble Nonionics [80]

Product	Company	Hydrophobe	Avg. moles EO	EO, wt%	HLB	Pour point (°C)
Neodol 91-6	Shell	C_9,C_{10},C_{11}	6.0	62.1	12.4	52
Neodol 91-8	Shell	C_9,C_{10},C_{11}	8.3	69.7	13.9	80
Neodol 1-7	Shell	C_{11}	7.0	64.3	12.9	58
Neodol 1-9	Shell	C_{11}	9.0	69.6	13.9	74
Neodol 23-6.5	Shell	C_{12}/C_{13}	6.6	60.0	12.0	43
Neodol 25-7	Shell	$C_{12}/C_{13}/C_{14}/C_{15}$	7.3	61.3	12.3	50
Neodol 25-9	Shell	$C_{12}/C_{13}/C_{14}/C_{15}$	8.9	65.6	13.1	74
Neodol 25-12	Shell	$C_{12}/C_{13}/C_{14}/C_{15}$	11.9	71.8	14.4	78
Neodol 45-7	Shell	C_{14}/C_{15}	7.0	58.2	11.6	45
Neodol 45-13	Shell	C_{14}/C_{15}	12.9	71.8	14.4	80
Alfonic 810-60	Vista	C_8/C_{10}	4.8	60	12	38
Alfonic 1012-60	Vista	C_{10}/C_{12}	5.4	60	12	37
Alfonic 1412-60	Vista	C_{14}/C_{12}	7.2	60	12	52
Novel 1412-70[a]	Vista	C_{14}/C_{12}	11	70	14	78
Brij35	ICI	Lauryl	23	85	16.9	—
Genapol 24-L-45	Hoechst-Celanese	C_{12}/C_{14}	6.3	58	11.6	45
Genapol 24-L-60	Hoechst-Celanese	C_{12}/C_{14}	7.2	61	12.2	60
Genapol 24-L-75	Hoechst-Celanese	C_{12}/C_{14}	8.3	65	12.9	75
Genapol 24-L-92	Hoechst-Celanese	C_{12}/C_{14}	10.6	70	14.0	92
Genapol 24-L-60N[a]	Hoechst-Celanese	C_{12}/C_{14}	7.0	60	12.1	60
Genapol 24-L-98N[a]	Hoecsh-Celanese	C_{12}/C_{14}	11.3	71	14.2	98
Surfonic L12-6	Huntsman	C_{10}/C_{12}	6	62	12.4	48
Surfonic L24-7	Huntsman	C_{12}/C_{14}	7	59	11.9	48
Surfonic L24-9	Huntsman	C_{12}/C_{14}	9	65	13.0	73
Surfonic L24-12	Huntsman	C_{12}/C_{14}	12	72	14.4	65
Surfonic L46-7	Huntsman	C_{14}/C_{16}	7	58	11.6	48

[a] Narrow range ethoxylates.

carbon and single polyoxyethylene adducts, namely, foaming, solubility, wettability, cloud point, and pour point. This section will deal with information and studies using commercially available products.

1. Physical Properties

(a) Cloud Point. Information concerning the physical and chemical properties for a selection of major commercial nonionic surfactants is given in Tables 4 and 5 [80]. For convenience, the list is divided into water-insoluble (oil-soluble) and water-soluble nonionics.

(b) Viscosity. At room temperature, most nonionic surfactants form a gel with the addition of water. Table 6 compares the viscosities of selected commercial polyoxyethylene alcohols with their alkylphenol and secondary alcohol counterparts at various surfactant concentrations in water. This information

TABLE 6 Viscosities of Aqueous Ethoxylate Solutions (Centipoise at 22°C)

Product	10	20	30	40	50	60	80
Neodol 91-6	3	13	63	173	187	144	80
Neodol 91-8	2	6	29	138	Gel	Gel	120
Neodol 1-5	30	48	58	71	1649	30350	54400
Neodol 1-7	3	14	109	Gel	Gel	235	87
Neodol 1-9	2	6	26	245	Gel	Gel	104
Neodol 23-5	282	4895	Gel	Gel	Gel	Gel	56500
Neodol 23-6.5	27	431	1620	Gel	Gel	37000[a]	Gel
Neodol 25-7[b]			960[c]	Gel	Gel	Gel	Gel
Neodol 25-9[b]			70[c]	Gel	Gel	Gel	Gel
Neodol 25-12[b]			71[c]	Gel	Gel	Gel	Gel
Neodol 45-7[b]			2530	Gel	Gel	Gel	Gel
Neodol 45-13[b]			80	Gel	Gel	Gel	Gel
Linear 1012 primary alcohol (5.2EO)			160	208	176[a]	37750[a]	201[a]
Random secondary alcohol (7EO)			88	179	205[d]	1940[a]	116
Nonylphenol (9EO)			290	Gel	Gel	3020	1080[a]
Octylphenol (9.5EO)			100	Gel	Gel	1640	456

[a]Fluid gel, by examination with polarized light.
[b]Measured at 25°C.
[c]Centistokes.
[d]Clear solution.

indicates the problems involved with handling and trying to formulate aqueous products using various surfactants

(c) *Gel Curves or Phases.* As can be seen in Table 6, concentrated solutions of polyoxyethylene alcohols can often form gels. The gelling characteristics are depicted in temperature vs. surfactant concentration plots, called *gel curves*. To avoid gel formation, the formulator must add the surfactant to, or dilute neat surfactant with, water that has been heated above the peak temperature displayed on the gel curve. A knowledge of gel curves is essential for the proper handling of these materials. A typical gel curve for Neodol 1-9 is shown in Fig. 12.

2. Surface Properties

The surface properties of commercially available products can be obtained from supplier information. The main properties of interest are wetting, foamability, and detergency.

(a) *Draves Wetting.* The Draves wetting test measures the ability of a surfactant to cause the solution to wet or spread evenly onto surfaces. The surfactant with the shortest wetting time generally works faster. Table 7

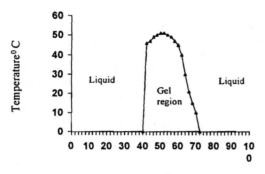

wt% NEODOL 1-9 in Water

FIG. 12 Gel curve for Neodol 1-9 ethoxylate [80].

displays the wetting times of some selected commercial surfactants at room (25°C) and elevated (60°C) temperatures.

(b) Foam Performance. Polyoxyethylene alcohols have moderate foaming characteristics, significantly less than their sulfated derivatives. However, foaming can be both an asset or a liability depending on the desired end use. Polyoxyethylene alcohols are actually used as foam breakers when used in formulations above their cloud points, e.g., with automatic dishwashing prod-

TABLE 7 Draves Wetting Times [79]

Product	Seconds at 25°C	Seconds at 60°C
Neodol 91-6	5	4
Neodol 91-8	10	5
Neodol 1-5	5	5
Neodol 1-7	5	—
Neodol 1-9	9	—
Neodol 23-5	20	13
Neodol 23-6.5	9	16
Neodol 25-7	11	21
Neodol 25-9	14	6
Neodol 25-12	35	11
Neodol 45-7	17	15
Neodol 45-13	48	11
Linear 1012 primary alcohol (5.2EO)	5	5
Random secondary alcohol (7EO)	6	9
Nonylphenol (9EO)	11	11
Octylphenol (9.5EO)	10	7
Linear 812 primary alcohol EO/PO nonionic	7	6
Branched tridecyl alcohol ethoxylate	5	7

TABLE 8 Dynamic Spray Foam Performance

Product	PSIG	Dynamic spray foam height, cm (24°C, 0.1 wt%, distilled water) Time, min						
		3	5	10	15	20	25	30
Neodol 91-6	10	—	33.2	47.0	53.5	>65	—	—
Neodol 91-6[a]	10	—	29.7	42.5	48.0	51.0	—	—
Neodol 91-6	15	0	8.0	30.0	36.0	42.0	52.0	53.0
Neodol 91-8	10	—	35.8	51.3	>60	>65	—	—
Neodol 91-8[a]	10	—	34.0	48.0	56.0	>60	—	—
Neodol 1-5	10	—	24.0	36.0	44.0	49.0	—	—
Neodol 1-5[a]	10	—	26.2	36.8	41.3	43.5	—	—
Neodol 1-7	10	—	32.5	47.0	56.5	>60	—	—
Neodol 1-9	10	—	35.0	52.0	>60	>60	—	—
Neodol 23-5	10	—	24.0	37.0	44.0	47.0	—	50.0
Neodol 23-6.5	10	—	33.0	46.0	53.0	58.0	—	>65
Neodol 25-7	10	—	19.5	37.5	46.7	50.8	—	—
Neodol 25-7[a]	10	—	27.2	40.7	46.9	50.3	—	—
Neodol 25-9	10	—	32.0	45.5	53.2	>60	—	—
Neodol25-9	15	0	9.0	3.0	44.0	52.0	53.0	55.0
Neodol25-12	10	—	36.5	51.0	>60	>65	—	—
Neodol 25-12[a]	10	—	32.3	45.8	51.5	56.0	—	—
Neodol 45-7	10	—	28.3	42.7	49.5	53.0	—	—
Neodol 45-7	15	0	0	13.0	26.0	37.0	44.0	47.0
Neodol 45-13	10	—	8.0	—	22.0	33.0	—	—
Nonylphenol (9EO)	10	—	29.3	42.6	49.0	52.1	—	—
Nonylphenol (9EO)	15	0	0	13.0	26.0	37.0	44.0	48.0
Octylphenol (9.5EO)	10	—	33.0	45.3	52.3	57.0	—	—
Secondary alcohol (5EO)	10	—	28.5	42.1	48.5	51.5	—	—

[a]Solutions containing 2% NaOH.

ucts. Therefore, knowledge of the foam profiles of various surfactants enables one to choose the correct application.

Foam characteristics are often measured using dynamic spray foam tests. Table 8 shows the dynamic foam heights for a variety of commercial ethoxylates determined using the Shell dynamic spray foam test in distilled water. This test method was designed to generate foam data under more realistic dynamic conditions. Foam is generated by injection of the surfactant solution through a spray nozzle under specific conditions, onto a glass column. Impingement of the sprayed circulating solution onto the glass column wall generates foam continuously and is measured as a function of time. Foam height is interpreted as a relative measurement.

V. APPLICATIONS

It is not possible to discuss all the uses of the polyoxyethylene alcohols in detail. However, the overall use of these products will be discussed in general terms, with the major areas of applications being addressed. At the end of this section, future trends for applications of polyoxyethylene alcohols will be discussed.

A. Household Surfactants

The household market is one of the largest uses for surfactants with the products being compounded as powders, liquids (or gels), pastes, or bars. There are a variety of uses that nonionic surfactants find in the home. Heavy-duty liquids or powders are commonly used in general home laundry [87–93] and light-duty liquids are used in hand dishwashing [94]. The polyoxyethylene alcohols are often used in combination with their sulfated derivatives for maximum efficacy. Such starter formulations can be found in commercial brochures [94].

Polyoxyethylene alcohols are also a key component as a defoaming aid for automatic dishwashing applications. Degreasers are also used both in the house and in the garage to remove oil spills and stains. Home carpets are often cleaned with the aid of prespotters that are a combination of surfactants including nonionics. Hard surface cleaning (e.g., floors, walls, countertops, tiled areas) is a problem that is facilitated by the use of nonionics such as Shell's Neodol 1 and Neodol 91 series of surfactants [95,96].

B. Industrial and Institutional Applications

1. Pulp and Paper

The pulp and paper industry uses polyoxyethylene alcohols in several stages of their manufacturing processes. Wetting agents are essential for the proper treatment of paper and box board, i.e., for rapid impregnation of solutions into the fibers. Deinking of magazines and newsprint requires significant usage of these alcohols in the need for recycling [97,98]. The deinking process utilizes surfactants such as polyoxyethylene alcohols for the dispersion and removal of various inks and coatings. The surfactants must be sufficiently effective to return the sheet brightness to essentially that of original newsprint quality paper [99,100]. An essential characteristic is low foaming for the deinking process. Often propoxylated polyoxyethylene alcohols are used for this purpose.

2. Textile and Wool Scouring

Conventional processes for applying chemicals to fabric in textile finishing mills require 60–100% wet pickup from aqueous solutions. Then the fabric must be dried to remove the water used in the process. An economic problem is encountered when large amounts of water must be handled and long drying times used. These problems can be minimized using polyoxyethylene alcohols to aid in the wet pickup process using foams that contain much less water [101].

Another application is the use of the nonionic surfactants to aid in the dyeing of the fabrics. This is a major application in the area of textile dyeing [102–107] and finishing of cotton textiles. The surfactants used must be effective in agglomeration of dye particles into micelles and deposition onto the textile.

3. Agricultural Emulsions and Spray Adjuvants

The polyoxyethylene alcohols are excellent emulsifiers for agricultural formulations. Most active ingredients for pesticides and herbicides are hydrocarbon-soluble materials and must be mixed with water prior to use. Consequently, the solution must be emulsified using a surfactant [108,109]. Modern liquid fertilizers also use surfactants to aid in the delivery of the phosphate salt to the root system. Surfactant are also used in spray adjuvants to enable the liquid pesticide or herbicide to wet and penetrate the plant leaves.

4. Enhanced Oil Recovery (EOR)

The recovery of crude oil from oil-producing properties that have declined in production over the years has been a profitable industry over the years. The use of surfactants to increase the production of oil from uneconomic drilling rigs has increased steadily over the years [110,111]. Interest is always the highest when the price of crude oil increases. The mechanism of action for surfactant-assisted recovery of crude oil has also been the subject of much academic and industrial investigation [112].

5. Future Uses

One attractive potential new application for polyoxyethylene alcohols and their sulfated derivatives is in the area of soil remediation and environmental cleanup [113]. Soil has become contaminated in many parts of the country through a combination of activities (excessive use of fertilizers, pesticides, chemical disposal, etc.). One promising new research area is the development of surfactants and formulations for the removal of contaminants from soil. With an increasing emphasis on environmental issues, soil remediation could become a new area for applications for polyoxyethylene alcohols.

REFERENCES

1. H. Adkins and R. Conner, J. Am. Chem. Soc. 53:1091 (1931).
2. H. Adkins and K. Folkers, J. Am. Chem. Soc 53:1095 (1931).
3. H. R. Arnold and W. A. Lazier, U.S. Patent 2,116,552 to Dupont (1938).
4. W. B. Johnston, U.S. Patent 2,347,562 to American Cyanamid (1994).
5. E. F. Hill, G. R. Wilson, and E. C. Steinle, Ind. Eng. Chem. 46(9):1917 (1954).

6. J. Burgers, Canadian Patent 672,715 to Procter & Gamble (1960).

7. V. Mills, German Patent 1,008,273 to Procter & Gamble (1957).

8. A. Abbey, British Patent 573,788 to Procter & Gamble (1945).

9. H. Corr, British Patent 921,477 to BASF (1963).

10. K. Negi, French Patent 1,375,472 (1964).

11. G. Harrison, Pet Int. App. 90/8127 to Davy McKee (1990).

12. M. Wilmott, U.S. Patent 5,138,106 to Davy McKee (1992).

13. M. Wilmott, U.S. Patent 5,157,168 to Davy McKee (1992).

14. Linear Alpha Olefins, Report No.12C, Process Economic Program, SRI International (1990).

15. P. J. Flory, J. Am. Chem. Soc. *62*:1561 (1940).

16. H. B. Fernald et al., U.S. Patent 3,482,000 to Gulf (1969).

17. R. T. Armstrong, Canadian Patent 831,550 to Gulf (1970).

18. H. B. Fernald et al., U.S. Patent 3,510,539 to Gulf (1970).

19. G. R. Lappin, et al. (eds.), *Alpha Olefins Applications Handbook*, Marcel Dekker, New York, 1989.

20. K. Ziegler, British Patent 944,935 (1963).

21. K. Ziegler, German Patent 1,191,815 (1966).

22. British Patent 1,037,866 (1966).

23. A. E. Harkins, Jr., U.S. Patent 3,655,809 to Ethyl (1972).

24. C. W. Lanier, U.S. Patent 3,789,081, to Ethyl (1974).

25. W. T. Davis, Canadian Patent 834,217, to Ethyl (1970).

26. W. T. Davis, U.S. Patent 3,395,292 to Ethyl (1967).

27. C. W. Lanier, U.S. Patent 3,686,250 to Ethyl (1972).

28. P. Kobetz, U.S. Patent 3,689,584 to Ethyl (1972).

29. Shell Higher Olefins Process—SHOP, Technical Bulletin SC:335-84, Shell Chemical Company, 1984.

30. E. F. Magoon et al. U.S. Patent 3,483,269 to Shell Oil (1969).

31. P. W. Glockner, U.S. Patent 3,530,200 to Shell Oil (1970).

32. H. J. Alkema et al., U.S. Patent 3,634,539 to Shell Oil (1972).

33. E. F. Magoon et al., U.S. Patent 3,592,866 to Shell Oil (1971).

34. R. Van Helden et al., U.S. Patent 3,728,414 to Shell Oil (1973).

35. P. A. Verbrugge, U.S. Patent 3,776,975 to Shell Oil (1973).

36. R. F. Mason et al., U.S. Patent 3,758,558 to Shell Oil (1973).

37. E. R. Freitas et al., "Shell Higher Olefins Process—SHOP," Paper No.27a, presented at the American Institute of Chemical Engineers, 85th National Meeting and Chemical Plant Equipment Exposition, Philadelphia, Pennsylvania, June 4–8, 1978.

38. A. E. O'Donnell et al., U.S. Patent 4,377,499 to Shell Oil (1983).

39. D. M. Singleton, U.S. Patent 4,503,280 to Shell Oil (1985).

40. E. F. Lutz, U.S. Patent 4,528,416 to Shell Oil (1985).

41. E. R. Freitas, Chem. Eng. *75*(1):73 (1979).

42. I. Wender, S. Metun, S. Ergun, H. W. Sternberg, and H. Greenfield, J. Am. Chem. Soc. *78*:5401 (1956).

43. I. Wender, H W. Sternberg, and M. J. Orchin, J. Am. Chem. Soc. *75*:3041 (1953).

44. H. Lemke, Hydrocarbon Process. Petrol. Refin. 45(2):148 (1966).

45. R. F. Heck and D. S. Breslow, J. Am. Chem. Soc. 83:4023 (1961).

46. L. Marko, G. Bork, G. Almasy, and P. Szabo, Brennstoff-Chem. *44*:184 (1959).

47. L. Kirch and M. Orchin, J. Am. Chem. Soc *80*:4428 (1958).
48. J. Falbe, *Carbon Monoxide in Organic Synthesis*, Springer-Verlag, New York, 1967, pp. 1–75.
49. L. H. Slaugh et al., U.S. Patent 3,239,566 to Shell Oil (1966).
50. L. H. Slaugh et al., U.S. Patent 3,239,569 to Shell Oil (1966).
51. L. H. Slaugh et al., U.S. Patent 3,239,570 to Shell Oil (1966).
52. L. H. Slaugh et al., U.S. Patent 3,239,571 to Shell Oil (1966).
53. C. R. Greene et al., U.S. Patent 3,274,263 to Shell Oil (1966).
54. Robert H. Schwaar and Kenneth E. Lunde, "Linear C12-C15 Primary Alcohols," Report No.163, Process Economics Program, SRI International, 1983.
55. French Patent 1,379,144 to Ethyl (1964).
56. W. T. Davis et al., French Patent 1,433,679 to Ethyl (1964).
57. Robert G. Muller, Linear Higher Alcohols, Report No. 27, Process Economics Program, SRI International, 1967.
58. T. Hellthaler et al., German Patent 552,886 to A. Riebecksche Montanwerke, (1934).
59. Japanese Patent 18-157177 to Kao Soap (1943).
60. Technology Newsletter, Chem. Week, p. 59, Jan. 28, 1967.
61. G. H. Twigg, The Mechanism of Liquid-Phase Oxidation, Special Supplement to Chem. Eng. Sci. *3*:5 (1954).
62. V. V. Veselov et al., Oxidation of Liquid Paraffinic Hydrocarbons to Alcohols in the Presence of Small Amounts of Boric Acid, Nauch-Issled. Inst. Sintetich, Zhirozamenitelei I Moyushchikh Sredstv 2:39 (1961).
63. V. V. Veselov et al., The Thermal Stability of Borates of Secondary High-Molecular-Weight Alcohols, Nauch-Issled. Inst. Sintetich, Zhirozamenitelei I Moyushchikh Sredstv 2:33 (1961).
64. F. R. Mayo and C. Walling, Chem. Rev. 27:351 (1940).
65. P. J. Garner and H. N. Short, U.S. Patent 2,587,990 to Shell Development Co. (1952).
66. K. L. Butcher and G. M. Nickson, J. Appl. Chem. (London) *10*:65 (1960).
67. R. B. Mosely et al., U.S. Patent 3,299,150 to Shell Oil (1967).
68. A. Mitsutani et al., French Patent 1,457,718 to Kurashiki Rayon (1966).
69. British Patent 1,049,043 to Kurashiki Rayon (1966).
70. K. C. Rottenberg, U.S. Patent 2,779,803 to Phillips Petroleum (1957).
71. E. Santacesaria, M. Di Serio, R. Garaffa, and G. Addino, Ind. Eng. Chem. Res. *31*:2419 (1992).
72. W. D. Satkowski and C. G. Hsu, Ind. Eng. Chem. *49*:1875 (1957).
73. N. Shachat and H. L. Greenwald, in *Nonionic Surfactants* (M. J. Schick, ed.), Marcel Dekker, New York, 1966, p. 22.
74. C. L. Edwards, unpublished results.
75. Michael Arne, Nonionic Surfactants, Process Economic Program Report No. 168, SRI International, 1984.
76. K. W. Dillan, J. Am. Oil Chem. Soc. *62*:1144 (1985).
77. H. L. Benson and Y. C. Chiu, Relationship of Detergency to Micellar Properties for Narrow Range Alcohol Ethoxylates, Technical Bulletin SC:443-80, Shell Chemical Company, 1980.
78. C. L. Edwards, U.S. Patent 5,057,627 to Shell Oil Co. (1991).
79. W. C. Griffin, J. Soc. Cosmet. Chemists *1*:311 (1949).

80. Neodol Ethoxylates and Competitive Nonionics Properties Guide, SC:569-94, Shell Chemical Company, 1994.
81. B. M. Finger et al., Detergent Alcohols: The Effect of Alcohol Structure and Molecular Weight on Surfactant Properties, Technical Bulletin SC:364-80, Shell Chemical Company, 1980.
82. E. B. M. Finger, G. A. Gillies, G. M. Hartuig, E. R. Ryder, Jr., W. M. Sauyer, and H. Stupel, J. Am. Oil Chemists Soc. 44:525 (1967).
83. H. L. Benson and K. H. Raney, The Requirements for Optimum Detergency, presented at New Horizons 1989 Detergent Industry Conference, Hershey, PA., October 29–November 1, 1989.
84. K. H. Raney and H. L. Benson, J. Am. Oil Chemists Soc. 67:722 (1990).
85. K. H. Raney, J. Am. Oil Chemists Soc. 68:525 (1991).
86. *McCutcheon's Volume 1: Emulsifiers and Detergents*, North American Edition, M. C. Publishing, Glen Rock, NJ, 1994.
87. L. Kravetz and D. Scharer, The Formulation and Performance of Heavy-Duty Liquid Laundry Detergents, Technical Bulletin SC:180-77, Shell Chemical Company, 1977.
88. D. W. Bisacchi, W. T. Shebs, and H. Stupel, No-Phosphate Liquid Laundry Detergents, presented at the American Oil Chemists Society, 47th Annual Fall meeting, September 18, 1973, Chicago, Illinois.
89. L. Kravetz and M. M. Wald, Detergency and Formulation Optimization of Unbuilt, Quarter Cup Heavy Duty Laundry Liquids, Technical Bulletin SC:344-18, Shell Chemical Company, 1981.
90. C. L. Merrill, The Optimization of Heavy Duty Liquids Containing Fabric Softener, Technical Bulletin SC:839-85, Shell Chemical Company, 1985.
91. L. Kravetz and K. F. Guinn, Effect of Surfactant Structure on Stability of Enzymes Formulated into Laundry Liquids, Technical Bulletin SC:814-91, Shell Chemical Company, 1991.
92. C. L. Merrill, Development of a Nonionic Surfactant-Based High Density Laundry Powder, Technical Bulletin SC:968-91, Shell Chemical Company, 1991.
93. L. Kravetz, Effects of Surfactant Structure on Detergency Performance of Laundry Powders Formulated with Zeolite 4A as Builder, Technical Bulletin SC:581-94, Shell Chemical Company, 1994.
94. Neodol Product Guide for Alcohols, Ethoxylates and Derivatives, Technical Bulletin SC:7-94, Shell Chemical Company, 1994.
95. Nonionic Surfactants in Hard Surface Cleaners, Technical Bulletin SC:441-80, Shell Chemical Company, 1980.
96. B. Lindman, S. Engstroem and J. Baeckstroem, J. Surf. Sci. Technol 4(1):23 (1988).
97. C. W. Schroeder, U.S. Patent 2,913,356 to Shell Development Co. (1959).
98. K. J. Lissant, U.S. Patent 3,069,306 to Petrolite Corporation (1962).
99. J. K. Borchardt and D. L. Wood, Preliminary Studies of the Deinking of Office Papers, Technical Bulletin SC:1430-92, Shell Chemical Company, 1992.
100. J. K. Borchardt and D. L. Wood, Surfactants in the Deinking of Old Newspapers, Technical Bulletin SC:1394-92, Shell Chemical Company, 1992.
101. NEODOL Ethoxylates in Textile Foam Finishing, Technical Bulletin SC:541-87, Shell Chemical Company, 1987.

102. H. B. Goldstein and H. W. Smith, Lower Limits of Low Wet Pickup Finishing, Textile Chemist and Colorist *12*:49 (1980).
103. B. L. Richardson, Foam Dyeing of Carpets Hits Its Stride, Textile World, *January 1981*, p. 55.
104. D. H. Ashmus, W. W. Rankin, and A. T. Walter, U.S. Patent 4,023,526 (1977).
105. H. Werdenberg, U.S. Patent 3,068,058 to Ciba (1962).
106. K. Knopf and E. Schollmeyer, Tenside Det. *24*(2):101 (1987).
107. W. Becker, D. Knittel and E. Schollmeyer, Tenside Det. *24*(5):264 (1987).
108. R. L. Sundberg and J. M. Cross, U.S. Patent 2,965,678 to General Aniline and Film Corp. (1960).
109. J. C. Wiedow, U.S. Patent 2,893,913 to Monsanto (1959).
110. K. Kosswig, Chemie in unserer Zeit *18*(3):87 (1984).
111. M. M. da Gama and K. E. Gubbings, Mol. Phys. *59*(2):227 (1986).
112. C. C. Mattax, Proc. World Pet. Congr. 11th, *3*:205 (1984).
113. C. C. West and J. H. Harwell, Environ. Sci. Technol *26*(12):2324 (1992).

4

Polyoxyethylene Esters of Fatty Acids

KURT KOSSWIG[*] Hüls AG, Marl, Germany

I. INTRODUCTION

Polyoxyethylene esters of fatty acids were among the first nonionic surfactants commercialized, when Schöller and Wittwer invented oxyethylation, i.e., the reaction of ethylene oxide with a proton-active substrate, in 1930 [1]. The

---

[*] Retired.

123

mono- and diesters of ethylene glycol and oligoethylene glycols had of course been known before, having been synthesized by esterification of the corresponding glycols with the acids or by analogous condensation reactions.*

Goldsmith [2] has given an excellent review on work reported up to 1942 on the preparation, properties, and uses of polyhydric alcohol esters of fatty acids, starting with a publication of A. Wurtz in 1859. In this review six synthesis methods are discussed: (1) direct esterification of hydroxyl groups with fatty acids or their anhydrides in the presence or absence of dehydrating agents; (2) esterification by means of fatty acid halides; (3) transesterification of triglycerides with polyhydric alcohols; (4) reaction of hydroxyalkyl halides with fatty acid salts accompanied by the formation of metal halides; (5) the partial hydrolysis of completely esterified polyhydric alcohols; and (6) the addition of ethylene oxide to compounds containing hydroxyl or carboxyl groups. Of these methods, direct esterification, transesterification, and addition of ethylene oxide to fatty acids are being carried out today to synthesize polyoxyethylene esters of fatty acids. While the esterification and transesterification reactions lead to the polyoxyethylene diesters in high yields when the appropriate molar ratios of starting materials are chosen, it is a key feature of these reactions that they are not suited to selective preparation of the monoesters, when the molar ratio of polyoxyethylene and acid is accordingly adjusted. Instead, the reaction product consists of a mixture of the desired monoesters together with considerable amounts of diesters and unreacted polyoxyethylene, as long as the polyoxyethylene is not introduced in great excess. This composition is explained by equal reactivity of both hydroxyl groups of the polyoxyethylene (glycol) as well as by the disproportionation of the polyoxyethylene monoester formed in a first step by transesterification:

$$2RCO(OCH_2CH_2)_xOH = RCO(OCH_2CH_2)_xOCOR + H(OCH_2CH_2)_xOH$$

The addition of ethylene oxide to a fatty acid yields a product mixture similar to that obtained in the esterification reaction of polyoxyethylene and fatty acid in a molar ratio of (1:1).

The preparation of pure polyoxyethylene monoesters, described below, is more demanding. At the same time the pure monoesters by themselves may disproportionate even in the absence of catalysts [7]. The pure monoesters therefore are of no relevance in industrial applications, despite the fact that they are, on the grounds of their chemical structure, the only true surface active species in the mixtures obtained by esterification or oxyethylation of fatty acids. Instead, the latter mixtures are produced industrially and used as such in a variety of applications. Today, quite a few products are obtained by esterification or oxyethylation of fats and oils, i.e., triglycerides, instead of the fatty acids originating therefrom, yielding products with a somewhat different

---

* See Refs. 1–6 for historical accounts and general overviews.

composition and containing chemically bound glycerol. Because of the close relation of these products to the true polyoxyethylene esters, they are also treated in this chapter.

II. SYNTHESIS AND PROPERTIES OF DEFINED POLYOXYETHYLENE ESTERS

A. Synthesis of Defined Polyoxyethylene Esters

The direct esterification of polyoxyethylene with fatty acids in a molar ratio of 2:1 to yield the diesters is a simple and straightforward reaction. It is the oldest method recorded [2]. Although the reaction already proceeds at room temperature, it is too slow to be of practical use. In general, the esterification is carried out at temperatures between 150 and 250°C. Acidic, amphoteric, and alkaline catalysts have been suggested for this reaction [2], sodium or potassium hydroxide or *p*-toluenesulfonic acid being preferred. To complete the conversion, water of reaction must be removed by vacuum or by a current of air or an inert gas. As the reaction mixture is often heterogenous in the initial stage, it has to be agitated. To lower the reaction temperature, carrier solvents such as toluene are often used. At the end of the reaction, unreacted polyoxyethylene may be removed by washing with water. Residual fatty acid may be extracted or precipitated as sodium salt, although some solvent for the diester such as dichloroethylene or acetone, which subsequently has to be evaporated, is requisite in this operation.

The formation of the polyoxyethylene monoester is the first step in the esterification reaction. Fatty acids are more readily miscible with the monoesters than with polyoxyethylenes and therefore they react more rapidly with the free hydroxyl group of the monoester than with one of the two hydroxyl groups of the polyoxyethylene. This is detrimental for the synthesis of monoesters. Consequently, high selectivities of monoesters in the esterification reaction can only be achieved if large excesses of polyoxyethylene over the fatty acid are employed, with molar ratios of polyoxyethylene to fatty acid of 6–12 being quite common [8,9]. After completion of the reaction, excess polyoxyethylene is washed out with water, preferably with a concentrated salt solution [2,10]. In this reaction the methyl esters of the fatty acids may be used instead of the free acids as starting material, preferably in the presence of an alkaline catalyst [11].

A more pretentious method of preparing monoesters is the esterification of fatty acids with polyoxyethylene in which one of the two terminal hydroxyl groups is protected. Boric acid was shown to be a good protective agent [12]. Some fatty acid monoesters of ethylene glycol [12], polyethylene glycol 600 or dodecaoxyethylene, POE-(12) [13], and monodisperse polyoxyethylene with 1–8 ethylene oxide units in the polyoxyethylene chain [8,14] were prepared. In this method, which can be accomplished as a one-pot reaction, first polyoxyethylene is esterified with boric acid at about 100°C in a molar ratio of 3:1, which leads to the tris(1-hydroxypolyoxyethylene)orthoboric acid ester. This in turn is esterified with fatty acid at 100°C in the presence of p-toluenesulfonic acid as catalyst. After esterification by removal of water in vacuo, the boric acid is hydrolyzed by addition of water. The polyoxyethylene fatty acid monoester can then be extracted, e.g., with trichloroethylene. Alternatively, the reaction mixture is washed with brine.

$$3H(OCH_2CH_2)_xOH + (HO)_3B \rightleftharpoons [H(OCH_2CH_2)_x]_3B + 3H_2O$$

$$3RCOOH + [H(OCH_2CH_2)_x]_3B \rightleftharpoons [RCO(OCH_2CH_2)_xO]_3B + 3H_2O$$

$$[RCO(OCH_2CH_2)_xO]_3B + 3H_2O \rightleftharpoons 3RCO(OCH_2CH_2)_xOH + (HO)_3B$$

The selective synthesis of polyoxyethylene monoesters or diesters of fatty acids by the addition of ethylene oxide to fatty acids in an oxyethylation reaction is not possible:

$$RCOOH + xCH_2CH_2O \rightarrow RCO(OCH_2CH_2)_xOH$$

As will be discussed in Sec. III.A, the ethylene glycol monoesters initially formed in this reaction quickly transesterify, leading to a mixture of mono- and diesters and ethylene glycol. The same is true for the polyoxyethylene esters formed by successive addition of ethylene oxide to the monoglycol ester formed in the first place. It is only with certain catalysts, such as amines [15] or amine oxides [16], that transesterification is suppressed until 1 mole of ethylene oxide has been added to 1 mole of acid. Beyond that equilibration occurs again.

The mono- and diesters of polyoxyethylenes can obviously be isolated from their mixtures by separation techniques. Ethylene glycol monolaurate and dilaurate for example, as well as diethylene glycol dilaurate, have been isolated from oxyethylated lauric acid by distillation [17]. Nevertheless, this method, as well as molecular distillation [18], is of limited use in the preparation of polyoxyethylene esters, especially their higher homologs.

B. Properties of Defined Polyoxyethylene Esters

Physical data of pure polyoxyethylene fatty acid mono- and diesters are scarce and moreover difficult to scrutinize, as information on the purity of the product and the polydispersity of the polyoxyethylene chain is lacking in most cases. Unambiguous data were published by Gerhardt and Holzbauer [14], who reported the melting points and refractive indices for the first eight homologous polyoxyethylene monoesters of lauric acid, as shown in Table 1.

Collected data on melting points of polydisperse polyoxyethylene esters of fatty acids follow a general pattern. The monoesters melt at a lower temperature than the diesters with the same polyoxyethylene chain length. Starting with the ethylene glycol esters of lauric through stearic acid, the melting points decrease

TABLE 1 Melting Points and Refractive Indices[a] $(n_D{}^{40})$ of Homologous Polyoxyethylene Monoesters of Lauric Acid

$C_{11}CO(OCH_2CH_2)_xOH$ $x =$	Melting point (°C)	$n_D{}^{40}$
1	28–30	1.4405
2	17–18.5	1.4431
3	15–16.5	1.4449
4	17–18	1.4462
5	19–21	1.4478
6	21–23	1.4495
7	26–27.5	1.4515
8	30–31	1.4522

[a]According to Ref. 14.

with increasing number of ethylene oxide units in the polyoxyethylene chain, with a minimum at an average chain length of about eight ethylene oxide units, increasing again to the level of the melting points of the monoethylene glycol esters when an average chain length of 12–20 ethylene oxide units has been reached. The polyoxyethylene esters of oleic acid have the lowest melting points and are liquid at room temperature. A comparison of the melting points of some fatty acid esters of ethylene and diethylene glycol may be instructive, taking the highest reported in each case [2]; see Table 2. A study of the phase behavior of the monomyristate, monopalmitate, and monostearate of ethylene glycol together with the esters of some other polyhydroxy compounds has shown that ethylene glycol monostearate is dimorphic [20].

TABLE 2 Melting Points of Fatty Acid Esters[a] of Ethylene and Diethylene Glycol

Ester		Melting point (°C)
Ethylene glycol	monolaurate	32
	dilaurate	54
	monomyristate	42
	dimyristate	64
	monopalmitate	52
	dipalmitate	72
	monostearate	58
	distearate	79
Diethylene glycol	monolaurate	18
	dilaurate	39
	monostearate	42
	distearate	58

[a]According to Ref. 2.

TABLE 3 Surface Tension[a] (mN/m) at 25°C of Polyoxyethylene
Monoesters of Fatty Acids at Three Concentrations

Fatty acid	Concentration (wt%)		
	0.05	0.1	0.2
Laurate	33.0	32.0	31.1
Myristate	35.3	34.8	34.5
Palmitate	38.5	38.1	37.5
Stearate	41.5	39.8	38.7
Oleate	38.3	36.7	35.6
Ricinoleate	39.4	38.5	38.0

[a]According to Ref. 13.

The fatty acid monoesters of polyoxyethylene are highly soluble in water at room temperature if their HLB values are 9 and higher; in the range of HLB values between 9 and 11 the solutions usually are cloudy; at higher HLB values they are clear (the lauric acid monoester of pentaoxyethylene, for example, has an HLB of 10.6).

Data on the surface activity of defined monoesters are scarce. Rao et al. report values for surface tensions of PEG 600 [polyoxyethylene-(12)] monoesters of some fatty acids at different concentrations [13], see Table 3.

Critical micelle concentrations (CMCs) cannot be extrapolated from these values, but one can surmise that these concentrations lie somewhat above 0.2% or roughly 0.04 mmol/L. Holzbauer and Gerhardt present a diagram from which CMCs between 0.05 and 1.10 mmol/L can be deduced for the lauric acid monoesters of ethylene glycol up to octaethylene glycol (octaoxyethylene) [19]. Unfortunately, these values are not very consistent. Aqueous solutions of polyoxyethylene monoesters are somewhat poor in foaming. In contrast, the monoesters of at least the lower fatty acids are good wetting agents. The wetting properties have been studied in some detail [8]. It was shown that monoesters with HLB values adjusted to about 10 are excellent wetting agents, comparing well with the corresponding polyoxyethylene ethers of fatty alcohols. The monoesters of polyoxyethylenes are good emulsifiers [13,19]. In practical use, the classical formula of Griffin for the hydrophilic-hydrophobic balance can be applied:

$$HLB = 20 \left(1 - \frac{\text{(saponification number of the ester)}}{\text{(acid number of the original acid)}}\right)$$

The colloidal properties of polyoxyethylene monoesters are impaired by diesters and free polyoxyethylene. A shortcoming is the fact that they are chemically unstable in the sense that they can disproportionate or that they can be hydrolyzed in aqueous medium.

III. SYNTHESIS OF POLYOXYETHYLENE FATTY ACID ESTERS IN INDUSTRIAL PRACTICE AND THE PROPERTIES OF PRODUCTS OBTAINED THEREBY

A. Oxyethylation of Fatty Acids and of Fats and Oils

The oxyethylation of fatty acids or of fats and oils, i. e., the addition of ethylene oxide to the mentioned substrates, is carried out in the presence of alkaline catalysts at temperatures between 120°C and 200°C, preferably 140–180°C, and advantageously at slightly elevated pressures of 1–5 bar gauge under vigorous stirring or mixing by circulation, with the usual safety precautions [21] being taken into account. The oxyethylation of fatty acids, fats, and oils today is carried out exclusively in a discontinuous batchwise manner. Continuous oxyethylations of fatty acids, fatty alcohols, and fatty amines have been described [22] but are not used in practice.

Catalysts are employed in concentrations from 0.01% to 2.0% by weight. Usually sodium or potassium hydroxide or carbonate are used. They are added to the substrate, with the water of reaction formed thereby being expelled by application of vacuum or by blowing a current of nitrogen through the mixture at elevated temperature prior to the successive addition of ethylene oxide. Of course, the alkali salts of the fatty acids themselves may be introduced directly as catalyst; furthermore, alkali metals or their alcoholates may be used to generate the carboxylate ions of the fatty acids as catalytically active species [23]. Besides these, other alkaline substances have been proposed, e.g., amines [15] and amine oxides [16].

The alkali-catalyzed oxyethylation of fatty acids has to be interpreted as the nucleophilic addition of a carboxylate ion to an ethylene oxide molecule in the first stage.

Wrigley et al. have elucidated the mechanism of this reaction, comparing the oxyethylation of fatty acids to that of fatty alcohols [24]. Some later publications concerning the kinetics of the reaction of stearic acid with ethylene oxide [25] have only confirmed the original findings and conclusions. The carboxylate ion is introduced either as a sodium, potassium, or ammonium salt, or it is generated in situ from an alkaline moiety and free carboxylic acid. The carboxylate ion approaches an ethylene oxide molecule yielding in a rather slow addition reaction an acyloxyethoxy anion, which in turn exchanges a proton with unreacted fatty acid in a fast reaction, the anion of the latter being less basic than the (acyloxy) ethoxy anion:

$$RCOO^- + CH_2CH_2O \rightarrow RCOOCH_2CH_2O^-$$

$$\underline{RCOOCH_2CH_2O^- + RCOOH \rightarrow RCOOCH_2CH_2OH + RCOO^-}$$

$$RCOOH + CH_2CH_2O \rightarrow RCOOCH_2CH_2OH$$

As long as unreacted acid is available in the reaction medium, the reaction path described above is the predominant one. Stockburger and Brandner suggested that the addition of the carboxylate ion to ethylene oxide might proceed via a loose addition compound of one molecule of fatty acid to one molecule of ethylene oxide analogous to an intermediary complex discussed by Woytech and Patat for the addition of ethylene oxide to phenols [26]. The acyloxyethanol or ethylene glycol fatty acid monoester formed in the first reaction step will not react with ethylene oxide present in the reaction mixture before all fatty acid is consumed because it is only then that the strongly basic and nucleophilic (acyloxy)ethoxy anion will not be protonated to an electrically neutral molecule immediately after it is formed but will nucleophilically attack an additional ethylene oxide molecule:

$$RCOOCH_2CH_2O^- + CH_2CH_2O \rightarrow RCOOCH_2CH_2OCH_2CH_2O^-$$

It is this reaction that runs into the normal oxyethylation reaction, yielding an assembly of homologs distributed in a normal way (Poisson distribution), but which in contrast to the oxyethylates of fatty alcohols does not contain any unreacted starting material when the degree of oxyethylation, i.e., the number of moles of ethylene oxide added to 1 mole of acid, exceeds 1.

The fact that the less nucleophilic and therefore less reactive carboxylate anion prevails in the reaction mixture in the starting phase has the peculiar effect that the rate of uptake of ethylene oxide is slow in this phase, as compared to that of the oxyethylation of fatty alcohols or as compared to the rate after all fatty acid in the mixture has been consumed, leaving only species with alcoholic groups to react. This phenomenon, which is also encountered in the oxyethylation of phenols, is exemplified in Fig. 1 [24].

As was already mentioned in the introduction to this chapter, the monoesters of polyoxyethylenes transesterify easily. Conversions in the oxyethylation of fatty acids therefore are not confined to the reaction sequence discussed above but also comprise transesterification, which leads to polyoxyethylene diesters and free polyoxyethylene in equilibrium with polyoxyethylene monoesters. Stockburger and Brandner have studied the oxyethylation of oleic acid in detail, determining the composition of the reaction product depending on the amount of ethylene oxide added [26,27]. Their findings are summarized in Fig. 2. This diagram shows impressively that the equilibrium reaction between polyoxyethylene monoesters on one side and polyoxyethylene diesters as well as polyoxyethylene on the other side occurs as soon as 1 mole of ethylene oxide has been added to 1 mole of acid, i.e., the point where all fatty acid has been consumed. The molar ratios of monoesters to diesters to free polyoxyethylene (polyethylene glycol, which is also called polyol) are found to be approximately 2:1:1, giving an equilibrium constant of about 4.

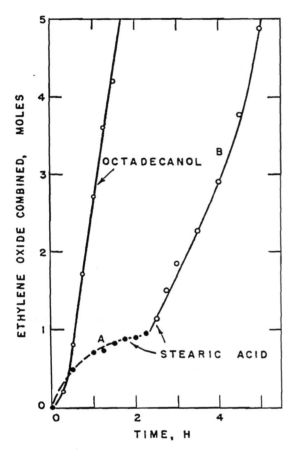

FIG. 1 Reaction rate of the oxyethylation of stearic acid as compared to that of octadecanol. Reaction conditions: 0.5% KOH, 160°C. (From Ref. 24, with permission.)

$$K = \frac{(\text{monoester})^2}{(\text{diester})(\text{polyol})} = 4$$

The diagram also shows that diester and free polyol formation starts at the beginning, yielding substantial amounts of these species even before all fatty acid has reacted. They originate from the esterification of ethylene glycol monoester primarily formed with free fatty acid to the diester and water, with the water reacting further with ethylene oxide to give polyoxyethylene (polyethylene glycol). In an uncatalyzed oxyethylation of fatty acids, which

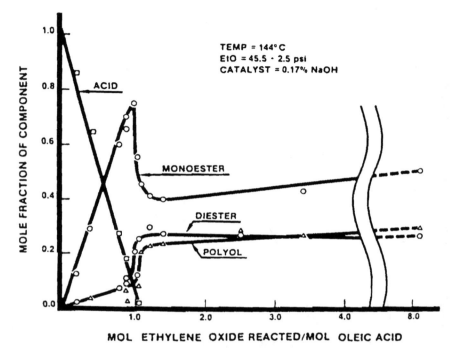

FIG. 2 Product composition in dependence of moles of ethylene oxide reacted with oleic acid. (From Ref. 27, with permission.)

proceeds very slowly, incidentally water of reaction does not react further with ethylene oxide [26].

Reactions occurring during the alkali-catalyzed oxyethylation of fatty acids, which in the end always lead to equilibrium mixtures, can be summarized as follows:

Initial formation of reactive species:

$$RCOOH + OH^- \rightleftharpoons RCOO^- + H_2O \qquad\qquad \text{very fast}$$

Starting phase:

$$RCOO^- + CH_2CH_2O \rightarrow RCOOCH_2CH_2O^- \qquad\qquad \text{slow}$$

$$RCOOCH_2CH_2O^- + RCOOH \rightarrow RCOOCH_2\ CH_2OH + RCOO^- \qquad \text{very fast}$$

Side reactions during the starting phase:

$$RCOO^- + RCOOCH_2CH_2OH \rightleftharpoons RCOOCH_2CH_2OCOR + OH^- \qquad \text{fast}$$

$$OH^- + CH_2CH_2O \rightarrow HOCH_2CH_2O^- \qquad\qquad \text{fast}$$

$$HOCH_2CH_2O^- + RCOOH \rightarrow HOCH_2CH_2OH + RCOO^- \qquad\qquad \text{very fast}$$

Consecutive phase after all fatty acid has reacted:

$$RCOOCH_2CH_2O^- + CH_2CH_2O \rightarrow RCOO(CH_2CH_2O)_2^- \qquad \text{fast}$$

$$RCOO(CH_2CH_2O)_2^- + CH_2CH_2O \rightarrow RCOO(CH_2CH_2O)_3^- \qquad \text{fast}$$

and so on, accompanied by proton exchange, which leads to a normal homolog distribution:

$$RCOO(CH_2CH_2O)_x^- + RCOO(CH_2CH_2O)_yH \rightleftharpoons$$
$$RCOO(CH_2CH_2O)_xH + RCOO(CH_2CH_2O)_y^- \qquad \text{very fast}$$

Side reactions during the consecutive phase via anionic intermediates:

$$2RCOO(CH_2CH_2O)_xH \rightleftharpoons RCOO(CH_2CH_2O)_xCOR + HO(CH_2CH_2O)H \quad \text{fast}$$

Water present in the reaction mixture, originating either from the catalyst or from the esterification of acyloxyethylene glycol with fatty acid in the starting phase, reacts via hydroxylation of ethylene oxide to ethylene glycol, which in turn reacts to polyethylene glycol, polyoxyethylene, as the oxyethylation reaction advances. Although transesterification, which renders polyoxyethylene, cannot be suppressed, the fraction of polyoxyethylene originating from the reaction water in the reaction mixture can be avoided if water is eliminated by distillation in the starting phase [28].

The transesterification of the monoesters of polyoxyethylenes to the diesters and free polyoxyethylene is hampered when sterically hindered acids such as neoacids are used instead of linear fatty acids [29].

Fatty acids most commonly used today are the well-known cuts of lauric and myristic acids, of palmitic and stearic acids, oleic acids, ricinoleic acid, and tall oils, which consist of linear fatty acids and rosin acids. The oxyethylation of brassylic acid has also been reported [30].

In the industrial production of polyoxyethylene esters of fatty acids today one can observe a trend to utilize fats and oils instead of the fatty acids as starting materials for the oxyethylation reaction. Of course, purified fats and oils, being triglycerides, do not possess a reactive proton to offer ethylene oxide a reaction site. Nevertheless, in the presence of an alkaline catalyst and optionally of water, the triglycerides are partially hydrolyzed, generating diglycerides, monoglycerides, or even glycerol on the one hand and salts of the fatty acids on the other [31]. These species may then be oxyethylated, giving a mixture of polyoxyethylene ethers of partial glycerides and of polyoxyethylene esters of fatty acids in a first step, whereas the salts or the polyoxyethylene esters of the fatty acids can esterify or transesterify in a second step, e.g.:

$$\begin{array}{l} RCOO-CH_2 \\ \ \ \ | \\ RCOO-CH \ + \ RCOO(CH_2CH_2O)_xH \ \rightleftharpoons \\ \ \ \ | \\ HO(CH_2CH_2O)_x-CH_2 \end{array}$$

$$\begin{array}{c} \text{RCOO–CH}_2 \\ | \\ \text{RCOO–CH} \quad + \quad \text{HO(CH}_2\text{CH}_2\text{O)}_x\text{H} \\ | \\ \text{RCOO(CH}_2\text{CH}_2\text{O)}_x\text{–CH}_2 \end{array}$$

Depending on the degree of saponification and of oxyethylation, the final reaction mixtures can contain:

Glycerol, monoglycerides, diglycerides, and triglycerides
Polyoxyethylene ethers of glycerol and of mono- and diglycerides
Esterified polyoxyethylene ethers of glycerol and of mono- and diglycerides
Polyoxyethylene monoesters and diesters of fatty acids and free polyoxyethylene

While these mixtures are far away from being well defined products, they have a wide area of application and are being used as emulsifying agents in the first place.

B. Esterification of Fatty Acids or Transesterification of Fats and Oils with Polyoxyethylene

Polyoxyethylene esters of fatty acids can be made by oxyethylation; industrially they are also produced by esterification or transesterification of polyoxyethylene (polyethylene glycol or polyol) with fatty acids or their methyl esters as described above (Sec. II.A), disregarding the fact that this reaction always yields mixtures of mono- and diesters of polyoxyethylene insofar as the fatty acid or its ester are not introduced in excess. This will be the procedure of choice when the diesters are the desired products. Transesterification may also be effected with triglycerides, leading to products similar in composition to products obtained by the oxyethylation of triglycerides discussed above [32].

It must be pointed out that it is common practice in industry to term products obtained by the reaction of polyoxyethylene with fatty acids or their esters in molar ratios around 1 or similar products manufactured by oxyethylation polyoxyethylene "monoesters," despite the fact that they are mixtures of monoesters, diesters, and free polyoxyethylene. This must be borne in mind the more so as common analytical values used to characterize the products such as the hydroxyl number or the saponification numbers will be identical for the pure monoesters and the 2:1:1 molar mixtures of monoesters, diesters, and free polyoxyethylene.

C. Polyoxyethylene Esters of Ricinoleic Acid and Castor Oil

Ricinoleic acid and its triglyceride, castor oil, may also be converted to the corresponding polyoxyethylene esters by the above-mentioned methods. Usually oxyethylation is preferred. The products obtained thereby are valuable emulsifiers. A special feature of these educts is a secondary hydroxyl group in allylic position to the double bond in the unsaturated fatty acid radical. This hydroxyl group can participate in the oxyethylation and transesterification reactions, augmenting the number of species in the final reaction product.

The alkali-catalyzed oxyethylation of ricinoleic acid leads to the expected mixture of monoesters, diesters, and polyoxyethylene. As side reactions, the dehydration of the ricinoleic acid at the site of the hydroxyl group and a Claisen condensation of the monoester to a 2-ricinoloxy-ricinoleic acid ester and polyoxyethylene were observed [33].

Müller presented a comprehensive study on the composition of oxyethylated castor oil with 40 moles of ethylene oxide per mole of triglyceride [34]. He could show that the product contained all of the species described above for the oxyethylation of triglycerides (III.A). In addition to these, he could identify higher esters, where the secondary hydroxyl group or the oxyethylated secondary hydroxyl group is esterified with ricinoleic acid. The degree of oxyethylation of the secondary hydroxyl group was found to be rather low.

D. Properties of Commercial Polyoxyethylene Esters of Fatty Acids

A great number of polyoxyethylene esters of fatty acids are offered by quite a few companies in the marketplace. These products are not strictly comparable, even if they have been assigned the same chemical names. The reasons for this are manifold: the composition of the starting fatty acid may be somewhat different; the chain length and the homolog distribution of the polyoxyethylene in the product, bound and free, may vary within certain limits; and the composition of the product with respect to monoesters, diesters, and polyol can differ as a result if specific reaction conditions of oxyethylation or esterification. While the so-called monoesters usually are produced by the oxyethylation of fatty acids, the diesters are synthesized by esterification.

Product characteristics and their performance of course depend decisively on the type of starting material, fatty acid (alternatively, its methyl ester in transesterification), or triglyceride.

Polyoxyethylene esters of fatty acids usually are described by their physical state, their melting or pour points, and their HLB values. Typical data of some esters are given in Tables 4–7, taken from the prospectus of a supplier [35]. In Table 8 data for some oxyethylated castor oils and hydrogenated castor oils are collected [36]. Hydrogenated castor oil and its derivatives are preferred for odor-sensitive applications because hydrogenated castor oil is practically odorless as compared to castor oil. Polyoxyethylene esters of fatty acids were studied in some detail in the 1950s. In this connection the good detergency of these products was an item of interest [37,38].

TABLE 4 Properties of Industrially Produced Ethylene Glycol and Diethylene Glycol Esters [35]

	Form	Melting point (°C)	HLB
Ethylene glycol			
monostearate	Flake	56–60	2.9
distearate	Flake	60–63	1.5
Diethylene glycol			
monostearate	Flake	44.5–47.5	4.3
distearate	Flake	42–48	2.8
for comparison:			
Propylene glycol			
stearate	Flake	33.5–38.5	3.4
distearate	Flake	36–38	2.2

TABLE 5 Properties of Industrially Produced Polyoxyethylene[a] Esters of Lauric Acid [35]

	Form	Melting point (°C)	HLB
Polyoxyethylene (4)			
laurate	Liquid	<5	9.3
dilaurate	Liquid	<9	5.9
Polyoxyethylene (6)			
laurate	Liquid	<8	11.4
dilaurate	Liquid	<13	7.9
Polyoxyethylene (8)			
laurate	Liquid	12	13.0
dilaurate	Liquid	18	9.7
Polyoxyethylene (12)			
laurate	Liquid	23	14.6
dilaurate	Soft solid	24	11.7
Polyoxyethylene (20)			
laurate	Soft solid	40	16.6
dilaurate	Soft solid	38	14.2
Polyoxyethylene (32)			
laurate	Wax	46	17.5
dilaurate	Wax	42	15.8
Polyoxyethylene (75)			
laurate	Wax	55	19.0
dilaurate	Wax	52	18.1
Polyoxyethylene (150)			
laurate	Wax	61	19.3
dilaurate	Wax	57	18.7

[a]Numbers in parentheses indicate number of moles of ethylene oxide bound per mole of ester.

TABLE 6 Properties of Industrially Produced Polyoxyethylene[a] Esters of Stearic Acid [35]

	Form	Melting point (°C)	HLB
Polyoxyethylene (4)			
stearate	Solid	31	8.1
distearate	Solid	34	4.8
Polyoxyethylene (6)			
stearate	Solid	28	10.3
distearate	Solid	32	6.9
Polyoxyethylene (8)			
stearate	Solid	32	11.7
distearate	Solid	36	8.5
Polyoxyethylene (12)			
stearate	Solid	37	13.5
distearate	Solid	39	10.7
Polyoxyethylene (20)			
stearate	Solid	41	15.7
distearate	Wax	40	13.3
Polyoxyethylene (32)			
stearate	Wax	47	16.9
distearate	Wax	45	14.6
Polyoxyethylene (75)			
stearate	Wax	56	18.7
distearate	Wax	51	17.6
Polyoxyethylene (150)			
stearate	Wax	61	19.1
distearate	Wax	55	18.4

[a] See footnote to Table 5.

An excellent review of properties of some polyoxyethylene esters obtained by the oxyethylation of lauric acid, sperm oil acid, stearic acid, and hardened fatty (palmitic/stearic) acid was given by Geipel et al. [39]. Their findings are summarized as follows: The pour point of the so-called monoesters first decreases with increasing number of moles of ethylene oxide added to 1 mole of fatty acid, reaches a minimum, and then increases again. This phenomenon is demonstrated in Fig. 3. The density of the esters increases with increasing content of ethylene oxide added. At 50°C the densities of the pure acids lie between 0.86 and 0.88 g/cm^3, increasing with increasing content of ethylene oxide added and approaching almost linearly the density of pure polyoxyethylene (1.10 g/cm^3). The refractive indices with increasing content of bound ethylene oxide also approach the refractive index of pure polyoxyethylene, n^{50}_D = 1.462, starting from the corresponding refractive indices of the pure acids. The cloud points depending on the degree of oxyethylation or content of

TABLE 7 Properties of Industrially Produced Polyoxyethylene[a] Esters of Oleic Acid [35]

	Form	Melting point (°C)	HLB
Polyoxyethylene (4)			
oleate	Liquid	<−15	8.2
dioleate	Liquid	<−15	5.0
Polyoxyethylene (6)			
oleate	Liquid	<−5	10.2
dioleate	Liquid	<−5	6.9
Polyoxyethylene (8)			
oleate	Liquid	<10	11.6
dioleate	Liquid	<7	8.3
Polyoxyethylene (12)			
oleate	Liquid	23	13.6
dioleate	Liquid	19	10.6
Polyoxyethylene (20)			
oleate	Soft solid	39	15.9
dioleate	Soft solid	37	13.2
Polyoxyethylene (32)			
oleate	Wax	45	17.0
dioleate	Wax	44	14.9
Polyoxyethylene (75)			
oleate	Wax	55	18.7
dioleate	Wax	49	17.7
Polyoxyethylene (150)			
oleate	Wax	59	19.1
dioleate	Wax	56	18.4

[a] See footnote to Table 5.

TABLE 8 Properties of Industrially Produced Polyoxyethylene[a] Esters of Castor Oil and Hydrogenated Castor Oil [36]

	Form	Pour point (°C)	HLB
Polyoxyethylene (14)			
hydrogenated castor oil	Liquid	7	8.6
Polyoxyethylene (22)			
castor oil	Liquid	5	10.8
hydrogenated oil	Liquid	5	10.8
Polyoxyethylene (26)			
castor oil	Liquid	9	11.8
Polyoxyethylene (32)			
castor oil	Liquid	12	12.6
Polyoxyethylene (176)			
castor oil	Liquid	50	18.1

[a] See footnote to Table 5.

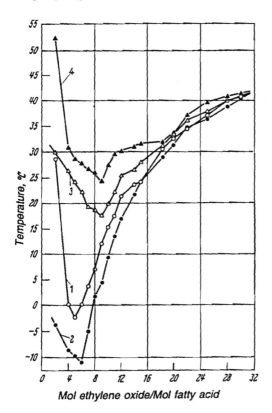

FIG. 3 Pour points of some polyoxyethylene esters of fatty acids in dependence of degree of oxyethylation. Curve 1: lauric acid; curve 2: sperm oil acid; curve 3: hardened fatty acid; curve 4: stearic acid. (From Ref. 39.)

bound ethylene oxide are shown in Figs. 4 and 5. The viscosities of the products increase in a typical way with increasing degree of oxyethylation. A scrutiny of the wetting power of the polyoxyethylene esters of different fatty acids in aqueous solutions depending on the degree of oxyethylation shows that the optimum wetting is reached by the lauric acid esters with 5 moles of ethylene oxide per mole and by stearic acid esters with 11 moles of ethylene oxide per mole. The dependency of such properties as cloud point or wetting power of the polyoxyethylene esters of fatty acids on their degree of oxyethylation is comparable to those of the polyoxyethylene ethers of fatty alcohols. This is also true for the foaming power. Like all surface active polyoxyethylene derivatives, their monoesters are poor foamers, especially those of a lower degree of

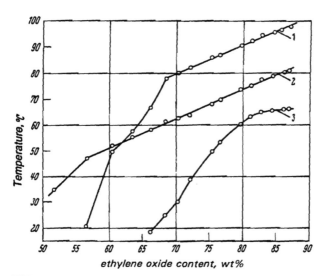

FIG. 4 Cloud points of polyoxyethylene esters of lauric acid in dependence of degree of oxyethylation. Concentration: 10%. Curve 1: dist. water; curve 2: 25% aq. solution of butyl diethylene glycol; curve 3: 10% aq. solution of sodium chloride. (From Ref. 39.)

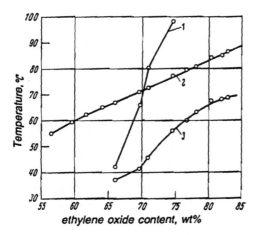

FIG. 5 Cloud points of polyoxyethylene esters of stearic acid in dependence of degree of oxyethylation. Legend as in Fig. 4. (From Ref. 39.)

oxyethylation that are insoluble in water. Aqueous solutions of polyoxyethylene diesters of fatty acids foam even less than the monoesters.

Almost no information was found in the literature about the reduction of surface tension by polyoxyethylene fatty acid esters synthesized by methods put forward in this section. In connection with adsorption studies of some poly-oxyethylene esters on fabrics and carbon black, curves of surface tensions over concentrations of polyoxyethylene-(30) and (80) (= PEG 1500 and 4000) stearic acid esters were published [40]. According to these curves polyoxyethylene-(80) monostearic and distearic acid esters reduce the surface tension of water to about 50 mN/m beyond a concentration of 0.05 mmol/L. The curves have two breaks, indicating two CMCs. The polyoxyethylene-(30) monostearic acid ester reduces the surface tension a little further to about 48 mN/m; this curve also has two breaks. Oxyethylated castor oil reduces the surface tension of water only slightly [41]. Oxyethylated hydrogenated castor oil gives vesicle dispersions in water [42].

E. Oxyethylation of Fatty Acid Methyl Esters

Discussing the alkali-catalyzed oxyethylation of triglycerides above the (Sec. III.A), it was pointed out that ethylene oxide only reacts with the substrate after a preceding partial hydrolysis of the triglyceride. A direct insertion of the ethy-lene oxide molecule into a ester bond does not take place with alkaline catalysts soluble in the reaction medium. It was only recently that the direct insertion of the ethylene oxide molecule into the ester bond was achieved with bifunctional catalysts, e.g., magnesium oxide modified by a aluminum oxide [43,44]. These catalysts, introduced into the reaction medium as fine powders, are not soluble therein and it is assumed that the reaction between the ethylene oxide and the ester proceeds on the surface of the solid catalyst. These catalysts have also been utilized to produce polyoxyethylene ethers of fatty alcohols with a narrow distribution homologs, the so-called narrow-range ethoxylates (NREs). In the oxyethylation of fatty acid methyl esters they yield the monomethyl ethers of polyoxyethylene monoesters in high yield with a normal distribution of homologs:

$$RCOOCH_3 + xCH_2CH_2O \rightarrow RCOO(CH_2CH_2O)_xCH_3$$

It is thus possible to produce true fatty acid monoesters of polyoxyethylenes, though terminally blocked with a methyl group. Such esters compare well in surface activity, wetting power, solubilization capacity, and detergency with the corresponding polyoxyethylene ethers of fatty alcohols, the fatty alcohol oxyethylates with a free terminal hydroxyl group. Due to the ester bond these products are susceptible to hydrolysis, although they are stable in aqueous solutions in a pH range from 3 to 9.

IV. ANALYSIS OF POLYOXYETHYLENE FATTY ACID ESTERS

The analysis of polyoxyethylene esters of fatty acids is rather straightforward if the products to be analyzed are of uniform composition, i.e., monoesters or diesters. These can be hydrolyzed, the polyoxyethylene and the fatty acids obtained thereby being identified and quantitatively determined by well-established methods. But in most cases, products encountered will be industrial products and will therefore be mixtures of monoesters, diesters, and free polyoxyethylene. The approved method for analyzing such mixtures is to separate first the polyoxyethylene therefrom by extraction with aqueous sodium chloride solution, leaving behind a mixture of monoesters and diesters free of polyoxyethylene, the former being characterized by saponification [45].

Schmitt has collected all relevant methods of analysis of polyoxyethylene esters of fatty acids [46]: specification (p. 56), characterization (p. 81), separation by distillation (p. 128), column adsorption chromatography (p. 150, an elegant method was overlooked [47]), thin-layer chromatography (p. 262), gas chromatography (p. 241), high-performance liquid chromatography (pp. 184, 198), superfluidal chromatography (p. 286), nuclear magnetic resonance spectroscopy (p. 82), and mass spectroscopy (pp. 319,322).

V. TOXICOLOGIC ASPECTS AND BIODEGRADATION

A. Toxicologic Aspects

The polyoxyethylene esters of fatty acids in general have a very low order of toxicity. They are nonirritant or of low irritancy to the skin and the mucous membrane. They can be characterized as mild products. This is especially true of the polyoxyethylene esters of ricinoleic acid and of castor oil [48].

B. Biological Degradation

A summary on the biodegradation of polyoxyethylene esters of fatty acids is given by Swisher [49]. The esters with polyoxyethylene chains as long as 20 ethylene oxide units are easily biodegraded. The rate of biodegradation is somewhat slow when the polyoxyethylene chain is as long as 50 ethylene oxide units. The polyoxyethylene esters of rosin acids seem to be biodegraded sluggishly.

VI. SYNOPSIS AND FUTURE OUTLOOK

With some exceptions the industrially more valuable fatty acid esters of polyoxyethylene are the monoesters because of their true amphiphilic character and the higher surface activity resulting therefrom. However, the synthesis of products with high contents of monoesters is cumbersome and questionable after all because the pure monoesters have a tendency to equilibrate to a mixture of mono- and diesters and free polyoxyethylene. In industrial practice, therefore, only such mixtures are used. They are produced by the esterification of polyoxyethylene (polyethylene glycol) with fatty acids or by the oxyethylation of these acids. For the use of the products in established areas of application, a trend can be observed to take the fats or oils as starting materials, leading to a somewhat different composition of the products that embody chemically bound glycerol.

Areas of application for the so-called monoesters, products in which the molar ratio of polyoxyethylene to fatty acids is approximately 1:1, are numerous. Their main use is as emulsifiers in metalworking, mold lubricants, and mold release agents; as emulsifying or dispersing agents in lubricating oils and dye-leveling auxiliaries in the textile and leather industries; as softeners for textiles or as emulsifiers in the preparation of fat liquors and greases for leather; as antistatic agents in the manufacture of textiles and plastics; or as emulsifiers in the preparation of plant protection agents, with the active substance being dissolved in an emulsified oil.

Because of their toxicologic and dermatologic harmlessness, the polyoxyethylene esters of fatty acids are particularly suited for use in cosmetic and pharmaceutic formulations, i.e., as emulsifiers in creams, lotions, and cleansing milks; as solvents, solubilizers, emulsifiers, or dispersants for volatile oils, scents, or perfumes; for colorants or special active ingredients. In this connection the polyoxyethylene esters of ricinoleic acid and particularly of castor oil play an important rôle. Besides their wide application in body care ad cosmetics, they are utilized as solubilizers or emulsifiers of hydrophobic drugs or vitamins for local or oral application.

In the above-mentioned applications, the (formal) monoesters of polyoxyethylene usually function as oil-in-water emulsifiers, which means that the HLB values of the products used are 9 and higher. The diesters of polyoxyethylene with comparable HLB values have more or less the same uses as the corresponding monoesters, although they are not so amply employed for those purposes. The diesters, on the other hand, have to peculiar applications in cosmetics and cleaning agents. Thee diesters of the lower polyoxyethylenes, and in particular the diester of stearic acid with ethylene glycol, serve as pearlescent agents and opacifiers in cosmetic creams, lotions, and shampoos. The diesters of the long chain polyoxyethylenes, e.g., polyoxyethylene-(30) distearate, are useful as in cosmetic formulations, although the fact that they are solids that have to be dissolved before application has induced a search for alternative thickeners [50,51]. The guiding concept in these attempts was the notion that the two hydrophobic residues on both ends of the polyoxyethylene chain would be integrated into two different micelles, thereby building up a three-dimensional network in the aqueous phase [52]. This concept is sometimes referred to as associative thickening.

Despite the very useful properties of the polyoxyethylene fatty acid esters, it must be borne in mind that these esters are in competition with industrial products of comparable structure, e.g., the esters of glycerol or polyglycerols and their polyoxyethylene ethers. The oxyethylated fats and oils treated in this chapter already constitute a connecting link between the polyoxyethylene esters and oxyethylated glycerides. Also worth mentioning are the esters of sorbitol and their polyoxyethylene ethers, allowing for variations or modifications in the polyoxyethylene chain by introducing, say, propylene oxide. Last but not least, the polyoxyethylene ethers of fatty alcohols can replace the esters in some applications.

ACKNOWLEDGMENT

The author thanks the following companies for providing product information and other literature: BASF, Croda, Henkel, Hüls, ICI Surfactants, Nikko Chemicals, PPG, Sino-Japan Chemical, Stepan, Taiwan Surfactants, Th. Goldschmidt, Witco, and Zschimmer & Schwarz.

REFERENCES

A. General Overviews

1. K. Lindner, *Tenside-Textilhilfsmittel-Waschrohstoffe*, Wissenschaftliche Verlagsgesellschaft, Stuttgart, 1964, pp. 882 ff.
2. H. A. Goldsmith, Chem. Rev. *33*:257–349 (1943).
3. R. Schneider, Fette Seifen Anstrichmittel *58*:549 (1956); *59*:876 (1957).
4. W. B. Satkowski, S. K. Huang, and R. L. Liss, in *Nonionic Surfactants* (M. J. Schick, ed.), Surfactant Science Series, Marcel Dekker, New York, 1967, pp. 142–174.
5. N. Schönfeld, *Surface Active Ethylene Oxide Adducts*, Pergamon Press, Oxford, 1969; *Grenzflächenaktive Äthylenoxid-Addukte*, Wissenschaftliche Verlagsanstalt, Stuttgart, 1976, pp. 25–28, 43–47; Ergänzungsband 1984, pp. 7–8.
6. R. A. Reck, in *Fatty Acids in Industry* (R. W. Johnson, ed.), Marcel Dekker, New York, 1989, pp. 201–215.

B. Special References

7. N. Parris and J. K. Weil, J. Am. Oil Chemists Soc. *56*:775 (1979).
8. J. K. Weil, R. E. Koos, W. M. Linfield, and N. Parris, J. Am. Oil Chemists Soc. *56*:873 (1979).
9. M. J. Astle, B. Schaeffer, and C. O. Obenland, J. Am. Chem. Soc. *77*:3643 (1959).

10. A. Kotzschmar and H. Metzger, German Patent 1,175,220 to Farbwerke Hoechst A. G. (1962).

11. R. Celades and C. Paquot, *Proceedings 4th International Congress on Surface Active Substances, CID,* Vol. 1 (F. Asinger, ed.), Brussels, 1964, pp. 249–255.

12. L. Hartman, J. Chem. Soc. (London) *1957*:1918 (1957).

13. T. C. Rao, Y. S. Ramasastry, R. Subbarao, and G. Lakshminarayana, J. Am. Oil Chemists Soc. *54*:15 (1977).

14. W. Gerhardt and H.-R. Holzbauer, Tenside Surf. Det. *12*:313 (1975).

15. E. Sung, W. Umbach, and H. Baumann, Fette Seifen Anstrichmittel *73*:88 (1971).

16. W. Umbach and W. Stein, Tenside Surf. Det. *7*:132 (1970).

17. A. N. Wrigley, F. D. Smith, and A. J. Stirton, J. Am. Oil Chemists Soc. *36*:34 (1959).

18. H. Szelag and W. Zwierzykowski, Seifen Öle Fette Wachse *121*:444 (1995).

19. H.-R. Holzbauer and W. Gerhardt, Tenside Surf. Det. *19*:67 (1982).

20. E. S. Lutton, C. B. Stewart, and A. J. Fehl, J. Am. Oil Chemists Soc. *47*:94 (1970).

21. For example, in Ref. 5 pp. 1–3.

22. W. Umbach and W. Stein, Fette Seifen Anstrichmittel *71*:938 (1969).

23. F. Wolf, G. Geipel, and K. Löffler, Tenside Surf. Det. *5*:270 (1968).

24. A. N. Wrigley, F. D. Smith, and A. J. Stirton, J. Am. Oil Chemists Soc. *34*:39 (1957).

25. M. Bareš, M. Bleha, B. Jeneralova, J. Zajíc, and J. Čoupek, Tenside Surf. Det. *12*:162 (1975); M. Bleha, M. Bareš, E. Votavová, J. Zajíc, and J. Čoupek, Tenside Surf. Det. *14*:123 (1977); M. Bareš, M. Bleha, M. Navratil, E. Votatvová, and J. Zajíc, Tenside Surf. Det. *16*:74 (1979).

26. G. J. Stockburger and J. D. Brandner, J. Am. Oil Chemists Soc. *43*:6 (1966).

27. G. J. Stockburger, J. Am. Oil Chemists Soc. *56*:774A (1979).

28. H. Grossmann, Tenside Surf. Det. *12*:16 (1975).

29. M. Coppersmith and R. C. Maggart, J. Am. Oil Chemists Soc. *46*:332 (1969).

30. T. K. Miwa, R. V. Madrigal, W. H. Talent, and I. A. Wolff, J. Am. Oil Chemists Soc. *45*:159 (1968).

31. V. Martin, Seifen Öle Fette Wachse *111*:51 (1985).

32. H. Pardun, Seifen Öle Fette Wachse *106*:65 (1980).

33. M. Bareš, J. Čoupek, S. Pkorný. J. Hanzalová, and J. Zajíc, Tenside Surf. Det. *12*:155 (1975).

34. K. Müller, Tenside Surf. Det. *3*:37 (1966).

35. Stepan Europe, Voreppe, *Stepan Line of Esters in Personal Care,* Leaflet, 1993.

36. PPG Industries, *Mapeg Polyethylene Glycol Esters,* Gurnee, Ill., Leaflet, 1994.

37. E. M. Stolz, A. T. Ballun, H. J. Ferlin, and J. V. Karabinos, J. Am. Oil Chemists Soc. *30*:271 (1953); A. T. Ballun, J. N. Schumacher, G. E. Kapella, and J. V. Karabinos, J. Am. Oil Chemists Soc. *31*:20 (1954).

38. W. B. Satkowski and W. B. Benett, Soap Chem. Specialties *33*(7):37 (1957).

39. G. Geipel, F. Wolf, and K. Löffler, Tenside Surf. Det. *5*:132 (1968).

40. V. G. Kumar, P. Aravindakshan, I. V. Nagarajan, and A. N. Bhat, Tenside Surf. Det. *29*:195 (1992).

41. C. Drugárin, P. Getia, and M. Vincze, Tenside Surf. Det. *17*:17 (1980).

42. M. Tanaka, H. Fukuda, and T. Horiuchi, J. Am. Oil Chemists Soc. *67*:55 (1990).

43. I. Hama, T. Okamoto, and H. Nakamura, J. Am. Oil Chemists Soc. *72*:781 (1995).

44. A. Behler, H.-C Raths, and B. Guckenbiel, Tenside Surf. Det., *33*:64 (1996).

45. J. D. Malkemus and J. D. Swan, J. Am. Oil Chemists Soc. *34*:342 (1957).
46. T. M. Schmitt, *Analysis of Surfactants*, Surfactant Science Series, Vol. 40, Marcel Dekker, New York, 1992.
47. R. Wickbold, Fette Seifen Anstrichmittel *74*:578 (1972).
48. BASF, Ludwigshafen, *Cremophor EL, Cremophor RH 40*, technical data sheets, 1987 and 1988.
49. R. D. Swisher, *Surfactant Biodegradation*, 2nd Ed. Surfactant Science Series, Vol. 18, Marcel Dekker, New York, 1987, pp. 427, 860–862.
50. W. Adam, Seifen Öle Fette Wachse *110*:599 (1984).
51. H. Meijer, Seifen Öle Fette Wachse *113*:135 (1987).
52. A. Behler, H. Hensen, H.-C. Raths, and H. Tesmann, Seifen Öle Fette Wachse *116*:60 (1990).

5

Polyoxyethylene Mercaptans

CHARLES L. EDWARDS Westhollow Technology Center,
Shell Chemical Company, Houston, Texas

I. INTRODUCTION

Alkyl and aromatic mercaptans (or thiols) can be reacted with ethylene oxide in a manner similar to alcohols or amines to produce surface active products.

However, in the case of mercaptans, there has been very little interest in polyoxyethylene mercaptans derived from aromatic sources. Due to the potential for odor development from impurities in the product, there are significantly fewer applications for polyoxyethylene mercaptans compared with their oxygen or nitrogen counterparts. However, the sulfur atom imparts unique properties to the polyether chain that in many cases show advantages relative to other non-sulfur-containing surfactants.

Polyoxyethylene mercaptans were first reported by I.G. Farben [1,2] in the 1930s and they were exclusively primary linear products. Later, emphasis was placed on using the less expensive tertiary branched chained hydrocarbons as the hydrophobe source (3). Over the years the tertiary polyoxyethylene alcohols found several unique industrial markets and have been commercially produced by GAF (Emulphogene LM series), Monsanto, and Rhone-Poulenc (Siponic SK and Alcodet series). This chapter begins with a discussion of the preparation of these materials and concludes with information concerning their physical properties and applications.

II. SYNTHESIS OF THIOL INTERMEDIATES

The majority of commercially available polyoxyethylene mercaptans are derived from tertiary branched hydrophobes leading to tertiary thiol polyoxyethylene derivatives. However, several recent reports have described the production of secondary thiol polyoxyethylene derivatives from the catalyzed addition of H_2S to linear α- or internal olefins. This section addresses laboratory and commercial procedures for the preparation of primary, secondary, or tertiary alkyl thiols. A brief discussion also includes the preparation of aromatic thiols. Additional information can be found in review articles describing the preparation and uses of thiols [4–7].

A. Aliphatic Primary Mercaptans

Aliphatic primary mercaptans are relatively easy to prepare using laboratory methods. However, they are more difficult to prepare commercially compared with their secondary and tertiary counterparts. A simple laboratory method for the preparation of primary mercaptans involves the reaction of sodium hydrosulfide with a primary halide, such as the bromide [1,2]:

$$CH_3(CH_2)_{10}CH_2Br + NaHS \longrightarrow CH_3(CH_2)_{10}CH_2SH + NaBr$$

This method produces the primary thiol in good yield but would be difficult to commercialize. A more recently reported method [8] that is more amenable to both laboratory and commercial operation involves the free radical addition of thiolacetic acid to a linear α-olefin:

$$CH_3\overset{O}{\overset{\|}{C}}SH \quad + \quad CH_2=CHR \quad \xrightarrow{\text{AIBN}} \quad CH_3\overset{O}{\overset{\|}{C}}SCH_2-CH_2R$$

$$CH_3\overset{O}{\overset{\|}{C}}SCH_2-CH_2R \quad + \quad H_2O \quad \longrightarrow \quad HSCH_2CH_2R \quad + \quad CH_3\overset{O}{\overset{\|}{C}}OH$$

In the presence of free radical initiators, H_2S and mercaptans add to double bonds by a free radical mechanism with the orientation being anti-Markownikoff [9–11]. This leads to the formation of primary thiols.

There have been a number of reports indicating that primary thiols could be prepared by the reaction of H_2S with linear α-olefins using free radical initiators with or without the presence of UV light [12–15]. Another study involved the investigation of the kinetics of the reaction of a primary alcohol with H_2S using an alkali-modified alumina catalyst [16].

B. Aliphatic Secondary Mercaptans

Secondary mercaptans have been produced using a variety of acidic catalyst systems. Various alkyl aluminum dichlorides [17], mixed cobalt/molybdenum oxides [18], ion exchange resins [19], and zeolite catalyst systems [20] have been reported as useful for the preparation of secondary mercaptans. Of these catalyst systems, the zeolite catalyst system [20] appears to be the most useful with respect to high yields and ease of operation.

C. Aliphatic Tertiary Mercaptans

The preparation of aliphatic tertiary polyoxyethylene mercaptans has been extensively reported [3,21–33]. The most common hydrophobe used in these preparations is either propylene tetramer or butylene trimers. Some of the earliest reports involved the use of silica-alumina catalysts [3] for the preparation of branched tertiary dodecyl mercaptans. Improvements in these catalyst systems have been reported more recently [21,20] using both acidified silica-aluminas as well as zeolites.

One of the most common types of catalyst systems used are the classical Friedel–Crafts catalysts. The most common of these are boron triflouride [22–26] and aluminum trichloride [27–33]. These homogeneous catalysts are quite active for the addition of H_2S to branched olefins. Since the mechanism of reaction involves carbonium ion intermediates, the addition of the H_2S to the olefin follows Markownikoff's rule:

$$RCH=C\overset{\displaystyle CH_3}{\underset{\displaystyle CH_3}{}} \quad + \quad H_2S \quad \xrightarrow{\text{acidic catalyst}} \quad RCH-\overset{CH_3}{\underset{SH}{\overset{|}{C}}}-CH_3 \atop \;\;\;\;\;\overset{|}{H}$$

Therefore, tertiary mercaptans are formed in preference to secondary mercaptans with little or no primary mercaptans being formed. Isomerization of the double bond occurs during hydrosulfurization, so that a mixture of mercaptan products is always produced even starting with pure olefin feedstock.

One significant problem encountered using any of these strongly acidic catalyst systems is the propensity for the branched hydrophobic chain to undergo acid-catalyzed chain degradation. Considerable cracking occurs in some of these processes resulting in the formation of low molecular weight olefins that can undergo further hydrosulfurization. Therefore, the crude product mixtures often contain significant quantities of low molecular weight thiols. These byproducts can, however, be stripped from the desired products and sent for disposal.

D. Aromatic Mercaptans

Aromatic mercaptans are seldom used as initiators for polyoxyethylation. However, the most common is alkylbenzenethiol, which is prepared by the acid-catalyzed addition of alkenes to benzenethiol (34).

III. PREPARATION OF POLYOXYETHYLENE MERCAPTANS

A. Laboratory Methods for Preparation

The most common laboratory method for the preparation of polyoxyethylene mercaptans is the reaction of the mercaptan with a substituted polyoxyethylene compound using stoichiometric quantities of sodium hydroxide. Monodispersed products can be obtained using this method as long as the appropriate polyoxyethylene derivative is used. This general method can be represented as follows:

$$RSH \quad + \quad X-(CH_2CH_2O)xCH_2CH_2OH \quad + \quad NaOH \quad \longrightarrow$$

$$RS(CH_2CH_2O)xCH_2CH_2OH \quad + \quad NaX \quad + \quad H_2O$$

The reason that this reaction proceeds in such high yield is that the mercaptan has a significantly higher acidity (pK_a = 12) relative to the polyoxyethylene alcohol derivative (pK_a = 16) [35]. When conducted at low temperature, the base reacts exclusively with the mercaptan to form the mercaptide anion, which undergoes an Sn_2 reaction with the substituted polyoxyethylene compound. No significant deprotonation of the alcohol occurs. This method has been used to prepare a variety of polyoxyethylene mercaptans in high yield [36,37].

B. Commercial Processes

Polyoxyethylene mercaptans are produced commercially by the base-catalyzed reaction of mercaptans with ethylene oxide:

$$RSH \quad + \quad n \quad CH_2\text{--}CH_2 \quad \xrightarrow{\text{catalyst}} \quad RS(CH_2CH_2O)nH$$

This reaction is analogous to the commercial production of polyoxyethylene alcohols discussed in Chapter 3. However, there are important differences between the polyoxyethylation of mercaptans and alcohols that affect the composition of the product and the process conditions used for their production. These differences result from the increased acidity, increased nucleophilicity (reactivity) [35], and reduced catalyst solubility of the mercaptans compared with the alcohols.

The reaction of mercaptans with ethylene oxide has been conducted using a variety of catalysts such as sodium methoxide [38], aqueous sodium hydroxide [39,40], methanolic sodium hydroxide [41], and the sodium mercaptide of the mercaptan to be polyoxyethylated [42]. Each of these reports indicates that the reactivity of the mercaptan with ethylene oxide is considerably higher than that for a comparable alcohol. This is due to the increased nucleophilicity of the mercaptide anion compared to the anionic alkoxide species. Because of this increased nucleophilicity, the addition of the first mole of ethylene oxide to the mercaptan is extremely exothermic and must be carefully conducted at lower temperatures and ethylene oxide partial pressures (e.g., 60°C, 10 psia EO) than those used for alcohol polyoxyethylations (e.g., 120–140°C, 25 psia EO). After the first mole of ethylene oxide has been added per mole of mercaptan, more conventional temperatures and pressures can be used to complete the reaction.

The product chain length distributions of polyoxyethylene mercaptans are also dramatically different from those of polyoxyethylene alcohols containing the same molar ratio of ethylene oxide to initiator. The mercaptan products have a much narrower chain length distribution due to the increased acidity of the mercaptan relative to the polyoxyethylene adducts. In this respect, the poly-oxyethylene mercaptans are similar to the polyoxyethylene alkylphenols (see Chapter 2), which also have narrower chain length distributions than compara-ble alcohols due to the increase acidity of the phenols relative to the adducts. A comparison of polyoxyethylene chain length distributions for a dodecyl mer-captan ethoxylate and a dodecyl alcohol ethoxylate is given in Table 1.

Another difference between the polyoxyethylation of the mercaptans relative to the alcohols is the difference in the solubilities of the catalytically active species. Although the potassium alkoxide is soluble in the alcohol to be reacted, the mercaptide anion is not soluble in the mercaptan initiator. During the early stages of catalyst drying (e.g., from 25°C to 100°C) the KOH is soluble in the

TABLE 1 Comparison of Polyoxyethylene Chain Length Distributions of Polyoxyethylene Mercaptan vs. Polyoxyethylene Alcohols [43,44]

Product	Anadet 912[a]	Neodol 23-9[b]
EO/RXH (mole/mole)	9.1	9.0
Adduct (wt%)		
EO 0	0.1	2
1	0.1	1
2	0.2	2
3	0.8	3
4	2.5	4
5	3.6	5
6	8.3	5
7	10.2	6
8	11.2	7
9	11.4	8
10	11.2	8
11	9.8	8
12	8.6	8
13	6.7	7
14	4.9	6
15	3.4	5
16	2.8	4
17	1.9	3
18	1.6	3
19	0.4	2
20	0.3	2

[a] Anadet is a trade name for a tertiary dodecyl thiol ethoxylate.
[b] Neodol is a trade name for an oxo-type alcohol-based material.

thiol. Continued heating at 130°C causes the formation of a heterogeneous mixture and evolution of water:

$$KOH + RSH \xrightarrow{RSH} RS^-K^+ \cdot H_2O \underset{+H_2O}{\overset{-H_2O}{\rightleftharpoons}} RS^-K^+ \downarrow$$

                                               Soluble   Insoluble

This could have significant consequences for commercial production because heterogeneous systems take longer to dry and also have a propensity to plug catalyst transfer lines.

This problem has been overcome by the observation that the thiolate hydrate was soluble in the thiol, thus suggesting that a thiolate alcoholate might also be soluble in the thiol. Addition of small amounts of an alcohol or a polyoxyethylene alcohol to the reaction mixture (approximately 1 mole of additive per mole of catalyst) was shown to produce homogeneous systems throughout the catalyst drying process [45]. In the process described the additive used was the 1-mole ethylene oxide adduct of the mercaptan to be polyoxyethylated:

$$RS^-K^+ \begin{array}{c} H \\ | \\ O \\ {}^{\cdots}CH_2 \\ CH_2 \\ S^- \\ {}^{\cdots}R \end{array}$$

Solubilized Catalyst Species

In this manner there would be no contamination of the product with an alternate polyoxyethylene alcohol.

A typical process for the polyoxyethylation of mercaptans is similar to that of alcohols, except for staged addition of the ethylene oxide at two different temperatures. The potassium hydroxide is added to the mercaptan containing the solubilizing additive. In the case of the example cited previously [45], this would involve using small amounts of the 1-mole ethylene oxide adduct of the mercaptan to be reacted (~1 wt% basis the mercaptan charge). The reaction mixture would be dried by nitrogen sparging or by vacuum drying for 1 h at 130°C or until a residual water level of <200 ppm is attained. Then the catalyzed thiol mixture is added to the reactor along with additional dried thiol sufficient for the desired initial charge. The reaction temperature is increased to ~80°C at which time all external heating is terminated. Ethylene oxide and nitrogen are introduced to the reactor, which results in an exothermic reaction ($\Delta H = 22$ kcal/mole) increasing the temperature to the desired final reaction temperature of ~120–130°C. Ethylene oxide is added at a rate sufficient to maintain reaction temperature and the heat of reaction is removed by circulating a portion of the batch reactor contents to a heat exchanger. Safety procedures require that limits be placed on temperature and pressure. The gas phase ethylene oxide concentration is continually monitored by sensors, and if the reaction temperature decreases for some reason, ethylene oxide feed is immediately terminated to prevent a buildup of vapor phase ethylene oxide concentration.

After all of the required amount of ethylene oxide is added, the residual ethylene oxide is consumed by maintaining reaction temperature until there is no change in the reactor pressure. This usually requires approximately 30–60 min. Because many customers require that residual ethylene oxide be <5 ppm in the product, a nitrogen sparge of the product is optionally performed if product "aging" is insufficient to reduce ethylene oxide concentrations to the desired levels. The reaction product is normally neutralized with acetic acid, but other homogeneous acids can be used.

Because polyoxyethylene mercaptans often have residual odor due to the presence of residual mercaptan or sulfur-containing volatile impurities, the batch reactor product can be stripped under vacuum at this time to reduce odor.

TABLE 2 Factors Affecting Selectivity in Polyoxyethylation of Tertiary Dodecyl Mercaptan [45]

Conditions employed			wt % TTE[b]	wt % PEG[c]
Drying without solubilizer[a] (1 h @ 130°C)	+	conventional temp. (155°C)	88	12
Drying with solubilizer[a] (1 h @ 130°C)	+	conventional temp. (155°C)	93	7
Drying with solubilizer[a] (1 h @ 130°C)	+	lower temp. (125°C)	98	2

[a]One mole ethylene oxide adduct of dodecyl mercaptan.
[b]TTE, tertiary dodecyl thiol ethoxylate.
[c]PEG, polyethylene glycol.

Additional methods to reduce odor have been shown to be effective, such as steam stripping of the product.

The improved process for the polyoxyethylation of mercaptans described previously [45] also involves the improvement of reducing odor and the amount of polyoxyethylene glycol (PEG) produced during reaction. It has been shown that the odor and PEG formation is a consequence of using high-temperature operations and heterogeneous catalyst mixtures due to base-catalyzed cleavage reactions as follows:

$$RSCH-CH_2-O(CH_2CH_2O)xH \longrightarrow RSCH=CH_2 + {^-O}(CH_2CH_2O)xH + base\text{-}H$$

base

This elimination reaction producing vinyl alkyl sulfide and PEG is a significant contributor to the production of odorous components and PEG in commercial

TABLE 3 Comparison of Polyethylene Glycol Impurities in Commercial Polyoxyethylene Mercaptans

Source	Shell	GAF	Alcolac
Trade name	Anadet 912	Emulphogene LM-710	Siponic SK
EO/RSH (mole/mole)	9.1	7.8	8.0
Hydrophobe			
wt% C10	—	5.0	a
wt% C11	3.4	28.0	a
wt% C12	96.6	67.0	a
%w PEG in product	2.4	4.4	12.0
Relative odor	Acceptable	Strong	Stronger

[a]No analysis available.

polyoxyethylene mercaptans. The effect of temperature and heterogeneous catalyst mixtures on selectivity, as well as a comparison of PEG levels in commercial products, is shown in Tables 2 and 3, respectively. This problem can be minimized through the use of solubilizers and low reaction temperatures [45].

IV. PROPERTIES OF POLYOXYETHYLENE MERCAPTANS

A. The "HLB Problem"

Care must be taken in comparing the physical properties and detergency of the polyoxyethylene mercaptans with polyoxyethylene alcohols having the same ethylene oxide/initiator molar ratio due to the impact of the sulfur atom in the chain. The effective length of hydrophobes containing sulfur (i.e., mercaptans) is greater than that of their alcohol counterparts. One study has found that polyoxyethylene products obtained from dodecyl mercaptan are more similar to products obtained from tetradecyl alcohol. This study has addressed this issue and found that a sulfur in a chain is equivalent to approximately 1.5 carbon atoms [46]. This is a result of the less polar sulfur atom "behaving" more like a carbon than an oxygen. Consequently, the HLB definition must be modified to take into account this difference. In general, more ethylene oxide must be added to mercaptans of a specific carbon number to prepare products more comparable to the same carbon number alcohol.

B. Solubility and Cloud Points

The solubility of various polyoxyethylene mercaptans has been reported [47] and compared with their polyoxyethylated alcohol, amide, and acid counterparts [48]. These materials are soluble in almost all common organic solvents, e.g., in toluene, ether, carbon tetrachloride, acetone, ethyl acetate, and ethanol. Products containing >3 moles of ethylene oxide per mole of mercaptan have limited solubility in hydrocarbons. The cloud points of a series of aqueous solutions containing a polyoxyethylene tertiary dodecyl mercaptan with an average of 7 moles of ethylene oxide per mole of mercaptan is shown in Table 4. These products were stripped to various levels (0–11.38 wt%) to remove lower molecular weight oligomers (mercaptan and low molecular weight EO adducts) in order to explore the effect of stripping on cloud point and turbidity [49]. As can be seen from Table 4, stripping enables the preparation of clear, one-phase, homogeneous, aqueous formulations of these polyoxyethylene mercaptans.

C. Viscosity

The polyoxyethylene mercaptans form gels with water over a relatively wide range of concentrations, as does their alcohol counterparts. For most

TABLE 4 Cloud Points for Stripped Polyoxyethylated Tertiary Dodecyl Mercaptan [49]

Sample[a]	% Stripped	Cloud point (°C)	Turbidity (% transmittance)
a	0	15.4	7.7
b	1.67	16.7	8.0
c	2.49	17.8	8.3
d	3.09	18.7	8.4
e	6.69	23.6	76.2
f	7.66	24.5	83.3
g	11.38	28.8	89.2

[a] a–g are samples of polyoxyethylated tertiary dodecyl mercaptans that had been stripped to various levels using a wiped film evaporator.

commercial mercaptan surfactants, gel formation does not occur until concentrations as high as 40 wt% in water are reached. However, their viscosities are generally lower than those of comparable polyoxyethylene alcohols and alkylphenols. This can be seen in Fig. 1 for Anadet 912 vs. Igepal CO 630. Additional data are published in another study [48].

D. Wetting

The polyoxyethylene mercaptans are excellent wetting agents with tertiary mercaptan hydrophobes showing much superior wetting compared with primary or normal mercaptans. This is illustrated in Table 5 [50].

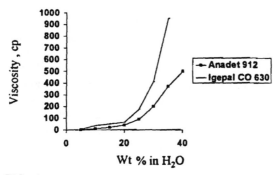

FIG. 1 A comparison of viscosities of a polyoxyethylene tertiary dodecyl mercaptan (Anadet 912) and a polyoxyethylene alkylphenol (Igepal CO 630) [37].

TABLE 5 Effect of Alkyl Chain and Ethylene Oxide Ratio on Wetting Efficiency of Polyoxyethylene Mercaptans [50].

Hydrophobe	EO mole ratio	Time to wet in Draves–Clarkson test, s conc., %	
		0.25	0.125
n-Dodecyl	5:1	31.3	67.9
	10:1	9.1	20.6
	15:1	39.1	110.7
t-Dodecyl	5:1	1.9	5.4
	10:1	3.6	11.2
	15:1	7.9	16.5
n-Tetradecyl	5:1	—	—
	10:1	140	180
	15:1	160	180
t-Tetradecyl	5:1	24.3	67.1
	10:1	4.4	9.6
	15:1	13.2	24.9

E. Foam Breaking

A unique property of the polyoxyethylene mercaptans is the chemical reactivity of the sulfur atom in the chain. The polyoxyethylene mercaptan can easily be oxidized to the sulfoxide or sulfone using sodium hypochlorite (bleach) and base. The former has excellent foaming characteristics, whereas the latter (the sulfone) has none. This feature has been exploited in breaking emulsions encountered when using polyoxyethylene mercaptans as surfactants in wool scouring and pulp and paper processing [51].

$$RSCH_2CH_2O(CH_2CH_2O)_xH + 2NaOCl \rightarrow RS\overset{\overset{\displaystyle O}{\|}}{\underset{\underset{\displaystyle O}{\|}}{C}}H_2CH_2O(CH_2CH_2O)_xH + 2NaCl$$

foamer   nonfoamer

Foams in the wastewater streams are a significant problem for many industrial cleaning operations. The mercaptan surfactants can be used as excellent emulsifiers for cleaning and then be easily disposed of by converting them to nonfoaming sulfones using bleach and base. This is a significant physical property and application advantage that the mercaptan derivatives have compared to polyoxyethylene alcohols or alkylphenols.

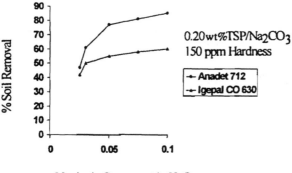

Nonionic Conc., wt% in H_2O

FIG. 2 A comparison of the effectiveness of soil removal using a polyoxyethylene tertiary dodecyl mercaptan (EO/RSH = 7) (Anadet 712) vs. a polyoxyethylene alkylphenol (EO/ROH = 7) (Igepal CO 630) in a hard surface cleaning test program [37].

F. Detergency

Polyoxyethylene mercaptans have been shown to be excellent detergents in a wide variety of applications [45, 49, 51–54]. They are particularly useful in hard surface cleaning and removing triolein fat in wool scouring. Results from tests in these areas are shown in Figs. 2 and 3 [37].

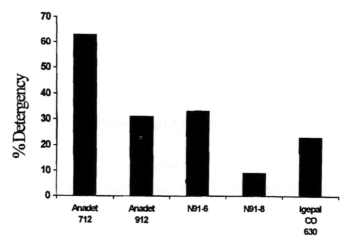

FIG. 3 Effect of nonionic surfactant structures on removal of triolein fat from wool at 35°C. The N91-6 is a C_9–C_{11} alcohol reacted with an average of 6 moles of ethylene oxide per mole of alcohol [37].

V. APPLICATIONS

A. Household

The polyoxyethylene mercaptans have found limited use in the household markets due to the concern for odor development in the formulations. The main household application is in shampoos, because the mercaptan-based products have excellent foaming, rinsability, and resistance to hard water [55–57]. A recent patent claims that their polyoxyethylene mercaptans have substantially reduced odor and when used with other materials produce a superior permanent wave lotion formulation [58]. Additional hair waving formulations containing polyoxyethylene mercaptans are claimed to have much improved performance [59].

B. Industrial and Institutional

There are quite a few industrial applications for the polyoxyethylene mercaptans, most of which do not have strict odor requirements. The main applications are found in metal cleaning, textile cleaning (wool scouring), pulp and paper, hard surface cleaning, agricultural chemical spray adjuvants, and emulsion stabilizers and polymerizations.

1. Metal Cleaning

Mercaptan-based surfactants are excellent for oil and grease removal from metal surfaces and materials [60]. The mercaptan derivatives are also effective as foam breakers in these formulations [61]. Improvements have been made in reducing any odor that may be associated with these products in the metal cleaning area [62]. Water-soluble lubricating additives for water-based metal working fluids have been developed with a key component being polyoxyethylene mercaptans [63].

2. Pulp and Paper

As mentioned previously the polyoxyethylene mercaptans are easily oxidized to the sulfone or sulfoxide using bleach, which makes them valuable surfactants for the breaking of emulsions that occur in the flotation de-inking process. The mercaptan products are particularly good at ink removal and dispersion, thus aiding in the recycling of newspaper and magazines.

3. Wool Scouring

As shown in Fig. 3, the polyoxyethylene mercaptans show superior removal of oil and grease from wool. Several studies have shown additional grease removal properties from various wool blends under a variety of conditions [35,64,65].

4. Miscellaneous Applications

The polyoxyethylene mercaptans have been shown to be useful for hard surface cleaning applications (see Fig. 2), agricultural spray adjuvants [66], and as emulsion stabilizers and in emulsion polymerization [67,68].

REFERENCES

1. H. Schuette and C. Schoeller, U.S. Patent 2,129,709 to I.G. Farbenindustrie A. G. (1938).
2. H. Schuette, C. Schoeller, and M. Wittwer, U.S. Patent 2,205,021 to I. G. Farbenindustrie A. G. (1938).
3. W. A. Schulze, J. P. Lyon, and G. H. Short, Ind. Eng. Chem. 40:2308 (1948).
4. C. M. Raynor, Contemp. Org. Synth. 1(3):191–203 (1994).
5. L. Field, Sulfur Rep. 13(2):197–277 (1993).
6. Y. Labat, Phosphorus Sulfur Silicon Rel. Elem. 74(1–4):173-94 (1993).
7. C. Forquy and E. Arretz, Stud. Surf. Sci. Catal. 41 (1988).
8. H. E. Fried, Shell Chemical Co., unpublished results.
9. F. W. Stacey and J. F. Harris Jr., Org. Reactions 13:150–376 (1963).
10. G. Sosnovsky, *Free Radical Reactions in Preparative Organic Chemistry*, Macmillan, New York, 1964.
11. C. Walling, *Free Radicals in Solution*, John Wiley and Sons, New York, 1957.
12. Jap. Patent 55009087 to Pennwalt Corp. (1980).
13. Fr. Patent 1582488 to United Kingdom Atomic Energy Authority (1970).
14. Shiro Ishida and Tokiyuki Yoshida, Japan Oils and Fats Co, Ltd. (1969).
15. G. F. Kite, U.S. Patent 3,397,243 to Gulf Research and Development Co. (1968).
16. L. I. Kiseleva, L. N. Khairulina, and A. V. Mashkina, Reaction Kinet. Catal. Lett. 53(1):73–78 (1994).
17. G. A. Tolstikov, F. Y. Kanzafarov, Y. A. Sangalov, and U. M. Dzhemilev, Neftekhimiya 19(3):425–429 (1979).
18. J. S. Roberts, Eur. Patent 354460 to Phillips Petroleum Co. (1990).
19. Emmanuel Arretz, Eur. Patent 329520 to Societe Nationale Elf Aquitane (1990).
20. H. E. Fried, Eur. Patent 122654 to Shell Internationale Research Maatschappij B. V., Neth. (1985).
21. P. F. Warner, U.S. Patent 3,534,106 to Phillips Petroleum Co. (1971).
22. R. T. Bell, U.S. Patent 2,531,602 to the Pure Oil Co. (1950).
23. J. F. Olin, U.S. Patent 2,434,510 to Sharples Chemicals, Inc. (1948).
24. J. L. Eaton and J. B. Fenn, U.S. Patent 2,468,739 to Sharples Chemicals, Inc. (1949).
25. Monsanto Chemical Company Has Developed a Simplified Process for Synthesizing Dodecyl Mercaptan, Petrol Management 35(13):212 (1963).
26. J. Y. Lee, U.S. Patent 4,612,398 to Ethyl Corporation (1986).
27. C. H. Gross, A. V. Belea, and R. Marcu, Romanian Patent 95629 to Institutul de Cercetari pentru Produse Auxiliare Organice, Medias, Rom. (1990).
28. W. Cleve, R. Lindenhahn, H. Martin, and R. Scheibe, Ger. Patent 160,222 to VEB Chemische Werke Buna (1983).

29. E. Pajda, C. Przybyl, Z. Kubica, and S. Chybowski, Pol. Patent 102,838 to Instytut Ciezkiej Syntezy Organicznej "Blachownia" (1980).
30. V. E. Mazaev and M. A. Korshunov, Organ. Soedineniya Sery. *1*:328–333 (1976).
31. M. A. Korshunov, V. E. Mazaev, and R. N. Kudinova, Prom. Sin. Kauch.-Tekh. Sb. *9–10*:3–5 (1969).
32. S. Regula and E. Kardos, Chem. Prum. *20*(11):519–521 (1971).
33. H. D. Steinleitner and W. Cleve, Ger. Patent 1,274,122 to VEB Chemische Werke Buna (1968).
34. E. A. Bartkus, E. B. Hotelliing, and N. B. Neuworth, J. Org. Chem. *25*:232 (1960).
35. J. March, in *Advanced Organic Chemistry: Reactions, Mechanisms and Structure*, McGraw-Hill, New York, 1968, pp. 220, 288.
36. B. R. Schwartz, U.S. Patent 2,588,771 to Hart Product Corporation (1952).
37. Shell Chemical Company, unpublished data (1983–1986).
38. J. F. Olin, U.S. Patent 2,565,986 to Sharples Chemicals, Inc. (1951).
39. L. T. Eby, U.S. Patent 2,570,050 to Standard Oil Development Co. (1951).
40. P. H. Schlosser and K. R. Gray, U.S. Patent 2,392,103 to Rayonier, Inc., (1946).
41. W. W. Crouch and R. P. Louthan, U.S. Patent 2,720,543 to Phillips Petroleum Co. (1955).
42. Br. Patent 693,966 to Standard Oil Development (1953).
43. Shell Chemical Company, unpublished data (1986–1988).
44. Neodol Product Guide for Alcohols, Ethoxylates and their Derivatives, Shell Chemical Co. Technical Bulletin SC:7–94 (1994).
45. C. L. Edwards, U.S. Patent 4,575,569 to Shell Oil Company (1986).
46. A. Sokolowski, Comun. Jorn. Com. Esp. Deterg. *21*:259–272 (1990).
47. R. A. Greff and P. W. Flanagan, J. Am. Oil Chemists Soc. *40*:118 (1963).
48. T. M. Kassem, Tenside Det. *21*(3):144–146 (1984).
49. C. L. Edwards, H. E. Fried, and W. Lilienthal, U.S. Patent 4,931,205 to Shell Oil Company (1990).
50. British Patent 675,993 to Monsanto Chemical Company (1952).
51. D. L. Wood and H. E. Fried, U.S. Patent 4,618,400 to Shell Oil Company (1986).
52. L. J. Mathias and J. B. Canterberry, Polym. Sci. Technol. *24*:359–370 (1984).
53. E. Schmitt and M. Quaedvlieg, Br. Patent 1,205,721 to Farbenfabriken Bayer A. G. (1968).
54. W. K. Griesinger, J. A. Nevison, and G. A. Gallagher, J. Am. Oil Chemists Soc. *26*:241 (1949).
55. O. Albrecht and E. Matter, U.S. Patent 2,992,994 to Ciba, Ltd. (1961).
56. R. L. Mayhew and J. M. Cloney, Am. Perfumer, *72*(4):83, 86, 88 (1958).
57. G. Kalopissis, P. Roussopoulos and C. Zviak, U.S. Patent 3,768,490 to Oreal S. A. (1965).
58. A. Savaides and L. Salce, U.S. Patent 5,350,572 to Shiseido Co., Ltd. Japan (1993).
59. S. Naito and K. Ohshima, Eur. Patent 235783 to Kao Corp. Japan (1987).
60. Br. Patent 993,226 to Cowles Chemical Co. (1965).
61. S. A. Bolkan, G. Byrnes, S. Dunn, A. Vinci, and A. E. Winston, WO Patent 9,609,366 to Church and Dwight Co, Inc. (1995).
62. S. Dunn and A. E. Winston, WO Patent 9,609,368 to Church and Dwight Co., Inc. (1995).

63. A. D. Schuettenberg, M. R. Lindstrom, B. R. Bonazza, H. F. Efner, R. L. Bobsein, and M. D. Herd, Eur. Patent 249,783 to Phillips Petroleum Co. (1987).
64. J. C. Harris, U.S. Patent 2,572,805 to Monsanto Chemical Company (1951).
65. Technical Data Sheet I-169, Monsanto Chemical Company, St. Louis, Missouri.
66. Br. Patent 705,117 to Monsanto Chemical Company, (1954).
67. A. B. Goel, U.S. Patent 4,696,992 to Ashland Oil Company (1986).
68. F. Vidal, J. Guillot, and A. Guyot, Polym. Adv. Technol. 6(7):473–479 (1995).

6

Polyoxyethylene Alkylamines

MICHAEL D. HOEY and JAMES F. GADBERRY Surfactants America
Research and Development, Akzo Nobel Central Research,
Dobbs Ferry, New York

I. INTRODUCTION

Strictly speaking, polyoxyethylene alkylamines should be considered as cationic surfactants. Indeed, in some references polyoxyethylene alkylamines are reviewed as cationic surfactants [1]. This notwithstanding, polyoxyethylene alkylamines made by reaction of >6 moles of ethylene oxide per mole of primary amine do exhibit solubility characteristics, e.g., high-temperature cloud point, of classical nonionic surfactants. Although adducts from propylene oxide and mixed adducts of ethylene oxide and propylene oxide are beginning to become more available, they will not be discussed. In this review we shall address only those surfactants that have attained significant commercial importance. Any omissions are a matter of the perspectives of the authors.

For ease of discussion we employ a nomenclature based on listings given in [2] wherein, for example, an adduct of tallowamine with 10 moles of ethylene oxide is called PEG-10 tallowamine. Nomenclature used by Chemical Abstract Services is too complex and unwieldy for our purposes, e.g., PEG-XX oleylamine, CAS Registry Number 26635-93-8, is listed as poly(oxy-1, 2-ethandiyl), α,α'-[9-octadecenylimino)di-2, 1-ethandiyl]bis[ω-hydroxy-, (Z)-, in the 9th Collective Index. This causes PEG-15 oleylamine and PEG-25 oleylamine, for example, to be indistinguishable under Chemical Abstracts nomenclature. As much as possible we tried to avoid reference to commercial trade names, however, some of the major trade names, along with their respective manufacturer, can be found in the Appendix.

II. SYNTHESIS

The traditional synthesis of polyoxyalkylene amines may be separated into two reactions [3,4]. In the first, an epoxide, such as ethylene oxide, is reacted with an amine to generate an amino alcohol (Scheme 1). The second reaction is growth of the polyoxyethylene chain through reaction of more epoxide with the hydroxyl group of the aminoethyl alcohol (Scheme 2). The most commercially common surfactants prepared this way are the ethoxylated fatty amines. Although any epoxide may be used in preparing a polyalkoxylated amine, the most important are made from ethylene oxide.

Alkoxylation of the amine proceeds through the nucleophilic attack of the amine upon the epoxide to produce the hydroxyethylamine (Scheme 3). This reaction continues until all of the primary and secondary amine has been consumed. Further alkoxylation of the product beyond this point usually requires a catalyst. Historically, sodium hydroxide and potassium hydroxide have been the catalysts of choice [3,4]. These bases will generate an alkoxide anion at the terminal hydroxyl on the growing polyoxyethylene chain, which then lengthens through addition of further oxyethylene groups. This type of chain growth might be expected to give an approximately symmetrical distribution of oxyethylene units if all of the hydroxyls present in the system are of equal nucleophilicity and have an equal chance to react with the base to generate an alkoxylate anion [5,6]. The distribution that is obtained typically does not agree with this assumption. Because of this a considerable amount of recent

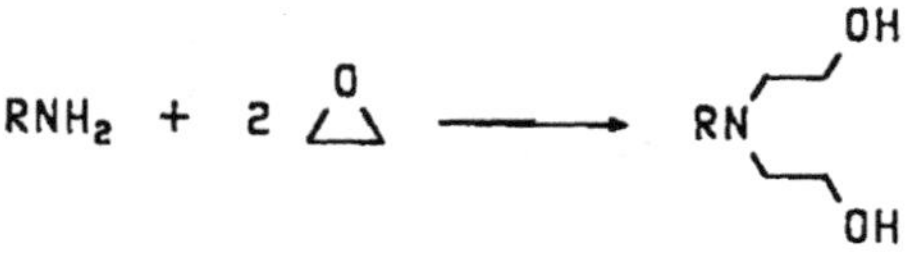

$$RNH_2 \ + \ 2 \ \triangle \ \longrightarrow \ RN \begin{array}{l} \diagup OH \\ \diagdown OH \end{array}$$

SCHEME 1

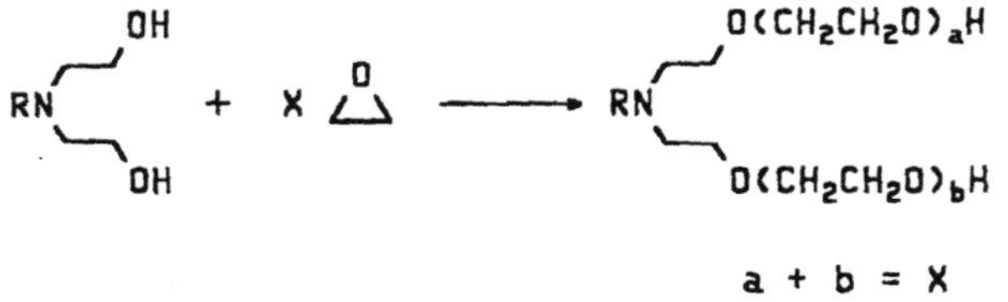

RN⟨OH/OH + X △O⟶ RN⟨O(CH₂CH₂O)ₐH/O(CH₂CH₂O)ᵦH

$$a + b = X$$

SCHEME 2

research activity has seized on this discrepancy to develop catalysts that will provide a narrower distribution than the one obtained with sodium and potassium hydroxides [7–10]. Much of this work is based on the theory that the growing polyoxyethylene chains can chelate with the sodium and potassium catalyst cation(s) (Scheme 4). These chelated cations would be able to stabilize an alkoxylate, thus giving the polyoxyethylene-cation-chelate-alkoxylate-anion a greater probability of reacting with more ethylene oxide and growing this polyoxyethylene chain. This would give a broad distribution with the longer, more reacted chain, tending to react more at the expense of the other alcohols in the reaction. Thus this difference in probability of reaction would skew the product distribution formed. By using a catalyst that contains different cations and thus will not chelate with the polyoxyethylene as sodium and potassium do, many of the new catalyst patents are expected to produce a product with a narrower distribution of chain length/mole equivalents of ethylene oxide reacted. Calcium [11–22], barium [23–30], strontium [12,22,25,26,28–31], aluminum [32–36], magnesium [37–39], and zirconium [40–42] have all been used as well as rare earth cations [43–48], bimetallic [49–54] and heterogeneous catalysts [46,55–61]. The new catalysts do generate a different distribution(s), but they also provide different costs to the producers who use them. Therefore one must balance the added costs with the real benefits that the new product distributions provide to the customer.

A complication that is observed with this synthesis of polyoxyethylene amines is the additional catalysis of the ethylene oxide polymerization by the tertiary amines [3,62]. This catalytic pathway involves the reaction of the tertiary amine with ethylene (or propylene) oxide to form a quaternary

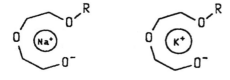

SCHEME 3

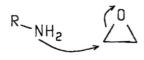

SCHEME 4

ammonium salt. These quaternary ammonium salts then can catalyze this poly-merization of ethylene oxide with an alcohol. The quaternary ammonium alkoxylate salt formed in situ can then catalyze other alcohols to react, or continue to react, at the anion formed from the quaternization. The quaternary ammonium salt may decompose during the reaction because the temperature of the ethoxylation is above the decomposition temperature of many quaternary ammonium salts. These decomposition products may be ethylene oxide and a tertiary amine, or they may be a polyoxyethylene glycol and a tertiary amine. Dioxane and other ring glycols may also be formed. Dioxane and polyoxyethy-lene glycols may also be formed by the reaction of water with the ethylene oxide. Unsaturated alkyl products may be formed via a Hoffman degradation. Many of these byproducts may continue to react/degrade and complicate the final product mix.

Another type of polyoxyethylene amine is typified by the Jeffamine series from Huntsman and the Tectronic series from BASF. The Jeffamine products are made by alkoxylating a short chain alcohol, or glycol, to obtain a polyalkoxylated alcohol. This alcohol is then aminated to generate an amine [63,64]. These products may be primary, secondary, or tertiary amines. This method has been described for a number of different reduction/amination cata-lysts [63–68]. The primary and secondary amines can also be alkoxylated by further reaction with an epoxide. This reaction may take place in the presence of water (5–15%) without forming undue amounts of glycols [69,70]. The Tectronic products are made by alkoxylating a small diamine, such as ethylene diamine. These products are all tertiary amines, and may be made of mixtures of ethylene oxide and propylene oxide.

Some other syntheses have been described in the patent literature. One such patent describes the reaction of a long chain epoxide, made from a long chain α-olefin, with an amine or ammonia [71]. Although this is a similar reaction to the traditional method described above, the size of the alkyl group on the epoxide will greatly affect the rate of this alkoxylation reaction as well as the solubility of the product amine.

Monoethoxylated amines have been prepared by reaction of an imine and an alkylene oxide to produce an oxazolidine [72]. This can then be cleaved with water to produce the monoethoxylated fatty amine.

III. REACTIONS OF POLYOXYETHYLENE ALKYLAMINES

Polyxyethylene alkylamines react as typical tertiary amines. These reactions include quaternization, oxidation, and formation of salts with inorganic and organic acids. Quaternization magnifies the cationic properties to such an extent that the products can no longer be considered nonionic. Oxidation is generally unknown except for the lower adducts, e.g., PEG-2 cocoamine. Amine oxides of higher adducts have not yet been exploited to any great extent.

The presence of terminal hydroxyl groups on polyoxyethylene alkylamines allows formation of esters with carboxylic acids. Though these products should have greater lipophilicity that the parent poloyoxyethylene amines, they have not been yet achieved significant commercial success.

IV. PROPERTIES

A. Physical

The best sources of information regarding the physical properties of poly-oxyethylene alkylamines are the product bulletins and brochures from commercial suppliers. This is especially important as small but important differences may exist for similar products from different suppliers. Solubility data and formulations may also be available. A good general source of information on products and suppliers is the *Handbook of Industrial Surfactants* [2].

Depending on fatty alkyl chain length and degree of ethoxylation, poly-oxyethylene alkylamines can vary from free-flowing liquids to pastes or solids. Generally, long saturated alkyl chains and high moles of ethoxylation give pastes to solids. The color of polyoxyethylene alkylamines varies from golden to dark brown depending on reaction catalyst, temperature, degree of ethoxylation, etc.

B. Conductivity and CMC

In a study of the conductivities of ethoxylated laurylamines and ethoxylated stearylamines, Riva and Ribe [73] found that lauryl derivatives have higher conductivities than the stearyl products at equivalent ethoxylation. Furthermore, conductivity of the ethoxylated laurylamines increases with increasing ethoxylation; increased ethoxylation was reported to little affect the conductivity of the stearyl derivatives. These results suggest that chain length of the hydrophobe plays an important role in the aggregation properties of these surfactants. This study also reports that critical micelle concentrations (CMCs) range from 0.4 to 1.0 mmol/L.

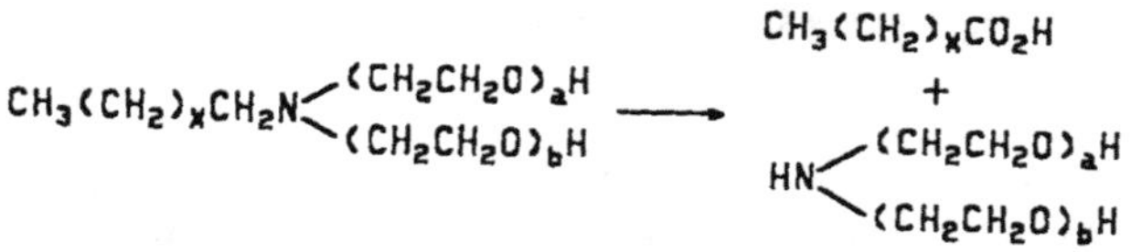

$$CH_3(CH_2)_xCH_2N \begin{cases} (CH_2CH_2O)_aH \\ (CH_2CH_2O)_bH \end{cases} \longrightarrow \begin{array}{c} CH_3(CH_2)_xCO_2H \\ + \\ HN \begin{cases} (CH_2CH_2O)_aH \\ (CH_2CH_2O)_bH \end{cases} \end{array}$$

SCHEME 5

C. Toxicity

Polyoxyethylene alkylamines are organic bases and therefore should be considered corrosive to skin if the adduct number of moles of ethylene oxide is <10. Nevertheless, adducts with >10 moles of ethylene oxide should be handled with care. All polyoxyethylene alkylamines should be considered as severely irritating and corrosive to eyes.

D. Biodegradation

Biodegradation of ethoxylated fatty amines has been studied by van Ginkel et al. [74] using a "prolonged" closed bottle test. The researchers propose that degradation of the polyoxyethylene alkylamines proceeds via initial oxidative cleavage of the fatty alkyl chain to form an easily degraded fatty residue and a less easily degraded polyoxyethylene secondary amine (Scheme 5). The polyoxyethylene secondary amines are not considered toxic to aquatic organisms. This is understandable because the polyoxyethylene secondary amines will have minimal surfactant properties.

V. APPLICATIONS

There is a wide range of applications for these polyoxyalkylene amines. For the purposes of this chapter, we will limit this discussion to polyoxyalkylene amines possessing an alkyl chain of more than six carbons. Also, due to the large number of patents in this area, we are limiting this discussion to U.S. patents. The list of patents is not exhaustive but rather is designed to be an overview.

Many of the applications described below owe their use to the polyoxyalkylene amine's affinity for surfaces. The use of these products to coat films or glass [75–77], to coat wool or cloth [78,79], to coat yarns [80–83], or as wettability aids [84] comes from the propensity for these amines to adsorb to surfaces from a solution. They are also used as solubilizers and emulsifiers in formulations [85–96], to stabilize dispersions [97], and as antistat additives [98–110]. They have been used as anticaking aids [111], dyeing auxiliaries [112–116], and, with borated derivatives, as friction-modifying additives [117–124]. They are included in herbicide formulations [125–131] and are used to degrease, or dye, leather surfaces [132–134]. They are included in many cleaning and detergent formulations for contributing antistatic behavior, antideposition properties, and dispersing abilities [135–144]. They have been used as additives in gasoline [145], drilling fluids [146–151], corrosion inhibitors [152–159], and plastics [160–165]. The great variety of uses seems to stem from the their ability to interact well with other surfactant types, and their surface absortivity. It is this effect that is used most often in the patents described herein.

VI. PROBLEMS TO BE SOLVED

Two major problems yet unsolved for polyoxyethylene alkylamines are dark color and formation of dioxane. Higher mole ethylene oxide adducts are typically brown to dark brown liquids. The exact nature of the color bodies is unknown but may be speculated to arise from the presence or formation of acetaldehyde, which may result in chromophoric condensation products. Improvement of the color of polyoxyethylene alkylamines should lead to an increased number of use possibilities, e.g., in personal care applications. Distillation of polyoxyethylene alkylamines is not practical.

Formation of dioxane, a suspected carcinogen, is becoming increasingly problematic as consumers of polyoxyethylene alkylamines impose more stringent requirements on suppliers. Removal of dioxane can add significantly to productions cycle times thereby increasing production costs. No published sources have addressed this problem in a meaningful manner.

Another similar issue that may or may not be important for some applications is the formation of PEGs simultaneously during preparation of polyoxyethylene alkylamines. Water (or hydroxide) must be present in order for PEGs to form. Therefore, sufficient drying of both the PEG-2 alkylamine and ethylene oxide should minimize PEG contamination.

APPENDIX: TRADE NAMES USED FOR ETHOXYLATED ALKYLAMINES

Trade name	Manufacturer
Accomeen	Karlsham
Alkaminox	Rhone-Poulenc
Chemeen	Chemax
Ethomeen	Akzo Nobel
Genamine	Hoechst
Hetoxamine	Hetere
Katapol	Rhone-Poulenc
Mazeen	PPG
Protox	Protameen
Teric	ICI
Trymeen	BASF
Varionic	Witco

REFERENCES

1. H. L. Sanders, J. B. Braunwarth, R. B. McConnell, and R. A. Swenson, J. Am. Oil Chemist Soc. 46:167 (1969).
2. C. Schoeller and M. Wittwer, U. S. Patent 1,970,578 to I. G. Farbenindustrie (1934).
3. E. Santacesaria, M. Diserio, R. Garaffa, and G. Addino, Ind. Eng. Chem. Res. 31:2413 (1992).
4. B. Bartha, L. Farkas, J. Morgos, P. Sallay, I. Rusznak, and G. Veress, J. Am. Oil Chemist Soc. 58:650 (1981).
5. J. Morgos, P. Sallay, L. Farkas, and I. Rusznak, J. Am. Oil Chemists Soc. 60:1905 (1983).
6. W. Hreczuch and J. Szymanowski, J. Am. Oil Chemists Soc. 73:73 (1996).
7. K. L. Matheson, T. P. Matson, and K. Yang, J. Am. Oil Chemists Soc. 63:365 (1986).
8. K. W. Dillan, G. C. Johnson, and P. A. Siracusa, Soap/Cosmetics/Chemical Specialties 34 (March 1986).
9. C. L. Edwards, U.S. Patent 4,396,779 to Shell Oil Company (1983).
10. J. H. McCain, and L. F. Theiling, Jr., U.S. Patent 4,453,022 to Union Carbide Corporation (1984).
11. R. J. Knopf and L. F. Theiling, Jr., U.S. Patent 4,754,075 to Union Carbide Corporation (1988).
12. R. J. Knopf and L. F. Theiling, Jr., U.S. Patent 4,820,673 to UOP (1989).
13. B. E. Leach, M. L. Shannon, and D. L. Wharry, U.S. Patent 4,835,321 to Vista Chemical Company (1989).
14. R. J. Knopf and L. F. Theiling, Jr., U.S. Patent 4,886,917 to Union Carbide Chemicals and Plastics Company (1989).
15. S. W. King and R. J. Knopf. U.S. Patent 4,902,658 to Union Carbide Chemicals and Plastics Company (1990).
16. C. F. Hauser, U.S. Patent 4,946,984 to Union Carbide Chemicals and Plastics Company (1990).
17. S. W. King and R. J. Knopf, U.S. Patent 5,104,575 to Union Carbide Chemicals and Plastic Company (1992).
18. S. W. King, U.S. Patent 5,114,900 to Union Carbide Chemicals and Plastics Company (1992).
19. S. W. King, U.S. Patent 5,120,697 to Union Carbide Chemicals and Plastics Company (1992).
20. T. S. Sandoval and P. A. Schwab, U.S. Patent 5,220,077 to Vista Chemical Company (1993).
21. K. Yang, G. L. Nield, and P. H. Washecheck, U.S. Patent 4,210,764 to Conoco, Inc. (1980).
22. K. Yang, G. L. Nield, and P. H. Washecheck, U.S. Patent 4,302,613 to Conoco, Inc. (1981).
23. K. Yang, U.S. Patent 4,239,917 to Conoco, Inc. (1980).
24. K. Yang, G. L. Nield, and P. H. Washecheck, U.S. Patent 4,376,868 to Conoco, Inc. (1983).
25. J. H. McCain and L. F. Theiling, Jr., U.S. Patent 4,453,023 to Union Carbide Corporation (1984).
26. A. J. Heuvelsland, U.S. Patent 5,010,187 to Dow Chemical Company (1991).
27. A. J. Heuvelsland, U.S. Patent 5,114,619 to Dow Chemical Company (1992).

28. R. A. Reck, in *Cationic Surfactants, Organic Chemistry* (J. M. Richmond, ed.), Marcel Dekker, New York, 1990, pp. 180–182.

29. M. Ash and I. Ash, in *Handbook of Industrial Surfactants*, Gower, London 1993.

30. D. E. Laylock and R. A. Sewell. U.S. Patent 5,374,705 to Dow Chemical Company (1994).

31. K. Yang, G. L. Nield, and P. H. Washecheck, U.S. Patent 4,223,164 to Conoco, Inc. (1980).

32. C. L. Edwards, U.S. Patent 4,721,816 to Shell Oil Company (1988).

33. C. L. Edwards, U.S. Patent 4,721,817 to Shell Oil Company (1988).

34. B. E. Leach, M. L. Shannon, and D. L. Wharry, U.S. Patent 4,775,653 to Vista Chemical Company (1988).

35. C. L. Edwards, U.S. Patent 4,825,009 to Shell Oil Company (1989).

36. C. L. Edwards, U.S. Patent 4,933,502 to Shell Oil Company (1990).

37. C. L. Edwards, U.S. Patent 4,375,564 to Shell Oil Company (1983).

38. C. L. Edwards, U.S. Patent 4,465,877 to Shell Oil Company (1984).

39. H. Nakamura, Y. Nakamoto, and Y. Fujimori, U.S. Patent 5,012,012 to Lion Corporation (1991).

40. S. W. King, U.S. Patent 4,727,199 to Union Carbide Corporation (1988).

41. S. W. King, U.S. Patent 4,873,017 to Union Carbide Corporation (1989).

42. S. W. King, U.S. Patent 5,025,094 to Union Carbide Corporation (1991).

43. C. L. Edwards, U.S. Patent 5,057,627 to Shell Oil Company (1991).

44. C. L. Edwards and R. A. Kemp, U.S. Patent 5,057,628 to Shell Oil Company (1991).

45. C. L. Edwards, U.S. Patent 5,059,719 to Shell Oil Company (1991).

46. R. A. Kemp, U.S. Patent 5,118,870 to Shell Oil Company (1992).

47. R. A. Kemp, U.S. Patent 5,208,199 to Shell Oil Company (1993).

48. R. A. Kemp and P. R. Weider, U.S. Patent 5,210,325 to Shell Oil Company (1993).

49. C. L. Edwards, U.S. Patent 4,665,236 to Shell Oil Company (1987).

50. C. L. Edwards, U.S. Patent 4,689,435 to Shell Oil Company (1987).

51. S. W. King, U.S. Patent 5,104,987 to Union Carbide Corporation (1992).

52. S. W. King, U.S. Patent 5,112,788 to Union Carbide Corporation (1992).

53. C. F. Hauser, S. W. King, R. J. Knopf, J. H. McCain, and C. A. Smith, U.S. Patent 5,112,789 to Union Carbide Corporation (1992).

54. S. W. King, U.S. Patent 5,118,650 to Union Carbide Corporation (1992).

55. S. W. King, U.S. Patent 4,873,017 to Union Carbide Corporation (1989).

56. S. W. King and R. J. Knopf, U.S. Patent 4,902,658 to Union Carbide Corporation (1990).

57. S. W. King, U.S. Patent 5,025,094 to Union Carbide Corporation (1991).

58. S. W. King, U.S. Patent 5,104,987 to Union Carbide Corporation (1992).

59. D. H. Champion and G. P. Speranza, U.S. Patent 5,110,991 to Texaco Chemical Company (1992).

60. S. W. King, U.S. Patent 5,136,106 to Union Carbide Corporation (1992).

61. M. Cuscurida, J. F. Knifton, and P. E. Dai, U.S. Patent 5,256,828 to Texaco Chemical Company (1993).

62. B. Bartha, L. Farkas, J. Morgos, P. Sallay, I. Rusznak, and G. Veress, J. Am. Oil Chemists. Soc. 62:824 (1985).

63. M. Cuscurida and E. L. Yeakey, U.S. Patent 5,103,062 to Texaco Chemical Company (1992).

64. J. F. Knifton and C. G. Naylor, U.S. Patent 5,344,984 to Texaco Chemical Company (1994).

65. J. M. Larkin and G. P. Speramza. U.S. Patent 4,827,038 to Texaco Chemical Company (1989).
66. J. M. Larkin and G. P. Speramza. U.S. Patent 4,847,417 to Texaco Chemical Company (1989).
67. K. R. Smith, U.S. Patent 4,888,447 to Ethyl Corporation (1989).
68. K. R. Smith, U.S. Patent 4,967,005 to Ethyl Corporation (1990).
69. M. Cuscurida and H. P. Klein, U.S. Patent 4,465,858 to Texaco Chemical Company (1984).
70. M. Cuscurida and W. P. Krause, U.S. Patent 4,479,010 to Texaco Chemical Company (1984).
71. R. Hoefer, U.S. Patent 4,689,366 to Henkel Komanditgesellschaft auf Aktien (1987).
72. H. Cho and D. Frank, U.S. Patent 4,677,208 to Akzona Incorporated (1987).
73. A. Riva and J. Ribe, Bol. Inst. Invest. Text. Coop. Ind. (Univ. Politec. Barcelona) *81*:31 (1982).
74. C. G. van Ginkel, C. A. Stroo, and A. G. M. Kroon, Tenside Surf. Det. *30*:213 (1993)
75. C. A. Johnson, U.S. Patent 4,039,716 to Owens-Corning Fiberglas Corporation (1977).
76. S. Razac and P. E. Eckler, U.S. Patent 5,409,574 to Rhone Poulenc Inc. (1995).
77. D. M. Regan, U.S. Patent 5,472,493 to Cabot Corporation (1995).
78 H. Abel, H. Hostettler, and R. Topfl, U.S. Patent 3,944,385 to Ciba-Geigy Corporation (1976).
79. H. U. von der Eltz, H. Maier, and K. Rostermundt, U.S. Patent 4,701,182 (1987).
80. K. C. Dardoufas and R. M. Marshall, U.S. Patent 4,103,068 to Allied Chemical Corporation (1978).
81. K. C. Dardoufas and R. M. Marshall, U.S. Patent 4,105,568 to Allied Chemical Corporation (1978).
82. K. C. Dardoufas and R. M. Marshall, U.S. Patent 4,210,700 to Allied Chemical Corporation (1980).
83. H. Dumm, G. Koener, M. Krakenberg, H. Rott, and G. Schmidt, U.S. Patent 4,434,008 to Th. Goldschmidt AG (1984).
84. R. E. Beetschen, B. P. Lynch, and A. A. Short, U.S. Patent 4,170,682 to Kellwood Company (1979).
85. B. J. Bishop and A. J. Moon, U.S. Patent 4,106,901 to Star Chemical, Inc. (1978).
86. T. Wershofen, U.S. Patent 4,803,012 to Henkel Komanditgesellschaft auf Aktien (1989).
87. H. M. Muijs and J. V. Schaik, U.S. Patent 5,061,386 to Shell Oil Company (1991).
88. I. Rivenaes, U.S. Patent 5,075,040 to Denbar, Ltd. (1991).
89. M. P. Breton, M. D. Croucher, K. M. Henseleit, B. Helbrecht, and F. M. Pontes, U.S. Patent 5,129,948 to Xerox Corporation (1992).
90. T. Wershofen, U.S. Patent 5,145,608 to Ecolab Inc.(1992).
91. D. W. Reichgott, U.S. Patent 5,282,992 to Betz Laboratories, Inc. (1994).
92. L. J. Barker, M. N. O'Connor, and R. G. Ryles, U.S. Patent 5,298,555 to American Cyanamid Company (1994).
93. R. M. Blanch and M. J. Kaszubski, U.S. Patent 5,310,772 to Allied Signal Inc. (1994).
94. L. J. Barker, M. N. O'Connor, and R. G. Ryles, U.S. Patent 5,376,713 to Cytec Technology Corp (1994).

95. S. O. Aston and J.R. Provost, U.S. Patent 5,403,358 to Imperial Chemical Industries PLC (1995).
96. R. M. Holdar, U.S. Patent 5,460,753 to NCH Corporation (1995).
97. M. J. Wisotsky, U.S. Patent 4,505,829 to Exxon Research and Engineering Co. (1985).
98. A. J. Castro and J. W. Stoll, U.S. Patent 4,147,742 to Akzona Incorporated (1979).
99. A. J. Castro and J. W. Stoll, U.S. Patent 4,210,556 to Akzona Incorporated (1980).
100. R. P. Pardee, U.S. Patent 4,232,072 to Ball Corporation (1980).
101. B. N. Hendy, U.S. Patent 4,268,583 to Imperial Chemical Industries Limited (1981).
102. A. J. Castro and J. W. Stoll, U.S. Patent 4,314,040 to Akzona Incorporated (1982).
103. W. Lybrand, U.S. Patent 4,391,952 to Bengal, Inc. (1983).
104. W. Lybrand, U.S. Patent 4,393,159 to Bengal, Inc. (1983).
105. W. Lybrand, U.S. Patent 4,393,176 to Bengal, Inc. (1983).
106. J. P. Duffy, U.S. Patent 4,417,999 to Witco Chemical Corporation (1983).
107. S. E. Gloyer, U.S. Patent 4,764,428 to Witco Chemical Corporation (1988).
108. B. D. Stevens, U.S. Patent 5,132,060 to the Dow Chemical Company (1992).
109. S. R. Seshadri, U.S. Patent 5,219,493 to Henkel Corporation (1993).
110. M. J. Incorvia, U.S. Patent 5,478,486 to Henkel Corporation (1995).
111. J. H. Finley, C. W. Lutz, and M. L. Pinsky, U.S. Patent 4,316,879 to FMC Corporation (1982).
112. T. Sato and S. Nagao, U.S. Patent 4,215,992 to Kao Soap Co., Ltd. (1980).
113. A. S. Canale and R. D. Canale, U.S. Patent 4,234,443 to Michael A. Canale (1980).
114. R. Fricker and R. Gross, U.S. Patent 4,362,530 to Sandoz Ltd. (1982).
115. U. Buchel and P. Muther, U.S. Patent 4,523,923 to Ciba-Geigy Corporation (1985).
116. A. Fredj, J. P. Johnston, and C. A. J. Theon, U.S. Patent 5,470,507 to the Procter and Gamble Co. (1995).
117. B. D. Busch and D. J. Yeavello, U.S. Patent 4,261,842 to Fremont Industries, Inc. (1981).
118. A. G. Horodysky, U.S. Patent 4,382,006 to Mobil Oil Corporation (1983).
119. A. G. Horodysky and J. M. Kaminisky, U.S. Patent 4,406,802 to Mobil Oil Corporation (1983).
120. A. G. Horodysky and J. M. Kaminisky, U.S. Patent 4,478,732 to Mobil Oil Corporation (1984).
121. A. G. Horodysky and J. M. Kaminisky, U.S. Patent 4,568,472 to Mobil Oil Corporation (1986).
122. A. G. Horodysky and J. M. Kaminisky, U.S. Patent 4,594,171 to Mobil Oil Corporation (1986).
123. W. T. Brannen, G. D. Burt, and R. A. McDonald, U.S. Patent 4,965,002 to Elco Corporation (1990).
124. K. P. Kammann, Jr., and W. R. Garrett, U.S. Patent 5,080,813 to Ferro Corporation (1992).
125. M. Grunert, and H. Tesman, U.S. Patent 4,356,020 to Henkel Komanditgesellschaft auf Aktien (1982).
126. J. L. Ahle, U.S. Patent 4,528,023 to Starffer Chemical Company (1985).
127. P. M. Ronning and G. A. Vandesteeg, U.S. Patent 4,556,410 to Minnesota Mining and Manufacturing Company (1985).
128. P. M. Ronning and G. A. Vandesteeg, U.S. Patent 4,557,751 to Minnesota Mining and Manufacturing Company (1985).

129. W. T. Hulshof, U.S. Patent 5,226,943 to Akzo N. V. (1993).
130. J. P. Bell, U.S. Patent 5,426,121 to Akzo Nobel N. V. (1995).
131. R. H. McIntosh and D. E. Triestram, U.S. Patent 5,474,739 to Interface, Inc. (1995).
132. F. Feichtmayr, D. Lach, and R. Streicher, U.S. Patent 4,272,243 to BASF Aktiengesellschaft (1981).
133. F. Landbeck, R. Leberfinger, H. Matschkal, and P. Rathfelder, U.S. Patent 4,470,825 to Schill and Seilacher GmbH and Co. (1984).
134. B. Danner, U.S. Patent 4,572,721 to Sandoz Ltd. (1986).
135. E. Daubach, H. Saukel, and L. Von Rambach, U.S. Patent 3,960,486 to BASF Aktiengesellschaft (1976).
136. J. T. Inamorato, U.S. Patent 3,959,157 to Colgate-Palmolive Company (1976).
137. R. M. Butterworth, J. R. Martin, and E. Willis, U.S. Patent 4,622,154 to Lever Brothers Company (1986).
138. R. M. Butterworth, J. R. Martin, and E. Willis, U.S. Patent 4,627,925 to Lever Brothers Company (1986).
139. E. P. Gosselink and Y. S. Oh, U.S. Patent 4,664,848 to the Proctor and Gamble Company (1987).
140. H. Andree, M. Berg, and W. Seiter, U.S. Patent 4,675,124 to Henkel Komanditgesellschaft auf Aktien (1987).
141. D. H. Leifheit, U.S. Patent 4,743,395 to the Drackett Company (1988).
142. R. P. Adams, U.S. Patent 4,741,842 to Colgate-Palmolive Company (1988).
143. G. Bosserhoff, H. Buchler, R. Puchta, and H. Wilsberg, U.S. Patent 4,931,063 to Henkel Komanditgesellschaft auf Aktien (1990).
144. S. Q. Lin, L. V. Salas, and L. S. Tsaur, U.S. Patent 5,429,754 to Lever Brothers Company, Division of Conopco, Inc. (1995).
145. M. Vataru, U.S. Patent 4,797,134 to Wynn Oil Company (1989).
146. J. J. Rayborn, U.S. Patent 4,645,608 to Sun Drilling Products Corp. (1987).
147. C. J. Davidson, U.S. Patent 4,828,724 to Shell Oil Company (1989).
148. K. A. Knitter and J. L. Villa, U.S. Patent 4,363,637 to Diamond Shamrock Corporation (1982).
149. M. J. Schick and E. L. Kelley, U.S. Patent 4,492,590 to Diamond Shamrock Chemicals Company (1985).
150. L. R. Norman and T. R. Gardner, U.S. Patent 4,324,669 to Halliburton Company (1982).
151. R. W. Jahnke, U.S. Patent 4,061,580 to The Lubrizol Corporation (1977).
152. F. W. Valone, U.S. Patent 4,420,414 to Texaco Inc. (1983).
153. B. Mosier, U.S. Patent 4,470,918 to Global Marine, Inc. (1984).
154. H. Haugen, U.S. Patent 4,224,170 to Texaco Inc. (1980).
155. R. N. Buggs, R. K. Gabel, and A. R. Syrinek, U.S. Patent 5,009,799 to Nalco Chemical Company (1991).
156. J. M. Brown, R. D. McBride, and J. R. Ohlsen, U.S. Patent 5,415,805 to Betz Laboratories, Inc. (1995).
157. J. M. Brown, J. J. Herrold, and H. A. Roberts, U.S. Patent 5,425,914 to Betz Laboratories, Inc. (1995).
158. R. Bohm, M. Hille, and R. Kupfer, U.S. Patent 5,421,993 to Hoechst AG (1995).
159. J. M. Brown, G. F. Brock, V. K. Mandlay, and J. R. Ohlsen, U.S. Patent 5,459,125 to BJ Services Company (1995).
160. H. K. Ficker and F. M. Teskin, U.S. Patent 4,684,683 to El Paso Products Company (1987).

161. H. K. Ficker and F. M. Teskin, U.S. Patent 4,692,489 to El Paso Products Company (1987).
162. K. G. McDaniel, U.S. Patent 4,760,100 to Arco Chemical Company (1988).
163. C. Bath and J. Trotoir, U.S. Patent 4,829,114 to Imperial Chemical Industries PLC (1989).
164. C. Breuer, U. Gelderie, and G. Holzer, U.S. Patent 5,030,686 to Isover Saint-Gobain (1991).
165. D. Augustin, C. Collette, and R. Parsy, U.S. Patent 5,089,546 to Atochem (1992).

7
Nonionic Surfactants Containing an Amide Group

ANNA LIF and MARTIN HELLSTEN* Central Research and Development, Akzo Nobel Surface Chemistry AB, Stenungsund, Sweden

I. INTRODUCTION

Nonionic surfactants containing an amide group are still a relatively small part of the total volume of nonionic surfactants, but they will probably be growing part. Their main advantage in comparison to other nonionic surfactants depends on the strong attraction between the amide groups, which leads to a dense packing at interfaces and thus an increased technical effect. Other advantages

---

* Retired.

In order to avoid the transamidation reaction as much as possible, Sandberg [6] was able to carry out the ethoxylation process at a significantly lower temperature than normal by using a tertiary or quaternary amine as catalyst. The reaction temperature can then be lowered from 140–150°C to 75–80°C. With this process light-colored amide ethoxylates may be produced, containing <0.5% esteramide.

Another way of eliminating esteramides from the ethoxylated products is given by Kado et al. [7] who treat the amide ethoxylates with monoethanolamine or monoaminoglycol ethers, thereby causing the reverse reaction, from III to I, in the reaction scheme given above.

Ethoxylated amides with a light color may also be produced at high ethoxylation temperature and alkali as catalyst if 0.01–0.5% (w/w) $NaBH_4$ is added to the fatty acid alkanolamide [21].

Instead of ethoxylating a fatty acid alkanolamide, it has been suggested [22] that an alkyl polyglycol ether amine be produced and reacted with a fatty acid ester according to the following:

$$R_1O(C_2H_4O)_nH + CH_2 = CHCN \rightarrow R_1O(C_2H_4O)_n(CH_2)_2CN \xrightarrow{H_2}$$

$$R_1O(C_2H_4O)_n(CH_2)_3NH_2 \xrightarrow{RCOOCH_3} RCONH(CH_2)_3(OC_2H_4)_nOR_1$$

where RCOOH is a fatty acid and R_1OH an alcohol containing one to four atoms.

Low-foaming surfactants, e.g., for machine dishwashing, may be produced according to this method.

C. Derivatives of Ethoxylated Amides

The ethoxylated fatty acid amides can be derivatized to anionic surfactants in the same manner as ethoxylated fatty alcohols. The only important difference consists of the numerous byproducts in the amides, e.g., the esteramides which may be completely or partly hydrolyzed during the reactions.

A process for the production of carboxymethyl acids of ethoxylated fatty acid alkanolamides is described in [8] whereby the alkanolamide is reacted first with ethylene oxide and then with a haloacetic acid in the presence of solid alkali. The haloacetic acid is used in less than stoichiometric amounts. The resulting surfactant mixture is claimed to have good foaming properties and mildness to the skin and is useful for the formulation of shampoos, skin cleansers, and dishwashing detergents.

A phosphate ester of ethoxylated alkanolamide may be prepared according to [9] by first treating an ethoxylated monoethanolamide based on coconut fatty acids with ethanol during 1 h at 80°C in order to eliminate the contents of esteramide and then to treat the product with $POCl_3$ and neutralize. The surfactants formed are suitable for the formulation of low-temperature liquid detergents and have a good foaming ability.

By using chlorosulfonic acid instead of $POCl_3$ in the process outlined above, sulfates of the ethoxylated ethanolamides may be produced [10]. The resulting surfactants are also claimed to be useful for low-temperature detergents and to have a good foaming ability.

D. Nonionic Surfactants Containing Both an Amide and a Carbohydrate Group

During the last 10 years much work has been done on the combination of a carbohydrate and a fatty acid via an amide group. Two different routes have become dominant:

1. A reducing sugar is reacted with a lower alkylamine and this product is then hydrogenated to the corresponding sugar amine. A fatty ester is then amidated with this amine to the N-acyl, N-alkyl sugar amide. So far glucose has been the most common reducing sugar used for this reaction and consequently the whole group of surfactants is often called N-alkylglucamides or just glucamides.
2. A reducing sugar is oxidized to the corresponding carboxylic acid and this acid is then amidated with a primary (or secondary) amine. The products are usually named after the sugar used, e.g., lactose is oxidized to lactobionic acid and then amidated to lactobionamide.

1. N-Alkyl Sugar Amines

A straightforward process for N-alkylglucamine is described in [12] and comprises the following steps: Glucose is reacted with a slight excess of CH_3NH_2 at 30–60°C with methanol as a solvent. After completion of this reaction the solvent is evaporated until no methylamine is present in the solution. The hydrogenation is then carried out at 50–90°C and a pressure of 250 psi using a supported nickel catalyst.

2. N-Alkyl Sugar Amides

An example of a possible process for the amidation of a fatty acid ester with N-methylglucamine is given in [13]: A mixture of methyllaurate and methanol + Na-methylate was treated at 80°C with a solution of N-methylglucamine in propylene glycol and the methanol was continuously evaporated from the reaction mixture resulting in a waxy N-lauryl, N-methylglucamide having a bright color.

A variation of this process is given in [14] whereby Na_2CO_3 or K_2CO_3 in an amount of 0.5–50% of the reactants (mol%) is used as a catalyst and 2–20% (w/w) of a preformed glucamide surfactant is used as a phase transfer agent. The advantage of this process is supposedly a very short reaction time (0.5–10 min) at the highest reaction temperature (160–180°C).

Another variation of the basic process is given in [15] whereby M_3PO_4, $M_4P_2O_7$, $M_5P_3O_{10}$, M_3-Citrate, M_2-tartrate, and M_n-basic aluminosilicates are mentioned as catalysts in an amount of 10–95% (w/w) based on the sum of the reactants. The resulting product seems to be aimed as an ingredient for a laundry detergent.

The possibility of balancing higher fatty acids by mixing the glucose with reducing di- or polysaccharides at the production of surfactants of the glucamide type has been observed in [16] where it is pointed out that the amidation should be carried out at 50–80°C to avoid cyclization and that the N-alkyl sugar amine should be derived from a glucose maltose mixture in ratio of 4:1 to 99:1 (w/w).

N-Alkylglucamides aimed for liquid hand dishwashing formulations should have a low content of fatty acid soaps and esters. Otherwise they will have a detrimental effect on foam stability, particularly in hard-water areas. The glucamides may be improved in this respect according to [17] by treating the final product with an amine, particularly an ethanolamine, thereby transforming the soaps and esters to alkanolamides, which normally will have a foam-stabilizing effect.

A continuous process for the amidation of fatty acid ester is described in [18] where the apparatus consists of a cascade of two stirred reactors. To the first reactor equimolar amounts on N-methylglucamine and C_{12-14}-fatty acid methyl ester were fed continuously together with 5 mol% of Na-methylate and 50 mol% of propylene glycol. The temperature was kept at 95°C and the mean reaction time was 1.9 h. The conversion in this first reactor was 82%. The effluent from this reactor was fed to a second one, which was kept at 90°C and a pressure of 22 mbar. The mean reaction time was 2.1 h and the overall conversion 94% based on the N-methylglucamine.

3. Formulation of N-Alkylamides

Due to the high melting point of the N-alkylglucamides they are normally delivered as solutions. As solvents have been suggested water, lower alcohols, and propylene glycol containing various amounts of water. In [19] the solubilizing effect of certain carboxylates, particularly Na-citrate, is pointed out. In an example, a 40% solution of C_{12-14}-akyl-N-methylglucamide is first dissolved in water at 60°C and then 2% of citric acid is added and neutralized to pH 7–9 with NaOH. The resulting solution has a viscosity of 1200 cPs at 31°C and remains stable at this temperature for at least 7 weeks.

A dry, flowable powder can be produced by treating a solution of N-lauroyl, N-methylglucamide, to which sodium chloride and silica has been added, with superheated steam [20]. The powder has no disagreeable odor or taste and may be used as an ingredient in toothpastes.

III. COMMERCIAL PRODUCTS

Table 2 lists some of the presently available fatty acid alkanolamide ethoxy-lates. The commercial fatty acid mono- and dialkanolamides are not included. The mean number of ethylene oxide units given includes the hydroxyethyl group(s) emanating from the alkanolamine. The hydrophile–lipophile balance (HLB) values are according to Griffin and the cloud points are valid for 1% of product in water unless other conditions are stated.

TABLE 2 Fatty Acid Amide Ethoxylates: Market Products and Producers

Trade name	Producer	EO units	HLB	Cloud point (°C)	Active matter (%)	Remarks
A. *Products derived from lauric acid CAS 26635-75-6*						
Amidox L-2	Stepan	3	—	—	100	
Amidox L-5	Stepan	5	—	—	100	
Mazamide L-5	PPG	6	—	—	100	Based on DEA
Unamide L-2	Lonza	3	—	—	100	
Unnamide L-5	Lonza	6	—	—	100	
B. *Products based on coconut or palm kernel oil acids, CAS 61791-08-0*						
Alkamide C2	Rhone Poulenc	—	—	—	100	
Amadol CMA 2 (80%)	Akzo Nobel	3	7.7	—	80	
CMA 5	Akzo Nobel	6	11	80[*]	100	[*] At pH 3
Empilan LP2	Albright & Wilson	—	—	—	100	
Empilan LP10	Albright & Wilson	—	—	—	100	
Empilan MAA	Albright & Wilson	6	9.6	—	100	
Ethylan CH	Harcros UK	—	—	80	100	
Ethylan CRS	Harcros UK	—	—	—	100	
Ethylan LM2	Harcros UK	—	—	—	100	
Eumulgin PC2	Pulcra SA	2	5.8	—	100	
Eumulgin PC4	Pulcra SA	4.5	8.8	—	100	
Eumulgin PC 10/85	Pulcra SA	10	14	—	100	
Eumulgin C4	Henkel	5	8.5	—	100	
Mazamide C5	PPG	6	—	—	100	
C. *Products based on oleic acid, CAS 31799-71-0 or 26027-37-2*						
Amadol OMA 2	Akzo Nobel	3	6.4	—	100	
OMA 3	Akzo Nobel	4	—	—	100	
OMA 4	Akzo Nobel	5	8.8	70[*]	100	[*] At pH 3

TABLE 2 Continued

Trade name	Producer	EO units	HLB	Cloud point (°C)	Active matter (%)	Remarks
C. Products based on oleic acid, CAS 31799-71-0 or 26027-37-2 (continued)						
Dionil OC	Hüls	3	—	67*	100	* In 25% BDG
Dionil SH 100	Hüls	6	—	—	100	
Dionil W 100	Hüls	14	—	60*	100	* In 10% NaOH
Ethomid 0/15	Akzo Nobel	5	8.8	80	100	
Ethomid 0/17	Akzo Nobel	7	—	—	100	
Lutensol FSA 10	BASF	—	12	80	100	
Rewopal 08	Rewo	9	—	—10	100	
Rewopal 015	Rewo	15	—	—	100	
D. Products based on hydrogenated tallow fatty acids, CAS 68155-24-8						
Ethomid HT/15	Akzo Nobel	5	8.8	—	100	
Ethomid HT/23	Akzo Nobel	13	13	55–77	100	
Ethomid HT/60	Akzo Nobel	50	18	—	100	
Schercomid HT/60	Scher	50	—	—	100	
E. Product based on rapeseed oil fatty acids, CAS 85536-23-8						
Aminol Tec N	Chem-Y	4	7.7	—	92	

IV. PHYSICAL AND AMPHIPHILIC PROPERTIES

In table 3 are given the foaming properties, wetting power, and HLB values for some of the commercial amide ethoxylates mentioned in Table 2. From the table it seems obvious that good foaming and foam stability characterizes the ethoxylated amides based on lauric and myristic acids and that these products also have a good wetting power independent of the polyglycol ether chain length in the range of 3–6 ethylene oxide units.

A. Polyoxyethylated Alkylamides

Heusch [36] studied the lyotropic phases of oleic amide heptaethylene glycol ether in water and interpreted the results depending on the formation of various hydrates with very different solubilities (Fig. 1).

Khan et al. [37] investigated a similar compound, an oleic amide nonaethylene glycol ether, with viscosity measurements, nuclear magnetic resonance, and direct imaging by the cryo-TEM technique and conclude that small, spherical micelles coexist with large spheroidal and thread-like micelles in dilute solutions. The solution has a high viscosity, which increases rapidly with the concentration to a maximum at 20% and then decreases again.

TABLE 3 Properties of Some Commercial Amide Ethoxylates

Product name	Foam height (mm)	Wetting power (s)	HLB	Fatty acids
Amadol CMA 2	113/108	22	7.7	Lauric/myristic
Cocodiethanolamide	90/85	27	8.5	Lauric/myristic
Empilan MAA	107/95	44	9.6	Lauric/myristic
CMA 5	125/122	20	11.1	Lauric/myristic
Ethomid	36/31	600	8.8	Oleic
OMA 4	20/18	280	8.8	Oleic
Lutensol FSA 10	95/90	110	12.1	Oleic
Ethomid HT/15	39/33	220	8.8	Stearic
Ethomid HT/60	84/32	600	17.7	Stearic
Aminol Tec N	16/13	500	7.7	Oleic/erucic

Foam height: Ross-Miles, 50°C, 0.05%, immediately and after 5 min. Wetting power: Draves, 20°C, 1 g/L. HLB: Griffin.

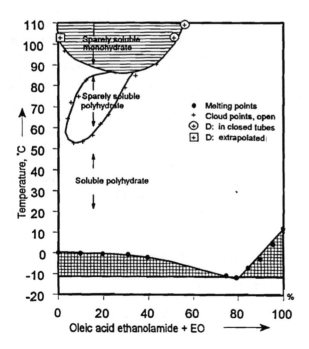

FIG. 1 Lyotropic phases of oleic amide heptaethylene glycol ether.

B. *N*-Acyl, *N*-Alkyl Sugar Amides

Pfannemüller et al. [23–28] conducted a comprehensive study to characterize the structures in water solution and the crystal structures of sugar amides synthesized from aldonic acid lactones of glucose and maltose and alkylamines or the similar compounds made from *N*-(2-aminoethyl)aldonamine and alkanoic acids where the acyl groups have been varied from C_6 to C_{20}.

Highly ordered aggregates are formed in water due to the strong hydrogen bonding between the amide and carbohydrate groups. This leads to the formation of stable monolayers suitable for the preparation of technical membranes (C_{12-20}-acyl groups).

With C_{6-10}-acyl chains, gels consisting of a loose network of regularly twisted helical strands are formed from dilute solutions. Lyotropic liquid crystalline phases are formed from more concentrated solutions, particularly for *N*-methylated samples with a C_{10}-acyl chain.

Fuhrhop et al. [29] made the very interesting observation that dihelical fibers, several micrometers in length, are formed in water solutions of octyl L- or D-glucamides, but not from the racemate. The single strands have the thickness of a bimolecular layer.

Since Hildreth in 1982 reported [30] that acyl-*N*-methylglucamides are useful for solubilization of membrane proteins from animal cells, other research groups have investigated the amphiphilic properties of these surfactants [31–34]. In Table 4 are given the CMCs for C_{8-9}- and C_{10}-acyl-*N*-methylglucamides, which in the biochemistry literature often are called MEGA-8, MEGA-9, etc. The CMC decreases only slightly with increasing temperature for MEGA-9 and MEGA-10 according to [32], whereas the CMC for MEGA-8 is practically independent of the temperature in the range 5–40°C.

It has long been known that carbohydrate surfactants can form thermotropic lamellar phases, but their lyotropic phases were recently investigated by Hall et al. [35] from whom the schematic phase diagram in Fig. 2 emanates. The diagram is somewhat rich in mesophases, whereas the corresponding diagram for MEGA-8 will show only a hexagonal phase existing between 0 and 39°C. The oleic acid *N*-methylglucamide, on the other hand, will show only a lamellar

TABLE 4 Critical Micelle Concentrations (CMCs) for Three *N*-Acyl, *N*-Sugar Amides

Surfactant	CMC (mmol/L)	
	Ref. 31, $T = 20$°C	Ref. 32, $T = 25$°C
MEGA-8	—	51
MEGA-9	25	16
MEGA-10	7	4.8

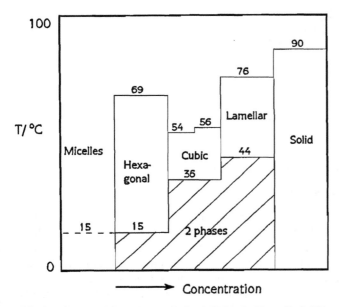

FIG. 2 Schematic phase diagram for MEGA-9. (From Ref. 35.)

phase. The phase behavior of *N*-acyl, *N*-methylglucamides is thus highly dependent on the size and structure of the hydrophobic group.

C. Dialkyl Disugar Amides

The concept of "gemini surfactants," i.e., surfactants containing two hydrophobic and two hydrophilic groups, has received much attention in recent years.
Eastoe et al. recently published two papers [173,174] regarding the amphiphilic
properties of dichain glucamides with the following structure:

$$C_nH_{2n}+1 \diagdown \quad \diagup CH_2NHCO(CHOH)_4CH_2OH$$
$$C$$
$$C_nH_{2n}+1 \diagup \quad \diagdown CH_2NHCO(CHOH)_4CH_2OH$$

As an acronym for the long names of these gemini surfactants, di-(Cn-Glu) has
been introduced.

In Table 5 the CMCs for the investigated four members of this series are
given at 35°C. This type of dichain glucamide is claimed to lower the interfacial
tension between oil and water even more than the monochain compounds and
thus be suitable as detergent ingredients [175].

TABLE 5 Critical Micelle Concentrations
for Four Dealkyl Sugar Amides

Surfactant	CMC (mmol/L)
di-(C5-Glu)	10.5
di-(C6-Glu)	1.35
di-(C7-Glu)	0.16
di-(C8-Glu)	0.04

V. FUNCTIONS AND APPLICATIONS

A. Polyoxyethylated Amides

Polyoxyethylated amides are used in broad range of applications, but nonetheless the total volume is limited. One reason could possibly be that the alkanolamides were introduced earlier and can be used in most of the same applications as the polyoxyethylated amides. However, for environmental and human toxicity reasons, the use of polyoxyethylated amides should be preferred, due to the very low amount of nitrosamines (see Sec. VI, "Environmental Properties") as well as the absence of diethanolamides in the products. The surfactant as a class has been known for a long time and many of the surfactant manufacturers can offer one or more polyoxyethylated amides. The alkanolamides will not be covered in this chapter.

1. Functions

The functions of polyoxyethylated amides are used in a wide range of different applications. Table 6 lists some functions and the corresponding applications. It can be seen in the table that the main functions of the polyoxyethylated amides are foam stabilization, thickening, and dispersion.

TABLE 6 Functions and Applications of Polyoxyethoxylated Amides

Function as:	Applications
Foam stabilizer	Liquid detergents, hard surface cleaners, general-purpose cleaners, hand cleaning gels, personal care products
Emulsifier	Cutting fluids, lubricants, rust inhibitors, plant protection agents, emulsifier for solvents and fuels
Suspending agent	Industrial degreasers, cleaning formulations, detergents
Antistatic agent	Polymers, fabric softeners
Solubilizer coupling agent	Personal care products, detergents
Fulling agent scouring agent	Textile-processing agents
Wetting agent	Low-temperature detergents, hand cleaners

For the low-oxyethylated amides thickening and foaming are the most important functions. An example of foaming is an alkaline foam cleaner composition reported to contain polyoxyethylated coconut fatty acid monoethanolamide giving copious stable foam [38]. Another example is a detergent composition with polyoxyethylated amide as a suds-boosting agent [39]. Concerning the polyoxyethylated amide as thickener, the thickening properties of polyoxyethylated rapeseed monoethanolamides with a maximum of 3 moles of ethylene oxide have been studied in different surfactant mixtures [40].

For the more highly oxyethylated amides suspending and emulsifying are very important functions. Examples are polyoxyethylated linseed oil fatty acid monoethanolamides with 7–18 moles of ethylene oxide that are used in different paint applications [41,42].

Concerning the other functions, an example is given of an antistatic effect when at polyoxyethylated (16-EO) oleic acid amide is used as a fabric softener [43].

2. Applications

(a) Personal Care. The use of different polyoxyethylated amides is not only restricted to shampoo and soap formulations, where the foaming and thickening properties are utilized. It can also be used in special applications such as lipstick [44] or in medicated face washing compositions in combination with Na selenite [45]. In the different compositions used as shampoo, soap, and other personal care products, the polyoxyethylated amides are used together with other surfactants and additives [46,47] and give varying properties to the compositions [48,49].

(b) Cleaning. In the cleaning area the polyoxyethylated amides are found in compositions both for hard surface and textile cleaning as well as in different fabric softener compositions.

The household hard surface cleaning agents cover dishwashing liquids and all-purpose kitchen liquid detergents [50]. The polyoxyethylated amides are used as suds-boosting agents [39], in combinations of anionic and nonionic surfactants [51], and in alkaline compositions [52]. The type of fatty acids used as a base for the amides differs between the various applications. A mixture of fatty acid carbon chain lengths and unsaturation in the polyoxyethylated amide, i.e., lauric, oleic, linoleic, and linolenic, is found to give superior detergency and a low melting point [53].

In laundry detergent composition the polyoxyethylated amides are used in combinations with soap [54], with Na-lauryl or cetyl sulfate [55], and with polyoxyethylated fatty C_{10-18}-alcohol [56]. The fatty acids contain both shorter (C_{10-12}) and longer (C_{18}) chains. In [55] both polyoxyethylated fatty acid monoethanolamide and diethanolamide can be used. The polyoxyethylated amides are also found in combination with an anionic surfactant and the enzyme lipase [57]. In a special application the detergent composition contains layered silicate in combination with the polyoxyethylated amide for improved washing power in the area of cosmetic staining, e.g., make-up, lipstick, and eyelash finish [58]. Also reported is a nonfoaming laundry detergent with superior laundering properties containing a polyoxyethylated rapeseed oil mono-ethanolamide with 5–7 moles of ethylene oxide [59].

The polyoxyethylated amide, based on C_{10}-fatty acid and up to 7 alkylene oxide units, are used in different fabric softener compositions in combination with cationic surfactants [60–62]. The polyoxyethylene (14–18 moles) oleic acid ethanolamide can be used as a textile softener [43]. Also in a combination of a cationic and nonionic softener the polyoxyalkylated (1–15 moles) C_{8-22}-fatty acid amide may be used [63]. Further, polyoxyethylated C_{10-16}-fatty acid amides are used in antistatic detergent compositions where cleaning and softening properties are combined in the same composition [64–66]. Polyoxyethylated oleic acid amide is also found in compositions with both detergent and softening properties (softergent) [67].

(c) Industrial Cleaning. Industrial cleaning and degreasing compositions are used in various applications, e.g., degreasing of metallic surfaces prior to painting or coating [68–71], degreasing to reduce corrosion [72], removal of deposits [73] or hydrocarbon oil [74] from metallic surfaces. Polyoxyethylated amides are also used in a composition with glycol *t*-butyl ether for cleaning of hard glass and ceramic surfaces [75].

The polyoxyethylated amides can also be used in cleaning and disinfecting compositions. For example, in combination with quaternary ammonium compounds (e.g., alkyltrimethylammonium chloride) for use in, say, the dairy industry [76,77]. In a disinfectant-cleaning composition polyoxyethylated coconut oil diethanolamide is added as a component that will reduce eye irritation [78].

A surfactant mixture of alkylglucoside and polyoxyethylated amide is used in washing and cleaning composition for wool, dishes, and so forth. The mixture can also be used as a flotation aid or in bore-flushing fluids [79].

Polyoxyethylated fatty acid amides with a short hydrophobic end group are useful in a wide range of washing and cleaning products, where low foaming is needed, and as emulsifiers [80].

(d) Miscellaneous. A very special application for polyoxyethylated amide is as a drag-reducing agent in cold water distribution systems [81–84]. The poly-oxyethylated amide forms thread-like micelles that will stabilize the laminar flow also at high flow rates and thus reduce the pump work. Heating of the cold water by the pumps will also be reduced.

Alkyd emulsions, in which the alkyd oil is emulsified in water and stabilized by surfactants, are interesting from an environmental point of view due to the lower need for organic solvents in the paint. The surfactants used are biodegradable polyoxyethylated amides, preferably based on linseed oil fatty acid and with 7–18 moles of ethylene oxide [41,42].

Compositions for the textile industry, concerning dyeing and finishing, are described in [85–88]. The compositions are low foaming and give good wetting.

Various special applications have also been found. For example, the polyoxyethylene amides are used as dye penetrants for detecting flaws in articles [89]. A coating for kaolin powderbased on polyoxyethylene amide has been effective [90]. Addition of polyoxyethylene amide to a lubricant for cold drawing of metals results in several improvements [91]. Polyethylene film for soil mulching in agriculture contains polyoxyethylated amide as an antistatic compound [92].

B. Glucamides

The use of glucamides has become strongly focused on cleaning formulations during the last 5 years. But there are also a lot of industrial and medical applications for glucamides, which have been patented since 1960.

1. Functions

The most important functions of the glucamides are foaming, thickening, and emulsifying.

A fatty acid polyhydroxyamide is said to give more foam than a conventional foam booster, e.g., amine oxide [93]. Also a mixture between a short chain (C_{8-10}) and long chain (C_{12-16}) fatty acid N-methylglucamide gives enhanced foaming [94]. A C_{6-22}-fatty acid N-methylglucamide should give excellent foam for toothpaste compositions [95].

The fatty acid (C_{2-18}) N-methylglucamides are used as thickeners for aqueous surfactant solutions [96]. The coconut oil–based glucamides are especially useful for thickening cosmetic and detergent compositions. They are said to be more effective in such compositions than fatty acid diethanolamides and alkoxylated fatty acids.

2. Applications

(a) Personal Care. Glucamides can be used in various personal care products, e.g., shampoos and other hair care formulations, soaps, and toothpastes.

There are a number of patents whereby glucamides are used in shampoos and other hair care formulations. A shampoo composition with glucamide is said to give rich foam properties and to be mild to the eyes [97]. A soap bar composition containing glucamide can give good lathering and a decreased tendency to crack [98]. A personal care or cosmetic detergent composition containing glucamide and vegetable oil adducts has been reported [99]. Addition of silicones to shampoos, hair rinses, etc., containing C_{12-14}-fatty acid alkyl-N-

methylglucamide improves their foaming capacity [100]. A personal cleansing composition comprising C_{12-18}-fatty acid polyhydroxyamide is very mild and has good conditioning benefits, stability, lathering, and rinsability [101]. The C_{12-14}-fatty acid–based glucamides are also used in combinations with cationic polymers giving excellent foaming, cleansing, and hair brightening [102]. Lauric acid N-methylglucamide is used in combination with monomeric cationic surfactants in a hair conditioner [103].

In toothpastes the glucamides are used as the sole surfactant [95] or in a mixture with sodium dodecyl sulfate [104]. The surfactants used show high mucosal compatibility, high foaming capacity, and improved flavor quality [105].

(b) Cleaning. Glucamides are also found in compositions for both hard surface and textile cleaning. There are also multipurpose compositions covering dishwashing and laundry cleaning (106–108) and, in some cases, personal care products [109–112]. The glucamides are combined with other nonionic or anionic surfactants—in one case with silicon added as foam control [107] and in another case with oxygen bleach [111].

The cleaning agents for hard surfaces in households cover chiefly dishwashing liquids. The different compositions have different benefits, such as enhanced foaming and wetting [94], removal of greasy deposits and tarnish [113], and high detergency and gentleness to skin [114]. There exist a wide range of patents concerning the advantages of dishwashing liquids. There are compositions with good cleaning and foaming properties that are mild to the skin [115–120]. There are also compositions that are low foaming [121,122] and high foaming [93,123], both types with good soil release performance. The addition of amine oxides to the glucamides improves the ease and rate of soil removal [124], and the addition of alkyl ether carboxylate provides good cleaning and grease removal [125,126].

The use of different detergent compositions comprises a broad variety of applications in the laundry cleaning area. There are compositions especially for removing oil and grease [127], with the help of lipase [128] or in combination with a polymeric dispersing agent for enhanced particulate clay soil removal, together with the grease and oil cleaning [129]. Low-foaming or controlled foaming detergent compositions are an advantage in automatic washing machines [130–136]. There are also compositions that are free of phosphate builders and alkylbenzenesulfonates [137,138], laundry compositions for handwash [139], compositions for color maintenance [140], compositions with enhanced soil release properties [141], compositions with bleaching agents [142], and compositions with added detergent enzymes [143].

Different types of builders are used, i.e., zeolite or layered silicate [144–146] or polycarboxylates [147]. To the detergent compositions with builder is also added brightener [148]. The production method for granular laundry detergent is also patented [149–151].

Glucamides can be used in compositions together with softening clay [152,153]. Other types of rinsing or after-treatment compositions for laundry have also been reported [154,155].

The aldobionamides when used in detergent compositions for washing fabrics in combination with anionic surfactants show good detergency [156, 157].

(c) Industrial Cleaning. The use of glucamides in the industrial cleaning area is limited. Only a few examples are found. Compositions containing mixtures of polyoxyethylene-polyoxypropylene ether and fatty acid *N*-alkylpolyhydroxy-amide can be useful as hard surface cleaners, dye dispersants, flotation aids, etc. [158]. Fatty acid *N*-alkylpolyhydroxyamides are used in rinsing agents for mechanical cleaning of hard surfaces [159].

A biodegradable C_{16-18}-fatty acid *N*-alkoxypolyhydroxyamide can be used in a composition for cleaning hard surfaces [160]. A combination of *N*-alkylpoly-hydroxyfatty acid amide and the C_{16-18}-fatty acid *N*-alkoxypolyhydroxyamide can be used as emulsifier for, say, paints, fabric softener, agricultural chemicals, cosmetics, and so forth.

(d) Miscellaneous. A C_{12}-fatty acid *N*-methylglucamide is used in a bio-degradable softener system for treating tissue paper [161]. Also C_{12-14}-fatty acid *N*-methylglucamide can be used in such systems [162].

The fatty acid polyhydroxyamides can be used as dispersants or wetting agents in pesticide formulations [163]. The amides can also be used in disper-sant composition for dyes and pigments [164]. Furthermore, glucamides based on saturated fatty acids can be useful as accelerating emulsifiers in emulsion polymerization of vinyl chloride [165].

The remediation of chlorocarbon-contaminated aquifers is performed by microemulsification of the chlorocarbons. The surfactants used are glucamides [166,167]. A mixture of alkylglucoside and glucamide, used as environmentally friendly synthetic leather or hide-greasing agents, is used instead of chlorinated hydrocarbons [168].

The *N*-alkylamide of lactobionic acid is used in a corrosion inhibitor for use in metal working media [169].

VI. ENVIRONMENTAL PROPERTIES

The renewable raw materials in polyhydroxy fatty acid amides (glucamides) and part of the polyoxyethylated fatty acid amides, and the fact that they are readily biodegradable, makes those two types of surfactants very attractive.

The biodegradability of a polyoxyethylated fatty acid amine has been reported [170]. A proposed route for biodegradation is given. The polyoxyethylated amides OMA 4, CMA 5, Amadol OMA 2, and Amadol CMA 2[1] are all readily biodegradable. The OMA amides are based on C_{18}-fatty acids and the CMA amides are based on C_{12-14}-fatty acids, and the numbers refer to the amount of ethylene oxide added.

The biodegradation of glucamides has been reported as well. An activated sludge study [171] and field studies [172] show that the concentration of glucamides is considerably reduced. In [173] the reduction is more than 99.6%. The reduction is assumed to be caused by biodegradation.

The nitrosamine content has been analyzed in Amadol OMA 2, Amadol CMA 2, and diethanolamide. The "total" nitrosamine content (as NO) was determined to 10–20 ppb in the Amadols and 100 ppb in diethanolamide. The former amount of nitrosamines is concerned low and the latter is considered high. High amounts of nitrosamines in the diethanolamides make them questionable for use in personal care products and light-duty detergents.

REFERENCES

1. E. Jungermann and D. Tabor, Polyoxyethylene Alkylamides, in *Nonionic Surfactants*, Surfactant Science Series Vol. 1, Marcel Dekker, New York, 1967, pp. 208–246.
2. H. Grossmann, Fette Seifen Anstrichm. *74*(1):58–63 (1972).
3. S. H. Fairheller, R. G. Bistline, A. Bilyk, R. L. Dudley, M. F. Kozempel, and M. J. Haas, J. Am. Oil Chemists Soc. *71*(1):863–866 (1994).
4. R. B. McConnell and D. W. Keuer, U.S. Patent 4085126 to Ashland Oil Co. (1978).
5. H. Grossmann, BRD Patent 2129425 to Chem. Werke Hüls AG (1972).
6. E. Sandberg, WO Patent 9208690 to Akzo Nobel Surface Chem. AB (1992).
7. K. Kado, K. Matsushita, T. Fujii, and K. Usuba, Jap. Patent 07102292 to Kawaken Fine Chemicals Co.
8. H. Terasaki, A. Fujiu, K. Isobe, T. Azuma, H. Nishikawa, and T. Imamura, Eur. Patent Appl. 960044 A2 to Kao Co., Japan (1996).
9. T. Fujii, H. Sekido, and K. Usuba, Jap. Patent 07197078 A2 to Kawaken Fine Chemicals Co. (1995).
10. T. Fujii and K. Usuba, Jap. Patent 07194958 A2 to Kawaken Fine Chemicals Co. (1995).
11. A. M. Naser, M. A. El-Azmirly, and A. Z. Gomaa, J. Oil Colour Chem. Assoc. *60*(1):18–22 (1977).
12. J. J. Scheibel, D. S. Connor, R. E. Shumate, and J. StLaurent, WO 9206984 to Procter and Gamble Co. (1990).
13. B. Strecker, H. Wolf, A. Oftring, H. H. Bechtolsheimerand, and D. Hertel, Ger. Pat. 4235783 A1 to BASF AG (1994).
14. D. S. Connor, J. Kao, J. J. Scheibel, and J. N. Kao, WO 9206071 A to Procter and Gamble Co. (1990).

---

[1] Trademarks by Akzo Nobel.

15. D. S. Connor, J. Kao, J. J. Scheibel, and J. N. Kao, WO 9206070 A to Procter and Gamble Co. (1990).
16. D. S. Connor, J. J. Scheibel, and R. G. Severson, WO 9206073 A (1990).
17. D. S. Connor, J. J. Scheibel, B. P. Murch, P. Bruce, M. N. Mao, E. P. Gosselink, and R. G. Severson, Jr., U.S. Patent 5188769 to Procter and Gamble Co. (1993).
18 R. Aigner, A. Fruth, H. Keck, U. Meyer, H. Seitz, J. Strauss, H. Stühler, M. Vervuert, G. Koch, and R. Vybiral, EP 63244 A2 to Höchst AG (1995).
19. Y. Fu, P. R. Foley, J. A. Dyet, P. G. Mather, R. Mermelsten, and B. J. Roselle, WO 9319146 A1 to Procter and Gamble Co. (1993).
20. W. Raehse, K. Paatz, W. Breitzke, W. Seispel, and H. Tesmann, DE 4340015 A1 to Henkel KgaA (1995).
21. S. Stüring and L. Kuhnen, BRD Patent Appl. 1806010 to Chem. Werke Hüls AG (1970).
22. B. Strecker, G. Oetter, A. Oftring, J. Perner, R. Baur, V. Schwendenmann, M. Kahmen, and W. Reif, BRD Patent Appl. DE 4336247 A1 to BASF AG (1995).
23. B. Pfannemüller and W. Welte, Chem. Phys. Lipids *37*:227–240 (1985).
24. V. Zabel, A Müller-Farnow, R. Hilgenfeld, W. Saenger, B. Pfannemüller, V. Enkelmann, and W. Welte, Chem. Phys. Lipids *39*:313–327 (1986).
25. B. Pfannemüller and I. Kühn, Makromol. Chem. *189*:2433–2442 (1988).
26. A. Müller-Farnow, R. Hilgenfeld, H. Hesse, and W. Saenger, Carbohydrate Res. *176*:165–174 (1988).
27. B. Pfannemüller, Starch/Stärke *40*:(12):476–486 (1988).
28. F. R. Taravel and B. Pfannemüller, Makromol. Chem. *191*:3097–3106 (1990).
29. J.-H. Fuhrhrop, P. Schnieder, J. Rosenberg, and E. Boekema, J. Am. Chem. Soc. *109*:3387–3390 (1987).
30. J. E. K. Hildreth, Biochem. J. *207*:363–366 (1982).
31. M. Hanatani, K. Nishifuji, M. Futai, and T. Tsuchiya, J. Biochem. *95*:1349–1353 (1984).
32. A. Walter, S. E. Suchy, and P. K. Vinson, Biochim. Biophys. Acta *1029*:67–74 (1990).
33. M. Shigeyoshi, K. Hirotaka, I. Yuuhichi, and Tsuyoshi, Bull. Chem. Soc. Jpn. *67*(9):2398–2402 (1994).
34. H. Shigeharu and S. Hideko, Langmuir *10*(11):4073–4076 (1994).
35. C. Hall, G. J. T. Tiddy, and B. Pfannemüller, Liquid Crystals *9*(4):527–537 (1991).
36. R. Heusch, Tenside Det. *21*(6):298–303 (1984).
37. A. Khan, A. Kaplun, Y. Talmon, and M. Hellsten, J. Coll. Interf. Sci. *181*:191–199 (1996).
38. U.S. Patent 3,639,283 to Grace & Co W R (1972).
39. G. J. Jakubicki, C. Schwarz, and A. J. Uray, EP Patent 487,170 to Colgate Palmolive (1992).
40. J. K. Smid and R. H. Van der Veen, EP Patent 386,826 to Kao Corp (1990).
41. G. Östberg, B. Bergenståhl, and M. Huldén, J. Coat. Technol. *66*:37 (1994).
42. G. Östberg, B. Bergenståhl, and M. Huldén. Colloids and Surfaces A *94*:161 (1995).
43. R. Brueckmann and T. Simenc, EP Patent 415,279 to BASF AG (1991).
44. DE patent 1,930,954 to Yardley of London Inc. (1969).
45. V. A. Yushchenko, E. A. Melnik and V. A. Drashchink, SU Patent 1,046,280 to Chem Ind Res Des and Biotech Res Inst (1983).
46. DE Patent 4,409,189 to Chem-Y Chem. Fab. GmbH (1995).

47. JP Patent 8,041,489 to Ajinomoto KK (1994).
48. DE Patent 2,846,639 to P. Jurgensen (1980).
49. A. R. Naik, EP Patent 552,032 to Unilever (1993).
50. G. Toninelli and G. Osti, EP Patent 109,022 to Mira Lanza Spa (1984).
51. JP Patent 59,078,297 to Asahi Denka Kogyo and Cope Clean KK (1984).
52. US Patent 3,644,210 to Chemed Corp. (1972).
53. JP Patent 59,210,998 to Sanyo Chem. Ind. Ltd. (1984).
54. JP Patent 56,038,399 to Asahi Denka Kogyo and Cope Clean KK (1981).
55. JP Patent 60,060,196 to Kanebo KK (1985).
56. B. K. Daurov, Z. H. E. Golovina, and L. K. Dzhelmach, SU Patent 1,011,682 to Chem Ind Res Des (1983).
57. J. Klugkist, EP Patent 341,999 to Unilever (1989).
58. H. Upadek, DE Patent 3,632,107 to Henkel KGaA (1988).
59. J. Przondo, Z. Kot, and Z. Kossinski, Pollena: Tluszcze, Srodki Piorace, Kosmet. 26:6 (1982).
60. R. M. Butterwort, J. R. Martin, and E. Willis, EP Patent 159,918 to Unilever (1985).
61. R. M. Butterwort, J. R. Martin, and E. Willis, EP Patent 159,919 to Unilever (1985).
62. R. M. Butterwort, J. R. Martin, and E. Willis, EP Patent 159,922 to Unilever (1985).
63. S. Billenstei, A. May, and H. W. Buecking, EP Patent 43,547 to Hoechst AG (1982).
64. A. G. Nikitenko, N. I. Golyatovsk, and B. I. Lesman, SU Patent 834,117 to Chem Ind Res Des (1981).
65. A. G. Nikitenko, N. I. Golyatovsk, and V. A. Yushchenko, SU Patent 883,168 to Chem Ind Res Des (1981).
66. A. G. Nikitenko, N. N. Lysova, and E. K. Ivanauskas, SU Patent 639,925 to Chem Ind Res Plan (1978).
67. JP Patent 60,096,695 to Sanyo Chem Ind Ltd. (1985).
68. A. Y. A. Nagina, A. I. Mikhalska, and E. I. Nechesova, SU Patent 732,370 to Atom Eng Reactor (1980).
69. A. Y. A. Nagina, A. I. Mikhalska, and E. I. Nechesova, SU Patent 785,350 to Atom Eng Reactor (1980).
70. M. P. Shemelyuk, M. M. Oleinik, and L. A. Gnutenko, SU Patent 745,925 to Car Ind Cons Techn (1980).
71. M. M. Oleinik, V. T. Protsishin, and A. V. Galkin, SU Patent 749,888 to Auto Ind Tech Inst (1980).
72. I. K. Getmanskii, M. M. Bebko, and V. D. Yakovlev, SU Patent 662,578 to I. K. Getmanskii (1979).
73. R. Baur, H. H. Goertz, H. W. Neumann, D. Stoeckigt, and N. Wagner, EP Patent 340,704 to BASF AG (1989).
74. M. Blezard and W. H. McAllister, EP Patent 84,411 to Albright and Wilson Ltd. (1983).
75. R. Osberghaus, K. H. Rogmann, and B. Frohlich, EP Patent 288,856 to Henkel KGaA (1988).
76. C. A. Beronio, B. T. G. Graubart, E. J. Sachs, and A. L. Streit, EP Patent 621,335 to Eastman Kodak Co and Reckitt & Colman Inc. (1994).
77. R. G. Alagezyan, B. V. Andriasyan, and A. G. Nikitenko, SU Patent 735,630 to Erev Zool Veter Ins. (1980).
78. B. M. Like, D. Smialowicz, and E. Brandli, U.S. Patent 4,336,151 to American Cyanamid Co. (1982).

79. B. Giesen, DE Patent 4,023,334 to Henkel KgaA (1992).
80. R. Baur, M. Kahmen, G. Oetter, A. Oftring, J. Perner, W. Reif, V. Schwendemann, and B. Strecker, DE Patent 4,336,247 to BASF AG (1995).
81. I. Harwigsson, A. Khan, and M. Hellsten, Tenside Surf., Det. *30*:174 (1993).
82. I. Harwigsson and M. Hellsten, J Assoc Off Chemists Soc *73*:921 (1996).
83. I. Harwigsson, Thesis "Surfactant aggregation and its application to drag reduction," Lunds University, 1995.
84. Y. Hu and E. F. Matthys, Submitted to J. Rheol.
85. A. Blum, H. Schwab, and Oppenlaend, DE Patent 2,643,804 to BASF AG (1978).
86. R. Brueckmann, A. Hohmann, and T Simenc, DE Patent 4,133,852 to BASF AG (1992).
87. E. Beckmann, R. Brueckmann, J. P. Dix, P. Freyberg, and E. Kromm, DE Patent 4,216,316, to BASF AG (1993).
88. L. V. Basova, I. N. Kokoreva, and E. N. Anishchuk, SU Patent 713,891 to L. V. Basova (1980).
89. G. B. Patent 1,265,476 to Ardrox Ltd. (1972).
90. U.S. Patent 3,301,812 to Philipp Minerals and Chem. Corp. (1967).
91. A. I. Soshko, O. M. Dzikovskii, and I. M. Shapoval, SU Patent 709,666 to Lvov Poly and Elektrostal Electrometal (1980).
92. V. I. Kuznetsov, I. V. Konoval, and R. G. Markovich, SU Patent 711,064 to V. I. Kusnetsov (1980).
93. J. M. Clarke, P. R. Foley, Y. Fu, and P. K. Vinson, WO Patent 95 20026 to Procter and Gamble (1995).
94. B. Fabry, DE Patent 4,400,632 to Henkel KGaA (1995).
95. B. Fabry, DE Patent 4,406,745 to Henkel KGaA (1995).
96. H. Kelkenberg, K. Engel, and W. Ruback, EP Patent 285,768 to Huels AG (1988).
97. E. D. Brock and A. L. Larrabee, WO Patent 92 05764 to Procter and Gamble (1992).
98. J. E. Kaleta and F. A. Pichardo, WO patent 92 13059 to Procter and Gamble (1992).
99. M. J. Giret, A. Langlois, R. P. Duke, and E. Jedla, WO Patent 93 21293 to Procter and Gamble (1993).
100. B. Fabry, DE Patent 4,435,383 to Henkel KGaA (1995).
101. C. M. Bellemain, M. J. M. Giret, and W. V. J. Richardson, WO Patent 96 03974 to Procter and Gamble (1996).
102. B. Fabry, DE Patent 4,422,404 to Henkel KGaA (1996).
103. H. Hensen, H. Tesmann, J. Kahre, R. Mueller, and W. Scholz, DE Patent 4,309,568 to Henkel KGaA (1994).
104. W. Breitzke and K.-H. Gantke, DE Patent 4,406,746 to Henkel KGaA (1995).
105. W. Breitzke and A. Behler, DE Patent 4,406,748 to Henkel KGaA (1995).
106. J. H. Collins and B. P. Murch, WO Patent 92 06160 to Procter and Gamble.
107. A. A. Fisk and A. Surutzidis, EP Patent 593,841 to Procter and Gamble.
108. D. S. Connor, B. P. Murch, and J. J. Scheibel, WO Patent 95 07341 to Procter and Gamble (1995).
109. B. Fabry and J. Kahre, DE Patent 4,430,085 to Henkel KGaA (1996).
110. M. Biermann, B. Fabry, U. Hees, and M. Weuthen, DE Patent 4,439,091 to Henkel KGaA (1996).
111. B. J. Roselle and D. T. Speckman, WO Patent 93 14183 to Procter and Gamble (1996).

112. D. S. Connor, Y. Fu, and J. J. Scheibel, WO Patent 95 07340 to Procter and Gamble (1995).

113. B. Burg, J. Haerer, K. Hill, and P. Jeschke, DE Patent 4,401,103 to Henkel KGaA (1995).

114. Y. Abe, T. Yasumasu, and T. Sakatani, JP Patent 7,053,988 to Lion Corp. (1995).

115. T. R. Rolfes and R. T. Richard, WO Patent 92 06156 to Procter and Gamble (1992).

116. Y. Fu and J. J. Scheibel, WO Patent 92 06157 to Procter and Gamble (1992).

117. M. H. K. Mao, WO Patent 92 06161 to Procter and Gamble (1992).

118. J. A. Dyet and P. R. Foley, WO Patent 92 06171 to Procter and Gamble (1992).

119. P. R. Foley, K. Ofusu-Asante, and K. W. Willman, WO Patent 93 05132 to Procter and Gamble (1993).

120. J. M. Clarke, P. R. Foley, D. L. Strauss, J. M. Van der Meer, and K. W. Willman, WO Patent 95 20024 to Procter and Gamble (1995).

121. K. Ofosu-Asante, WO Patent 95 20025 to Procter and Gamble (1995).

122. K. Ofosu-Asante, WO Patent 95 20028 to Procter and Gamble (1995).

123. D. S. Connor, Y. Fu, and J. J. Scheibel, WO Patent 95 07337 to Procter and Gamble (1995).

124. J. M. Clarke, P. R. Foley, D. L. Strauss, J. M Vander Meer, and K. W. Willman, GB Patent 2,292,562 to Procter and Gamble (1996).

125. K. Ofosu-Asante, WO Patent 94 05755 to Procter and Gamble (1994).

126. K. Ofosu-Asante, WO Patent 95 06106 to Procter and Gamble (1995).

127. G. M. Bailley and S. Powell, WO Patent 92 22629 to Procter and Gamble (1992).

128. M. S. Showell and A. M. Wolff, WO Patent 93 23516 to Procter and Gamble (1993).

129. B. P. Murch, WO Patent 92 06153 to Procter and Gamble (1992).

130. J. Boutique, A. Surutzidis, R. J. Jones, and A. A. Fisk, EP Patent 709,450 to Procter and Gamble (1996).

131. R. J. Jones and P. J. M. Baets, EP Patent 709,451 to Procter and Gamble (1996).

132. A. C. Huber and R. K. Panandiker, U.S. Patent 5,288,431 to Procter and Gamble (1994).

133. J. Boutique, D. S. Connor, Y. Fu, B. P. Murch, J. J. Scheibel, and A. Surutzidis, U.S. Patent 5,318,728 to Procter and Gamble (1994).

134. G. M. Baillely and T. E. Cook, WO Patent 92 06150 to Procter and Gamble (1992).

135. B. P. Murch, S. W. Morrall, and M. H. K. Mao, WO Patent 92 06162 to Procter and Gamble (1992).

136. J. P. Morelli, D. Lappas, S. L. Randall, R. K. Panadiker, J. Boutique, and C. E. Housmekerides, WO Patent 96 12000 to Procter and Gamble (1996).

137. M. H. K. Mao and B. P. Murch, WO Patent 92 06159 to Procter and Gamble (1992).

138. J. Boutique and Y. Fu, WO Patent 94 24246 to Procter and Gamble (1994).

139. F. Figueroa and R. Jarrin, WO Patent 96 01306 to Procter and Gamble (1996).

140. S. G. Cauwberghs and I. M. A. J. Herbots, EP Patent 576,778 to Procter and Gamble (1994).

141. E. P. Gosselink and R. Y. L Pan, WO Patent 92 06152 to Procter and Gamble (1992).

142. F. E. Hardy and B. P. Murch, WO Patent 92 06155 to Procter and Gamble (1992).

143. T. E. Cook, M. H. K. Mao, R. K. Panandiker, and A. M. Wolff, WO Patent 92 06154 to Procter and Gamble (1992).
144. S. W. Morrall and B. P. Murch, WO Patent 92 06151 to Procter and Gamble (1992).
145. M. C. Addison and M. A. J. Moss, WO Patent 94 03554 to Procter and Gamble (1994).
146. G. M. Baillely, S. C. Ersolmaz, M. A. J. Moss, and G. A. Sorrie, WO Patent 94 03572 to Procter and Gamble (1994).
147. S. L. Honsa and M. H. K. Mao, WO Patent 92 06164 to Procter and Gamble (1992).
148. S. L. Honsa, WO Patent 92 06172 to Procter and Gamble (1992).
149. A. L. Chisholm and K. M. A. Schamp, EP Patent 694,608 to Procter and Gamble (1996).
150. P. France, S. Ongena and C. Wilkinson, EP Patent 709,449 to Procter and Gamble (1996).
151. R. L. Tadsen and G. W. Bufler, WO Patent 92 06170 to Procter and Gamble (1992).
152. A. C. Convents, A. Busch, and A. C. Baeck, AU Patent 9,211,048 to Procter and Gamble (1993).
153. A. Convents, A. Busch, and J. A. Pretty, EP Patent 522,206 to Procter and Gamble (1993).
154. Fr. Patent 1,586,913 to Henkel and Cie GmbH (1969).
155. Fr. Patent 1,550,144 to Henkel and Cie GmbH (1968).
156. V. Au, U.S. Patent 5,516,460 to Lever Bros Co (1996).
157. V. Au, WO Patent 95 27770 to Unilever PLC (1995).
158. B. Fabry and I. Wegener, DE Patent 4,327,327 to Henkel KgaA (1995).
159. B. Burg, B. Fabry, J. Haerer, U. Hees, P. Jeschke, and M. Nejtek, DE Patent 4,323,253 to Henkel KGaA (1995).
160. D. S. Connor, Y. Fu, and J. J. Scheibel, WO Patent 95 07256 Procter and Gamble (1995).
161. L. N. Mackey, S. Ferershtekhou and J. J. Scheibel, U.S. Patent 5,354,425 to Procter and Gamble (1994).
162. A. V. Warner, L. N. Mackey, A. Wong, J. J. Franxman, B. A. Goldslager, T. J. Klofta, and D. V. Phan, WO Patent 95 16824 to Procter and Gamble (1995).
163. R. H. Garst, WO Patent 96 16540 to Henkel Corp. (1996).
164. C. Ullrich, I. Wegener, and B. Fabry, DE Patent 4,229,442 to Henkel KGaA (1994).
165. A. Fischer and R. Vybiral, DE Patent 4,237,434 to Hoechst AG (1994).
166. E. Arenas, J. R. Baran, Jr., G. A. Pope, W. H. Wade, and V. Weerasooriya. Langmuir 12:588 (1996).
167. J. R. Baran Jr., G. A. Pope, W. H. Wade, and V. Weerasooriya. Environ. Sci. Technol. 30:2143 (1996).
168. B. Fabry, J. Kahre, and I. Wegener, DE Patent 19,504,643 to Henkel KGaA (1996).
169. K. Gerling, H. Rau, P. Schwarz, K. Uhlig, and K. Wendler, EP Patent 726,335 to Solvay Deut GmbH (1996).
170. C. G. van Ginkel, C. A. Stroo, and A. G. M. Kroon. Tenside Surf. Det. 30:3 (1993).
171. M. Stahlmans, E. Matthijs, E. Weeg, and S. Morris, SÖFW J. 13:794 (1993).

172. E. Matthijs, G. Debaere, N. Itrich, P. Masscheleyn, A. Rottiers, M. Stahlmans, and T. Federle, Water Sci. Technol. *31*:321 (1995).
173. J. Eastoe and P. Rogueda, Langmuir *10* 4429–4433 (1994).
174. J. Eastoe and P. Rogueda, Langmuir *12* 2701–2705 (1996).
175. D. S. Connor, Y. Fu, and J. J. Sceibel, WO Patent 9519953 to Procter and Gamble (1995).

8

Polyol Ester Surfactants

JEREMY J. LEWIS Research and Technology, ICI Surfactants, Everberg, Belgium

I. INTRODUCTION

The purpose of this chapter is to give a practical insight into the chemical synthesis, processing methods, and composition of polyol fatty acid ester surfactants. These materials are among the most common surfactant types because esterification is one of the easiest chemical reactions to carry out. The ready availability of polyols such as glycerol, sorbitol, sugars, and ethoxylated feedstocks, in addition to a wide variety of fatty acid materials, ensures that a multiplicity of products can be tailored for many application needs. However, control over processing conditions, and hence over the final product composition, can give rise to very subtle variations in performance. In addition, as all commercially available materials are mixtures and exact processing pathways can be difficult to interpolate, precise structure–activity relationships for any particular application must be treated with a great deal of caution. Following an overview of current products and processes, a short survey of the emerging technology of enzyme esterification is also given.

This introduction outlines some of the basic ideas behind the concept of a fatty acid polyol ester. These include what type of structure is required for surfactancy effects, the composition of commercially available fatty acids, the complex nature of products arising from simple polyol esterification methods, and major families of materials currently being manufactured. Lastly, this introduction is summarized in a series of guidelines that provide the basic rules for this area of chemistry prior to consideration of the individual variations displayed by each individual family of product.

A. Structural Considerations

The simplistic view of these surfactant types can be readily understood. Fatty acids are used as the building blocks for many oleo chemical products, of which the commonest are the appropriate alkali metal salts, or soaps (Fig. 1). In these materials, a simple paraffinic chain 10–20 carbons long is functionalized on one end as a carboxylic acid group. If the carboxylic end is neutralized with, for example, sodium hydroxide, the acid is converted to its sodium salt.

$$RCO_2H \quad + \quad NaOH \quad \longrightarrow \quad RCO_2Na \quad + \quad H_2O$$

Fatty acid   Soap

$(R=CH_3(CH_2)_n,$ where $n=8{-}16)$

FIG. 1 Soap as a model surfactant system.

Soaps of this type have a good balance between water solubility (the salt of the carboxylic acid) and water insolubility (the oily hydrocarbon chain). The water solubility of a surfactant is known as its hydrophilicity, whereas its water insolubility is known as hydrophobicity. This balance leads to the collection of these molecules at an interface (e.g., between oil and water), such that both solubility properties can be satisfied at the same time. For oil and water, the hydrocarbon part is soluble in the oil, whereas the carboxylic acid salt portion is soluble in the water. As the water-soluble portion is usually smaller in size than the hydrocarbon chain, it is usually referred to as the headgroup. However, this tendency to collect at, and to alter the physical properties of, the interface between two different phases leads directly to numerous surfactancy effects. A more detailed introduction to surfactancy can be found elsewhere [1]. However, the key requirement in surfactancy applications is the balance between the water and oil solubility (the hydrophile–lipophile balance, or HLB requirement) for the system under study.

The HLB system was proposed as a way in which the relative water or oil solubility of nonionic surfactants could be gauged on a scale from 1 to 20. Figures toward the low end at the scale (1–5) indicated relatively oil-soluble materials, useful as water-in-oil emulsifiers. Figures toward the high end of the scale (14–20) indicated water-soluble materials, useful as oil-in-water emulsifiers. Materials with associated intermediate values could be used as wetters and detergents. Detergency remains the major outlet in tonnage terms for surfactant use.

This balance of HLB properties for soaps is easy to understand. If the soap has a hydrocarbon chain, or alkyl group, of less than 10 carbons long, the molecule is likely to be too water-soluble to allow it to collect at an interface and hence to perform effectively. Alternatively, if the alkyl chain is too long (more than 20 carbons), the material may become too water-insoluble and may preferentially collect in the oil phase rather than at the interface.

It is not a large step to view the replacement of the alkali salt part of a soap with another hydrophilic headgroup, joined through an ester linkage to the carboxylic group, to create molecules that are ester surfactants. As an example, ethylene glycol can be reacted with fatty acids in such a way as to provide a monoester product, with a hydrophilic headgroup containing a hydroxyl group, and a polar ester function (Fig. 2). Such a headgroup is likely to be much less hydrophilic than the salt of the carboxylic group due to the lack of strong ionic charges of the latter. Nevertheless it has the ability to form a hydration sphere in water due to hydrogen bonding with the terminal hydroxyl. Similarly, soap is used in its strongly ionic alkali metal salt form, rather than as the neutral carboxylic acid, as the latter is relatively oil-soluble.

The chemistry of commercially available ester surfactants is therefore the product of two main elements—a particular fatty acid, and a compound with a

$$CH_3(CH_2)_{10}CO_2H \quad + \quad HOCH_2CH_2OH \quad \longrightarrow$$

$$CH_3(CH_2)_{10}CO_2CH_2CH_2OH \quad + \quad H_2O$$

FIG. 2 An idealized fatty acid ester surfactant derived from ethylene glycol.

number of hydroxyl groups, a polyol. Some general points regarding the composition of feedstocks and the esterification reaction are worth considering prior to some more detailed discussion of individual products based on common polyol feedstocks.

B. Fatty Acids

Figure 2 depicts a reaction between pure lauric acid and ethylene glycol. While this can indeed be carried out, it is highly unlikely that any commercial product is based on such a fatty acid feedstock. Fatty acids are sourced from naturally occurring fats and oils, which not only contain a mixture of carbon chain lengths but also be fully saturated (no double bonds), or which can occur with various degrees of unsaturation (up to three double bonds in the chain). These naturally sourced acids also display another interesting feature in that they are normally mixtures of only even-numbered carbon atom chains. Other sources of fatty acids are available, e.g., from the oxidation of paraffin feedstocks, but these remain much less widespread in their occurrence.

Returning to lauric fatty acid, this is sourced by the hydrolysis of coconut oil. However, in addition to this material, significant quantities of caprylic, caproic, myristic, palmitic, stearic, and oleic acids are also found in the crude acid feedstock. Lauric acid material derived from simple hydrolysis of coconut oil is therefore usually a broad mixture of homologs and, although grossly impure, can be referred to commercially as lauric acid. The use of such broad mixtures can be seen as resulting from two complementary factors. First, it does make better commercial sense to use all of the raw material readily available, instead of discarding some and consequently increasing the price of the remainder. Second, most surfactancy effects appear to work better with a mixed alkyl chain, rather than with a single pure carbon chain homolog (a fortunate coincidence!). A list of typical commercially available fatty acids, together with their chemical compositions, is given in Table 1.

This table illustrates a number of important points. Commercial lauric acid can be seen as the hydrolysis product of coconut oil and indeed sometimes is referred to as coconut oil fatty acid (COFA). Tallow fat shows a wide range of saturated and unsaturated species, but by the use of common processing methods, these two types of fatty acid can be economically separated [2]. This gives rise to commercial stearic and oleic acid compositions, marked as derived from

TABLE 1 Typical Alkyl Chain Distributions for Commercially Available Fatty Acids

Carbon numbers	Single isomer fatty acids		Fatty acid isomers (%)						
	Systematic name	Trivial	Coconut oil	Tallow fat	Palm oil	Lauric (coconut oil)	Stearic (tallow)	Oleic (tallow)	Palmitic (palm oil)
C8	Octanoic	Caprylic	7	—	—	7	—	—	—
C10	Decanoic	Capric	6	—	—	6	—	—	—
C12	Dodecanoic	Lauric	48	—	—	48	—	—	—
C14	Tetradecanoic	Myristic	18	2	1	18	2	—	2
C16	Hexadecanoic	Palmitic	9	25	43	9	48	4	93
C18	Octadecanoic	Stearic	3	19	5	3	48	9	5
C16:1	9c-Hexadecenoic	Palmitoleic	—	3	—	—	—	6	—
C18:1	9c-Octadecenoic	Oleic	7	44	41	8	2	71	—
C18:2	9c, 12c-Octadecadienoic	Linoleic	2	6	11	1	—	9	—
C18:3	9c, 12c, 15c-Octadecatrienoic	Linolenic	—	1	—	—	—	1	—

"tallow." Again, commercial grades can either be rich in their nominal isomer (oleic), or much less pure than might be anticipated by their name (stearic). Finally, once separation of saturated and unsaturated material has been achieved, fractional distillation can give rise to fairly pure cuts of fatty acids (compare the commercial grade of palmitic acid with the alkyl chain distribution of the palm oil from which it was derived).

While the above description of common ester fatty acid feedstocks means that care must be exercised when considering the exact nature of the product resulting from it, it is worth remembering that to a large degree an alkyl chain of carbon length $C_{10}-C_{20}$, whether saturated or not, will produce essentially the same hydrophobic characteristics necessary for surfactant performance. The lower alkyl chain lengths, such as those found in COFA, and unsaturated alkyl chains will tend to confer liquidity on ester products at room temperature, whereas long and fully saturated chains tend to give a more solid, waxy character.

C. Polyol Esterification

Fatty acid esterification of a material containing a number of hydroxyl groups (a polyol) may at first sight be a simple way of converting two readily available feedstocks (such as lauric acid and ethylene glycol) to a product that should have utility as a surfactant. As shown in Fig. 2, the reaction product appears to have the right type of molecular structure (hydrophobic tail and hydrophilic head), and variation of the fatty acid and polyol type could be used to fine-tune application performance.

However, this simple view of the chemistry must be tempered by an appreciation of the complexities that such an apparently simple chemical conversion may generate. Two examples will be used to illustrate this idea further— ethylene and propylene glycol esters.

1. Esterification of Ethylene Glycol

Esterification of an acid with an alcohol is an equilibrium reaction in which the final product ester is reacting with the byproduct water at the same rate at which the raw materials are being converted to the product materials:

$$RCO_2H + R'OH \Leftrightarrow RCO_2R' + H_2O$$

The equilibrium constant for this reaction is given by the following equation, which is known as the law of mass action [3]:

$$K = [\text{ester}] [\text{water}] / [\text{acid}] [\text{alcohol}]$$

Thus, in order to force the equilibrium to give the greatest amount of product, esterification reactions are usually carried out such that water is removed from the reaction medium. However, this is not sufficient to yield only the desired

$$RCO_2H \quad + \quad HOCH_2CH_2OH \quad \longrightarrow$$

1 Mole   1 Mole

$$HOCH_2CH_2OH \quad + \quad RCO_2CH_2CH_2OH \quad + \quad RCO_2CH_2CH_2O_2CR$$

Minor   Major   Minor

FIG. 3 The common product mixture derived from simple reaction of equimolar quantities of a fatty acid and a polyol.

monoester. Reaction of 1 mole of ethylene glycol with 1 mole of fatty acid results in a mixed product composed of monoester, diester, and free ethylene glycol, as either of the two hydroxyl groups on the glycol is as statistically likely to react to give the ester function (Fig. 3). The molar ratio of product is therefore about 50:25:25, respectively. Only the monoester part of this mixture fits into the simplified description of a surfactant outlined above, and it is only this which is likely to be useful. The free ethylene glycol will be too water-soluble and the diester will be too oil-soluble to be of use. Simple esterification with equimolar amounts of starting materials is therefore an inefficient way to produce polyol ester surfactants, particularly the monoester materials, which are those of primary surfactant interest.

2. Esterification of Propylene Glycol

An added complication to the above consideration of statistical likelihood of polyol esterification is that if the polyol contains both primary and secondary hydroxyls, a further bias in the ratio of esters formed will occur. The esterifiability of a primary hydroxyl is approximately three times that of a secondary hydroxyl [4]. Thus, when propylene glycol is esterified with a molar equivalent of fatty acid, the final product composition consists of free glycol, 1-monoester, 2-monoester, and diester. However, the majority of the monoester is found at the 1 position (Fig. 4). When considering the final product composition of a fatty acid polyol ester, therefore, awareness must be given to the nominal ratio of fatty acid to polyol being described and to the ratio of primary to secondary hydroxyl groups in the polyol structure.

D. Major Polyol Ester Surfactant Families

The major families of the polyol ester surfactants are given in Table 2. This table gives a simple overview of ester manufacture in the United States for 1990 [5]. It is included to illustrate the point that more than 80% of the needs in this area are accounted for by only a relatively few product types and that this forms the basis for the rest of the chapter. While many other ester types have been

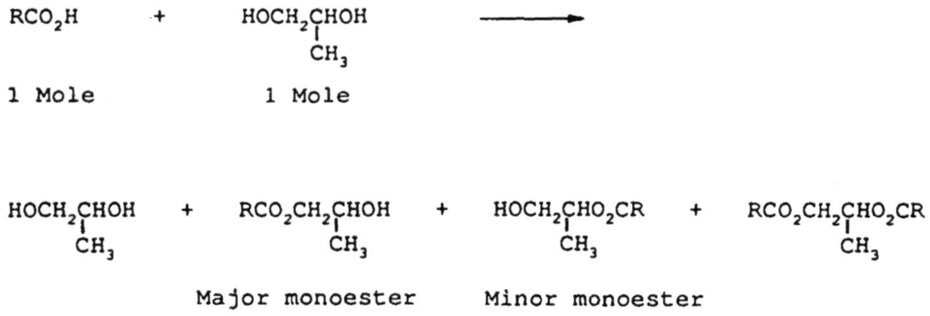

$$RCO_2H \quad + \quad HOCH_2\underset{\underset{\displaystyle CH_3}{|}}{C}HOH \quad \longrightarrow$$

1 Mole   1 Mole

$$HOCH_2\underset{\underset{\displaystyle CH_3}{|}}{C}HOH \quad + \quad RCO_2CH_2\underset{\underset{\displaystyle CH_3}{|}}{C}HOH \quad + \quad HOCH_2\underset{\underset{\displaystyle CH_3}{|}}{C}HO_2CR \quad + \quad RCO_2CH_2\underset{\underset{\displaystyle CH_3}{|}}{C}HO_2CR$$

Major monoester   Minor monoester

FIG. 4 Extension of esterification product complexity is due to the disproportionate reactivity of primary and secondary hydroxyl groups.

produced and commercialized over the years, this short overview seeks to illustrate the main industrial processes, limitations, and product compositions as a practical guide to the area.

The chemistry of sucrose esters is also discussed, although the volume of product manufactured and sold is very small. However, it remains an area in which the superficial appeal of a renewable resource surfactant is heavily outweighed by the reality of processing and product shortcomings.

For more details about individual product materials, available ranges, and manufacturing suppliers, the reader is referred to some standard reference works [6,7].

E. Introductory Guidelines

This introduction provides the basis for the discussion of the complex polyol ester chemistry that follows. However, the main reason for the above remarks has been to present the remainder of the chapter within the context of a few simple observations.

TABLE 2 Major Families of Polyol Fatty Acid Ester Surfactants[a]

Family	%
Glycerol	29
Glycol or polyethylene	20
Ethoxylated fats and oils	16
Anhydrosorbitol	11
Ethoxylated anhydrosorbitol (polysorbates)	9
Others	15

[a](U.S., by production, from Ref. 5.)

1. Fatty acid feedstocks used for the production of commercial surfactant polyol esters are always mixtures of homologs with varying degrees of unsaturation.
2. Esterification of a polyol will almost inevitably lead to a mixed product when the number of moles of fatty acid used is less than the total number of hydroxyls on the polyol.
3. If the polyol contains both primary and secondary hydroxyls, esterification will predominantly occur with the former.
4. Polyol monoesters tend to be the most desired forms of product sought, as they give the best balance of hydrophobic and hydrophilic properties.

Bearing these ideas in mind, the rest of this chapter will seek to clarify how the final surfactant ester composition is affected by the structure of the polyol considered and the stoichiometry of its reaction with a fatty acid. Although there are many chemical ways to produce much more well-defined, single-isomer species of great purity, it should be remembered that such species are really only of academic interest, as commercial competition and application utility over many years has established the performance baseline as gross mixtures, or materials that have undergone some simple processing modification. Introduction of innovative techniques to overcome the inelegant established processes can only be acceptable in the marketplace if the cost of the innovation does not lead to the newer products becoming noncompetitive.

II. GLYCOL ESTERS

Simple ethylene and propylene glycol esters have been mentioned in the introduction, but they are only specific examples of the much broader classes of glycols made from two common industrially important raw materials—ethylene oxide and propylene oxide. These feedstocks both feature the highly strained oxirane (epoxide) ring structure and can easily be polymerized under basic conditions to form either homo- or copolymers. The sodium hydroxide catalysis of this oxyethylation or oxypropylation leads to the polymer structure (Fig. 5). This is more fully covered in Chapter five, but is mentioned here in order to contrast the two ways in which a fatty acid glycol ester may be prepared. This can be done either by alkoxylation of a fatty acid by reaction with ethylene or propylene oxide, or by the esterification of a preformed glycol.

Before continuing this synthetic appraisal, it is worth remembering that what is ideally sought from this chemistry is the polyoxyalkylene monoester, in order to give the best surfactancy effect in the product. Unreacted polyoxyalkylene chains, or the diester compounds, are much less efficient at delivering these effects.

The reaction conditions for the two synthetic approaches to glycol esters can take place under very different catalysis. Alkoxylation is usually base-

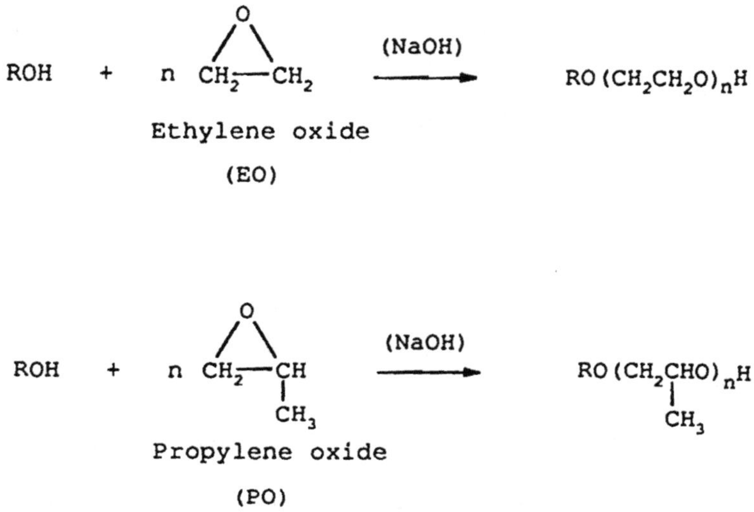

$$ROH \quad + \quad n \; CH_2\!\!-\!\!CH_2 \quad \xrightarrow{\text{(NaOH)}} \quad RO(CH_2CH_2O)_nH$$

Ethylene oxide

(EO)

$$ROH \quad + \quad n \; CH_2\!\!-\!\!CH \quad \xrightarrow{\text{(NaOH)}} \quad RO(CH_2CHO)_nH$$
$$\overset{|}{CH_3} \qquad\qquad\qquad \overset{|}{CH_3}$$

Propylene oxide

(PO)

FIG. 5 Oxyethylation and oxypropylation polymerization.

catalyzed, whereas esterification can be base-, acid-, or amphoterically catalyzed. However, the two different approaches lead to essentially the same general product mixture. This will be exemplified for polyoxyethylene glycol esters rather than for polyoxypropylene materials, which are far less common. This is due to the polyoxyethylene moieties adding hydrophilicity to the hydrophobic fatty acid feedstock, giving rise to surfactancy effects, whereas polyoxypropylene addition would only add a hydrophobic effect to a hydrophobic feedstock, thereby rendering it more oleaginous.

A. Oxyethylation Process

Oxyethylation of a fatty acid is carried out using a catalytic amount of base and proceeds by the addition of ethylene oxide to the anion of the fatty acid [Eq. (1), Fig. 6]. Once all of the carboxylate has been turned into an oxyethylated compound, further addition of ethylene oxide may occur [Eq. (2), Fig. 6 $n > 1$). This process represents a simple way to control the balance of hydrophilic and hydrophobic properties, as the hydrophilicity can be controlled by the degree of oxyethylation, and the hydrophobicity by the selection of the desired fatty acid feedstock. It should be noted that the basicity of the product monooxyethylate anion in Eq. (1) is higher than that for the carboxylate anion and that this results in all of the carboxylic acid being converted to a monooxyethylene derivative before any further reaction takes place. However, as the oxyalkylation reaction proceeds, there is an additional reaction taking place [Eq. (3), Fig. 6]. The

1. RCO_2H + EO $\longrightarrow$ $RCO_2CH_2CH_2OH$

2. $RCO_2CH_2CH_2OH$ + $(n-1)\,EO$ $\longrightarrow$ $RCO_2(CH_2CH_2O)_nH$

3. $RCO_2(CH_2CH_2O)_nH$ + $RCO_2(CH_2CH_2O)_nH$ $\longrightarrow$

$RCO_2(CH_2CH_2O)_nCOR$ + $HO(CH_2CH_2O)_nH$

FIG. 6 Sequential oxyethylation and then interesterification of fatty acids.

oxyalkylate anion end of the polyoxyethylene chain may react with another ester function to give a transesterification reaction. As this is a ready reaction, and also reversible, the product from fatty acid oxyethylation is a typical mixture of free glycol, mono- and diesters.

B. Esterification Process

Esterification of polyoxyethylene glycol can be carried out under a variety of typical conditions, such as by using p-toluenesulfonic acid at 140°C under a reduced pressure of about 30–50 mm Hg [8]. The reduced pressure allows the water of reaction to be stripped out of the product, and so helps to displace the equilibrium in favor of the desired ester product. Although water could be removed simply by increasing the temperature to drive it off more quickly, this will generally result in an inferior color and odor of product. Other techniques have also been used to aid water removal, including the addition of an azeotroping solvent such as toluene, but the occurrence of low levels of organic solvents in the final product is undesirable for most end-use applications.

Once again, due to the nonselectivity of the esterification reaction, whereby the ester can form with equal likelihood on either end of the glycol, the product consists of a mixture of free glycol, mono- and diesters. To some degree, the composition can be biased toward a greater monoester content by using an excess of glycol in the reaction mixture. Reaction of 3 moles of ethylene glycol with 1 mole of palmitic acid leads to a product containing almost 70% by weight of monoester [9]. A further increase in starting glycol content can drive the monoester content close to 100%. For a monostearate ester of propylene glycol, a molar ratio of 10:1 for the glycol to stearic acid has been reported as optimum conditions, using an ion exchange resin at 100–110°C. The practical drawback to these monoester-biased reaction strategies is whether such a crude product composition (rich in monoester, but containing a lot of free glycol) is either commercially or end-use effective. This must be balanced against materials in which extra processing is done to remove the excess polyol, if possible, or in which the simple mixed glycol, mono-and diester product (the cheapest option) remains cost and end-use competitive.

Glycol esters can also be synthesized using materials such as fatty acid chlorides, by carrying out the reaction in solvents such as pyridine [10]. However, although such a strategy is simple to carry out on a laboratory scale, where materials may be prepared for further experimentation and study, this approach is far too expensive to be used on a commercial scale. In addition, toxic solvents such as pyridine are generally unacceptable to most potential application uses, and quite involved purification of the final product is normally required.

III. GLYCEROL ESTERS

Glycerol esters are the largest volume group of fatty acid partial esters of commercial significance. This should not be surprising, as the hydrolysis of fats and oils with water to give fatty acid feedstocks results in glycerol as the byproduct. That is to say, most fats and oil are simply particular mixtures of triesters of glycerol (triglycerides). As both the fatty acids and glycerol have major uses outside the surfactants area [11], the hydrolysis of triglycerides is an industrially important process in itself (Fig. 7). To form the required monoester compounds, which have the greatest utility as surfactants, there are two possible synthetic routes. These are illustrated in Fig. 8.

A. Esterification of Glycerol

Esterification of glycerol with fatty acids is normally carried out at temperatures of 200–250°C, with a sparge of inert gas such as nitrogen through the mixture to aid the removal of water [12]. The catalysts normally used are either basic (sodium, potassium, or calcium hydroxide; or potassium carbonate) or acidic (such as p-toluene sulfonic acid). By the use of alkali metal salts of fatty acids, the resultant soap helps the emulsification of glycerol and the fatty acids. The esterification reaction allows the manufacturer to exercise some control over the final product properties by controlling the distribution of fatty acids homologs in the feedstock. As described in Sec. I, this can be as wide or as narrow as desired, and is essentially dictated by the cost of the uniqueness of the homolog blend versus the value of the product to the end user.

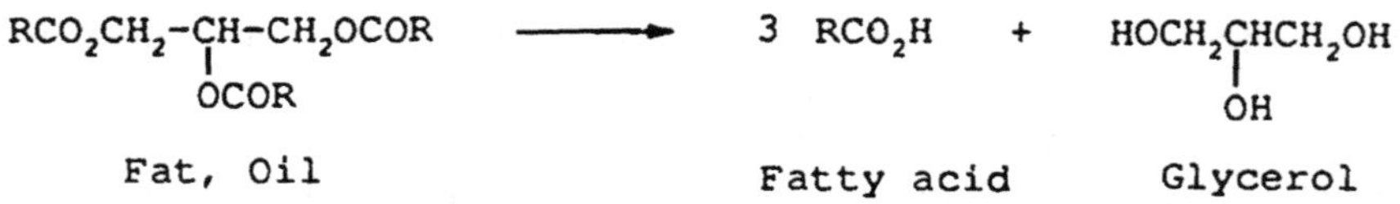

$$RCO_2CH_2-\underset{\underset{OCOR}{|}}{CH}-CH_2OCOR \longrightarrow 3\ RCO_2H\ +\ HOCH_2\underset{\underset{OH}{|}}{CH}CH_2OH$$

| Fat, Oil | Fatty acid | Glycerol |

FIG. 7 Simple hydrolysis of triglyceride compounds.

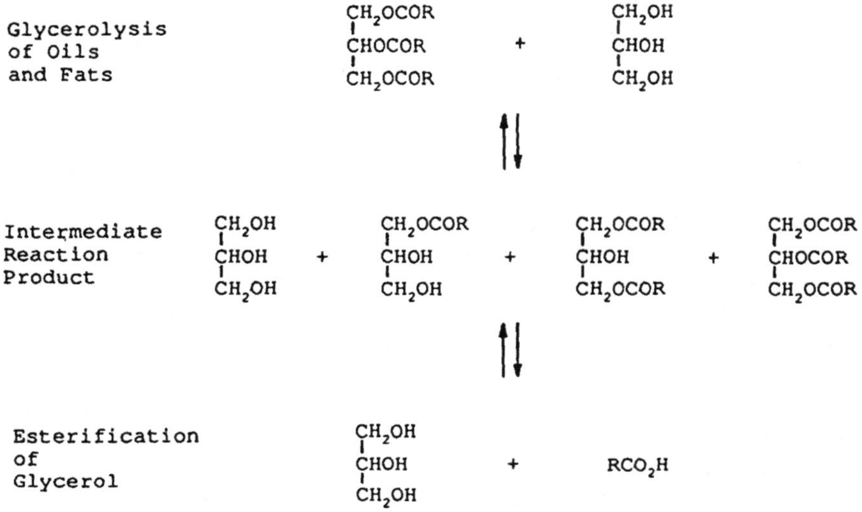

Glycerolysis
of Oils
and Fats

$$
\begin{array}{c}
CH_2OCOR \\
CHOCOR \\
CH_2OCOR
\end{array}
\quad + \quad
\begin{array}{c}
CH_2OH \\
CHOH \\
CH_2OH
\end{array}
$$

$$\upharpoonleft\downharpoonright$$

Intermediate
Reaction
Product

$$
\begin{array}{c}
CH_2OH \\
CHOH \\
CH_2OH
\end{array}
\; + \;
\begin{array}{c}
CH_2OCOR \\
CHOH \\
CH_2OH
\end{array}
\; + \;
\begin{array}{c}
CH_2OCOR \\
CHOH \\
CH_2OCOR
\end{array}
\; + \;
\begin{array}{c}
CH_2OCOR \\
CHOCOR \\
CH_2OCOR
\end{array}
$$

$$\upharpoonleft\downharpoonright$$

Esterification
of
Glycerol

$$
\begin{array}{c}
CH_2OH \\
CHOH \\
CH_2OH
\end{array}
\quad + \quad RCO_2H
$$

FIG. 8 Reaction products derived from glycerolysis of oils and fats, or by the esterification of glycerol.

B. Glycerolysis of Fats and Oils

Instead of the hydrolysis of fats and oils with water, reaction with glycerol (an interesterification known as glycerolysis) can also give a product that is similar to that obtained by direct esterification. This processing technology has been well reviewed by Sonntag [13]. The method has the advantage that no mass is lost during the reaction stage (water does not need to be liberated), but it has the disadvantage that account has to be taken of the distributions of fatty acid alkyl chain homologs in the product, which are dictated by the nature of the triglyceride used. Catalysts and temperatures are as for the simple esterification method. In practice, this method of glycerol ester synthesis is much more common than that of direct esterification.

Overall, however, if the distributions of the fatty alkyl chain homologs are comparable, then the product arising from either method will be comparable—a mixture of free glycerol, mono-, di-, and triester. If a base catalyst is used, then residual soap will also be present. The simple reaction of stoichiometric quantities of fatty acid and glycerol will therefore give a very impure product, with a typical monoester content of about 42–48%. This type of product has found many application outlets due to the economical way in which it is made.

In addition, it is possible to push up the amount of monoester in the product by again using an excess of the polyol feedstock, glycerol. If 10 moles of glycerol is used, the monoester content can approach 100%. Again, the trade-off for the operation of such a process becomes the acceptability of the excess polyol in the product, or the extra cost associated with the separation of the excess from the desired ester, with the consequently decreased utilization of the reactor volume.

One final aspect of these esterification procedures is also worth noting. This is the low solubility of glycerol in either the starting fatty acids or the resultant glyceride esters. At room temperature, the solubility is only around 5%, whereas at 250°C it is closer to 50%. Obviously, the higher temperatures not only make the esterification reaction proceed faster but also allow a greater degree of contact to take place between the glycerol and the fatty acid or triglyceride. The drawback to this is that as the reaction mixture cools, glycerol may split out of the ester layer. This physical loss of polyol from the ester layer means that the equilibrium set up between the forward-and-backward reaction of the esterification results in a disproportionation of the monoester content back into free glycerol and triglyceride. For this reason, after the esterification reaction has been carried out, the catalyst usually is neutralized. When the product is then cooled, the reversion to free glycerol and triglyceride is minimized.

Although the basic chemistry leads to gross mixtures of ester products, there remains a commercial demand for higher purity grades of glycerol fatty acid monoesters. This need is met by the use of a technique known as "molecular distillation" [14]. The crude reaction mixture from the esterification reaction is fed into a short path distillation still, and given that the different degrees of esterification give rise to a stepwise change in molecular weight (depending on how many fatty acid groups are attached to the glycerol moiety), fractions of relatively high purity can be obtained. Monoesters of approximately 95% purity are made in this way. In addition, the byproducts of this treatment (free glycerol and the di- and triglycerides) can be recycled back to the reaction stage. Typical compositions for the two main types of commercially available glyceride monoesters are given in Table 3. It should also be noted that the diester compounds, although very hydrophobic, still retain some useful surfactancy characteristics.

There remains, however, one further compositional detail which adds to the complexity of glycerol ester understanding. Glycerol has two dissimilar hydroxyl environments, and the variation in the proportions of the ester groups at these two different positions may be sufficient to give some subtle end-use variation. Current nomenclature is to describe monoesters as being joined to glycerol at the 1 or 2 position, on the primary or secondary hydroxyl, respectively. Earlier descriptors that have been used (and can still be found) are the α or β positions (Fig. 9). Once again, if consideration is given to how a glycerol ester may act as a surfactant, with a water-soluble headgroup and an oil-soluble tail, it is clear that the two positional isomers give rise to two different shapes of

TABLE 3 Composition (%) of Standard and Molecularly Distilled Monoglycerides

Feature	Glycerol monostearate	
	Standard grade	Molecularly distilled
Monoglyceride content	46	92
Diglyceride content	40	5
Free glycerol	1	1

headgroup (Fig. 10). If small changes in surfactant packing at an interface are critical to a particular application, then this effect may well be important. However, for most end uses this effect is not vital. As Sonntag has pointed out, use of monoglycerides for the baking of bread is done at a temperature that probably establishes an equilibrium between the two forms in any case!

In addition, as there is an equilibrium between the formation of an ester at either the 1 or 2 position, and this is temperature-dependent, then there is a difference in the composition of a product at reaction temperature when compared to that at ambient. However, before the hot product is cooled down, the catalyst is neutralized, and so the achievement of equilibrium between the 1 and 2 isomers is very slow at normal temperatures. This, coupled with the fact that many of the useful monoglyceride products are solids melting at 50–60°C, means that nonequilibrium mixtures of isomers may be frozen into the commercial grades.

The above description of the basic methodologies for the production of glycerol esters commonly sold is supplemented in the open literature [15] by a number of synthetic strategies that result in different reaction or product characteristics. Should specific well-defined products be required in order to gain a better understanding of structure-activity phenomena at a molecular level, or in preparation for further synthetic transformation, then such methodologies should be carried out. Effects of solvents, use of fatty acid chlorides, and protection followed by deprotection have all been studied extensively and their consequences for ester ratio and product composition reported. However, for the vast majority of industrial operations the financial penalty involved usually precludes their use.

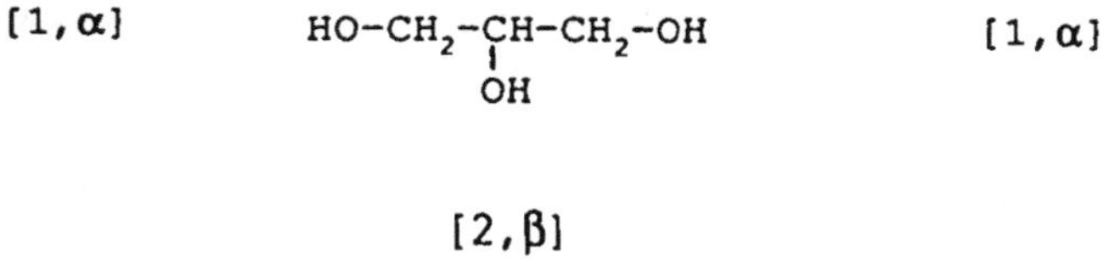

[1, α]   HO–CH$_2$–CH–CH$_2$–OH   [1, α]
                              |
                             OH

[2, β]

FIG. 9 Positional naming for the different glycerol hydroxyl environments.

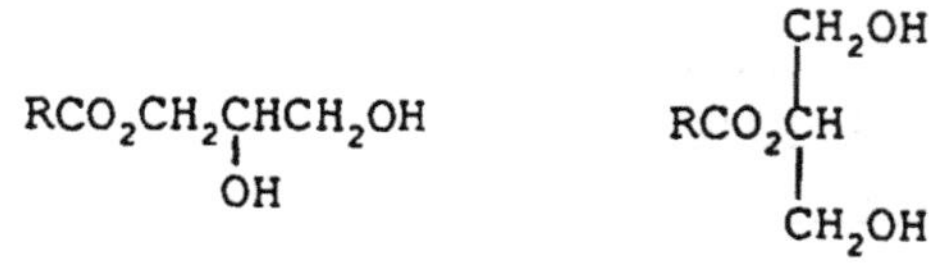

$$CH_2OH$$
$$|$$
$$RCO_2CH_2CHCH_2OH \qquad \qquad RCO_2CH$$
$$|\qquad\qquad\qquad\qquad\qquad\qquad\qquad |$$
$$OH \qquad\qquad\qquad\qquad\qquad CH_2OH$$

FIG. 10 Different shapes of headgroup arising from the different esterification sites of glycerol.

IV. POLYGLYCEROL ESTERS

Polyglycerol fatty acid esters are normally produced by a two-step procedure [16]. The first step involves the controlled reaction of glycerol into a polymeric form, and then this polymer is reacted with the appropriate amount of fatty acid to achieve a balance of hydrophobic and hydrophilic properties.

The simplest way of carrying out the polymerization step is to react glycerol at about 260–270°C in the presence of an alkali catalyst, such as 1% sodium hydroxide. Studies have been carried out for the optimization of this process [17] and acid-catalyzed procedures are also known [18]. The condensation of two glycerol molecules is illustrated in Fig. 11, with the dominant reaction being the formation of an ether linkage at the expense of two primary hydroxyl groups, and the concomitant formation of 1 mole of water.

Although the diglycerol compound is shown in Fig. 11, the presence of two primary hydroxyl groups at either end of this molecule allows further reaction to continue. Products formed from the condensation of up to 30 glycerol units are known, although most commercially available materials feature 2–10 units [7]. Traces of air must be excluded from the process during reaction, as oxygen will cause acrolein to form, giving rise to high colors and odors. The polymerization itself can be followed by the viscosity, refractive index, or hydroxyl value of the reaction mixture. Viscosity does rise markedly during polymer formation, as each additional glycerol unit adds an additional hydroxyl group to the final polymer in addition to the increase in molecular size. Hydrogen bonding between the multitude of hydroxyls in the final product can therefore lead to a very viscous product.

Esters of polyglycerol can be made at temperatures above 200°C, with or without the addition of further catalyst. The easiest process is simply not to neutralize the sodium hydroxide present from the polymerization reaction and

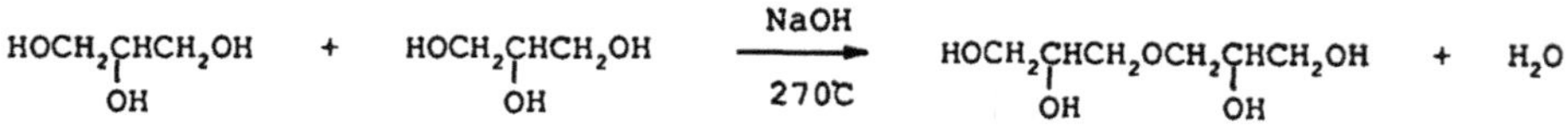

$$HOCH_2CHCH_2OH \quad + \quad HOCH_2CHCH_2OH \quad \xrightarrow[\;270°C\;]{NaOH} \quad HOCH_2CHCH_2OCH_2CHCH_2OH \quad + \quad H_2O$$
$$|\qquad\qquad\qquad\qquad\quad |\qquad\qquad\qquad\qquad\qquad\qquad |\qquad\qquad\quad |$$
$$OH \qquad\qquad\qquad\qquad OH \qquad\qquad\qquad\qquad\qquad OH \qquad\quad OH$$

FIG. 11 Polymerization of glycerol.

then to add the desired fatty acid. Treatment of the reaction mass after esterification is dependent on the degree to which ester formation has been carried out. If less than one third of the total number of hydroxyl groups has been reacted, the mixture may split into two layers when cooled. The lower free polyol layer may then be separated out and recycled for further reaction. If more than one third has been reacted, no settling may be required, and the products can be bleached and deodorized using methods normally used for fats and oils.

The addition of one more hydroxyl group with each additional glycerol unit means that a large range of polyglycerol esters can be made, and the potential range of HLB that can be spanned is quite large. For decaglycerol, the HLB range for various esters can be range of 3–16 for fatty acids with a carbon chain length of 10–18.

The ideal polyglycerol structure therefore appears to be quite simple. It would seem to be a linear glycerol polymer, usually 2–10 units long, with a variable amount of fatty acid attached as esters. Three potential problems with this idealized picture can be predicted from the preceding discussion, namely, that all fatty acids used will be mixtures of homologs, that there will be an equilibrium between species such as free polyglycerol, mono-, di-, and higher esters, and the location of the acyl groups distributed over all the hydroxyls on the polymer backbone. The additional complication that arises is from the polymerization reaction itself, as the polymer formed is not entirely linear.

The polymerization of glycerol is seen as proceding through the condensation of two primary hydroxyl groups. However, condensation of a primary hydroxyl with a secondary hydroxyl is possible, although the condensation of two secondary hydroxyls is very unlikely. The structures resulting from these three condensations is given in Fig. 12. The dominant nonlinear isomer is the 1,2'-diglycerol, which can be present in amounts of up to 15%. The presence of this 1,2' isomer is usually inferred, as the molecule readily undergoes a facile internal condensation to give a substituted dioxane structure. Although such a cyclic polymerization product has been detected in early attempts to analyze polyglycerol [19], it is only comparatively recently that Indian workers [20] have not only quantified this species but demonstrated how the amount is affected by the temperature at which the reaction is carried out.

Again, the essential finding is that the polymerization reaction mainly takes place through the primary hydroxyl groups. In addition, the cyclic dimeric species only tend to appear when the reaction temperature is 260°C or more. However, as the polymerization reaction itself is fairly slow below 250°C or less, cyclic dimer is a usual consequence of most commercial processes, although not to any major extent.

Some exploration of more highly reactive monomeric species (epichlorhydrin [21] or glycidol [22]) as potentially more selective ways of generating much better defined glycerol polymers has been carried out. These have not

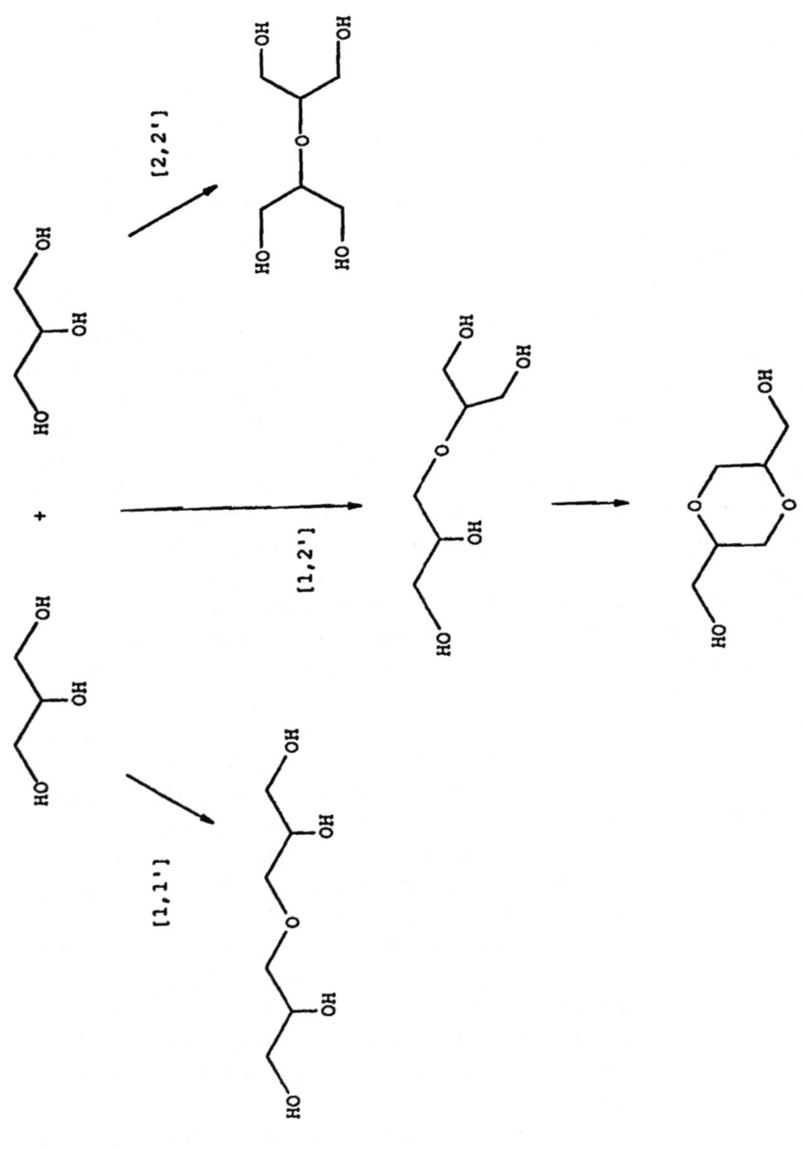

FIG. 12 Possible isomers of diglycerol resulting from reaction at either the primary or secondary hydroxyls.

resulted in commercially viable products, however, as these starting materials or newer processes have been more inherently expensive to realize than the established routes. Even the use of polyglycerol esters over polyoxyethylene esters (Sec. II) suffers from this problem.

V. SORBITAN ESTERS

The reaction of fatty acids with sorbitol results in a group of products known as sorbitan esters. Commercial quantities of the sorbitol feedstock are derived from the hydrogenation of glucose, and the material is classified as a sugar alcohol. Both sorbitol and its stereoisomer mannitol (Fig. 13) are used for the production of surfactants, although sorbitol has the much higher volume usage. From the structures shown in the figure, the initial impression might be that the ester chemistry of these materials should resemble a more complex version of the glycerol ester theme. That is, although both feature primary and secondary hydroxyls, the presence of six hydroxyl functions on these sugar alcohols would add a degree of complexity above that for glycerol. However, the further degree of complication is not this feature but rather is due to the property that sorbitol exhibits of undergoing internal dehydration during normal esterification reactions. This internal dehydration (or anhydrization) leads to the formation of substituted furan ring systems. It is also possible for such single-furan-ring systems to anhydrize further to give a fused bicyclic structure. The single anhydrization product is known as a sorbitan and the dual-ring system as isosorbide.

The ability of sorbitol to cyclize first to 1,4-sorbitan and then to isosorbide was originally demonstrated by the isolation and characterization of both of these compounds. However, more recent studies by Bock and coworkers [23] have shown that isomers of sorbitan other than the 1,4 material were also possible. These are shown in Fig. 14. The more usual anhydrisation pathway appears to be through the 1,4- or 3,6-sorbitans, which can both give rise to isosorbide.

FIG. 13 Fischer projection drawings of sorbitol and mannitol.

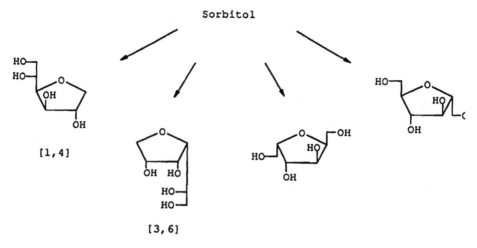

FIG. 14 Anhydrosorbitol (sorbitan) isomers arising from the dehydration of sorbitol.

However, a certain quantity of the 2,5 and 5,2 isomers can also be formed, both of which cannot be anhydrized further. The dominance of the 1,4- and 3,6- isomer pathways has been put down to the preference for secondary hydroxyl groups to displace primary hydroxyls during the anhydrization step [24]. Beyond this, steric arguments have been advanced to rationalize the fact that the 1,4 isomer can easily be isolated as a crystallizable solid, whereas the 3,6 isomer is difficult to detect. Effectively, the argument is used to explain that while cyclization to the 1,4 isomer is relatively quick, the onward reaction to isosorbide is slow [25,26]. For the 3,6 isomer, formation of the sorbitan is slow, but the onward reaction is quick. The intermediate is therefore only ever present in very low quantities (Fig. 15).

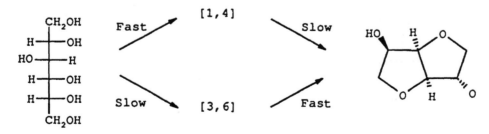

FIG. 15 Formation of 1,4–3,6-dianhydrosorbitan (isosorbide) via either 1,4- or 3,6- sorbitan.

The practical outcome of the above is that commercially available sorbitan esters do not contain sorbitol as the central polyol but rather mixtures of sorbitol that has been anhydrized to various degrees. This probably includes most of the above species illustrated in Fig. 14, but also possibly others in small percentages. However, given the long history of use of these products and their good toxicological profiles, they may be considered to be very useful and safe materials whatever their exact composition.

The compexity of the underlying chemical and compositional considerations can be contrasted with the essential simplicity of the processing conditions. Sorbitol and the appropriate fatty acid are reacted together at temperatures between 200°C and 280°C, in the presence of either acid or base catalysis [27]. Hydroxyl and acid values of the product are monitored until they fall into the required specification range, and then the reaction mixture is cooled. Recently, Ropuszinski and Sczesna reported work relating to the effect of various catalytic regimes on both the esterification and anhydrization reactions [28]. The systems they studied include p-toluenesulfonic acid, phosphoric acid, and sodium hydroxide. The acid catalysts appear to promote anhydrization over esterification, whereas the basic conditions give the opposite precedence. In addition, combinations of caustic catalysis (to initially promote esterification), followed by acidification of the reaction mixture (to promote anhydrization), as well as acid followed by base, were studied. The work implies that it may be possible to control the sorbitan ester reaction by controlling the two distinct chemical steps more exactly. As an extension of this work, it was later reported [29] that the rate of sorbitol anhydrization was dependent on the particular sodium phosphate species used as catalyst. At 230°C, trisodium phosphate had no catalytic effect. Disodium phosphate was moderately catalytic, but the monosodium salt was faster still.

The amount of academic study of these products has been relatively limited due to two main factors. First, the products resulting from these reactions are extremely complex mixtures, and hence the effect of process changes on the final product composition is likely to remain poorly understood. Analysis of these mixtures, even by recent high-performance liquid chromatography (HPLC) or supercritical fluid chromatography (SFC) methodologies, has still proved to be a difficult task [30,31]. Structure–activity relationships therefore may prove very difficult to determine. Second, the raw materials are relatively cheap, and the methods for their processing to sorbitan esters have been known for 50 years. The possibility of further optimization may be unnecessary, as formulators know how to optimize around the standard commercial materials.

An economic process for the production of sorbitol monoesters, on the other hand, does remain a research target. While the simple processing of sorbitol into fatty acid esters is accompanied by anhydrization and this internal etherification results in the loss of hydroxyl functionality, the hydrophobicity of these materials will remain high. If a sorbitol monoester could easily be made, this material would have a much greater water affinity and could be a much more versatile surfactant compound.

The reaction of equimolar amounts of fatty acid chloride and sorbitol in pyridine has been known for many years. However, the starting acid chloride is relatively expensive, and running the reaction in a toxic solvent such as pyridine also adds to the costs by requiring a scrupulous cleanup of the product. More recently, a process has been disclosed that involves running a simple sorbitan ester process featuring very quick heat-up and cool-down phases of the reaction [32]. Although some anhydrization takes place, the final product does retain much of the hydroxyl functionality of the sorbitol. A mixed sorbitol-glycerol-polyglycerol ester product has also been produced by the reaction of a triglyceride with sorbitol in the presence of 4–10% of various polyglycerols [33].

More highly substituted sorbitol esters do seem easier to produce with little anhydrization. For example, sorbitol tetraoleate can be made by the reaction of sorbitol with methyl oleate in the presence of a basic catalyst at temperatures of 120–155°C at various reduced pressures [34]. However, these types of product have very little surfactant activity. The proposed use for these compounds is as low-calorie butter or margarine substitutes, which is based on the idea that the greater the degree of ester functionality, the more difficult it is for the polyester fat to be broken down in the human gut.

In contrast to the highly complex picture of sorbitan/sorbitol chemistry, there has been some interest in the esterification of pure isosorbide. This diol can be prepared in moderate yields by driving the anhydrization of sorbitol through the sorbitan stage until the cyclization reactions are complete. A yield above 70% has been reported when using sulfuric acid at 135°C, or 66% when using hydrochloric acid [35]. This cyclization of optically active sorbitol produces a molecule featuring a cis-fused junction, and a V-shaped molecule with one hydroxyl inside (exo) and one outside (endo; Fig. 16).

While at first sight it might be expected on steric grounds that esterification of isosorbide should preferentially take place at the 2-exo hydroxyl, it has been

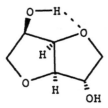

FIG. 16 Possible hydrogen bonding may explain the different reactivity of the 5-endo and 2-exo hydroxyls of isosorbide.

found that the 5-endo hydroxyl function is usually favored under kinetic conditions. Stoss and coworkers [36] found that the acylation with acetic anhydride at room temperature in the presence of lead oxide could give yields of up to 70–85% of the 5-endo acyloxy group. In contrast, if the same reaction is carried out and is followed by equilibration of the mixture with potassium hydroxide, then the 2-exo group is found in the product in yields of 50–90%. This migration of the acyl group shows that the ester at the 2-exo position as the thermodynamically favored outcome. One explanation put forward for this behavior is that the 5-endo hydroxyl can be seen as hydrogen-bonded to the ether oxygen of the adjacent ring, which enhances the reactivity to esterification [37].

The production of esters from mannitol has followed along the lines of those from sorbitol. Original work on these fatty acid esters was conducted at about the same time as that for sorbitol [38], but as mannitol is a more expensive feedstock for these types of products, these materials have a much smaller use in the marketplace. Studies into the cyclization of mannitol to give mannitan and isomannide have been carried out alongside the key work that has been conducted for sorbitol anhydrization [23,25].

VI. SUCROSE ESTERS

Sucrose would appear to be a very interesting feedstock for the production of polyol esters. It is a disaccharide with eight hydroxyl groups, and an additional three ether linkages, making it a very attractive hydrophilic group (Fig. 17). It is produced worldwide in million tonne quantities, and is therefore very cheap and abundant.

Unfortunately, the highly polar nature of sucrose results in several problematic attributes. It is a high-melting solid (melting point 160–180°C) and has a tendency to caramelize rather than melt. Given these physical constraints, the usual processing method for the manufacture of sucrose derivatives is to use a solvent for the reaction. This leads to a further constraint. There are only a few solvents that can dissolve this compound effectively. These include water as well as such highly polar aprotic solvents as dimethylsulfoxide (DMSO) and dimethyl formamide (DMF).

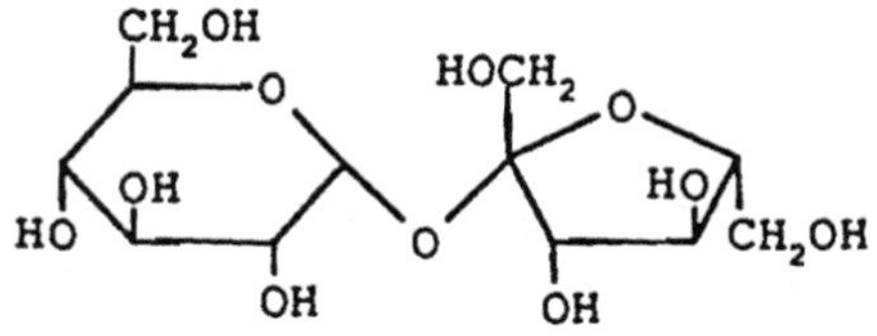

FIG. 17 Structure of sucrose.

The attraction of surfactant monoesters derived from sucrose is the higher water solubility of these compounds when compared with glycerol or sorbitan esters, due to the greater abundance of hydrophilic groups. This should result in a wider applicability of this type of material, given the cheapness of the raw materials, and the ready biodegradability and low toxicity.

The physical processing restrictions remain the main problem. Early work in surfactant ester preparation was carried out by reacting fatty acid methyl esters with sucrose in pyridine as the preferred solvent [39]. The reaction temperature was 100°C and sodium methoxide was used as the catalyst. Removal of methanol from the reaction vessel drives the preparation to the desired product, and this was complete in about 4 h. Purification of the desired product from the toxic and flammable solvent is an obvious drawback, as the production of a sucrose ester would appear to be ideal for use in applications such as food emulsification. Heterocyclic amines, DMF, and DMSO have also been reported as solvents for the reaction but suffer from the same drawback [40]. The usual catalysts for the reaction include alkali metal hydroxides and carbonates. Yields of up to 50% of sucrose ester are disclosed.

Osipow overcame the need for a highly polar but undesirable solvent by using a method whereby reaction took place in a transparent emulsion [41]. By bringing two immiscible liquids together in a very fine emulsion, it was shown that reaction between the two phases was possible, as if they were dissolved in a solvent. In order to achieve the intimate contact required, the droplet size needed to be about one-quarter of the wavelength of light, rendering the emulsion transparent. A suitable emulsifying agent was also required. For one specific example, an emulsion of sucrose, sodium oleate, and propylene glycol was formed after heating to 130°C. The propylene glycol was then removed by distillation over 5 h, and after an extractive workup, more than 50% of the desired ester had been formed.

This approach was later modified to use water as the solvent instead of propylene glycol [42]. The emulsifying agent was a combination of soap and previously prepared product at a level of 10–40%. The product mass contained approximately equal quantities of sucrose mono- and higher esters. Again, separation of the desired products from the rest of the reaction mass remains a potentially laborious process. This is particularly necessary when soap is used as one of the emulsifiers, as the surfactant properties of this anionic compound could radically interfere with those of the true sucrose derivatives. In addition, the taste of soap is certainly undesirable for emulsifiers used for food applications, currently a key outlet for these materials.

In 1973, Tate and Lyle published a patent that exemplified the transesterification of a triglyceride with sucrose, in a process that did not employ a solvent [43]. Reaction was carried out at 110–140°C in the presence of a basic catalyst such as potassium carbonate. An emulsifier, present in 5–10% by weight of the reaction mass, was also preferred. One of the advantages of this method is that the triglyceride ester is used as the acyl group source for transfer to sucrose. The byproducts of the reaction (di- and monoglycerides) also confer useful surfactant properties to the final mixture. Two of the disadvantages of this method are the longer reaction times required (up to 16 h) and the necessary restriction on the use of these products into applications where impure sucrose esters can be useful. These were exemplified as cosmetics, agricultural chemicals, and detergent powders.

Another interesting method disclosed for overcoming the physical incompatibility problems of the reactants is by use of a high-temperature worm shaft reactor [44]. This describes the use of a single-screw extruder with a fatty acid methyl ester, sucrose, and soap at 170–180°C. The soap catalyst was used at levels of 10–20%. This stage of the reaction provided only about 10% conversion to products, and the extruded product was then reacted further in a conventional reactor at 150°C under reduced pressure. The effectiveness of this technique was not explicitly demonstrated, however, as disappearance of the methyl ester was monitored, but the appearance of the required ester products was not.

Use of molten soap to provide a reaction medium into which both methyl ester and sucrose could be added has also been described. In this way, the overwhelming presence of soap as a solvent and reaction catalyst allows the buildup of product to happen gradually, and so helps facilitate the main reaction [45].

Preparation of sucrose polyesters has tended to be a much easier reaction to carry out than the production of the more interesting monoesters. Procter and Gamble have introduced a sucrose polyester (more than five fatty acid ester groups per molecule of sucrose) as a low-calorie alternative to butter and margarine [46]. The high degree of esterification allows the compound to pass through the human intestine unchanged. However, once the polyester has been formed, further reaction with sucrose can give a product that has a much lower degree of substitution. Polyesters with three to eight ester groups have been reacted to give products with about half the total number of esters per sucrose. Again, extractive purification adds to the complexity of the workup procedure.

The main production and use of fairly pure sucrose esters takes place in Japan. This is usually carried out utilizing a highly polar aprotic solvent for the synthesis step, followed by extensive product cleanup. Methods for the economic purification of sucrose esters continue to appear in the patent literature. Recently, an aqueous extractive method has been reported by Dai-Ichi, which is carried out after a conventional ester preparation in DMSO [47]. The reaction mixture is neutralized and a sediment precipitated out of solution. The sediment is triturated with acidified water, and fatty acid is added to the aqueous washings. The fatty acid forms micelles with the very water-soluble esters. About 85–95% of these esters can be recovered in this way. The resultant solid is neutralized, redissolved in water, and spray-dried to give a highly purified version of the sucrose monoester.

The Japanese market has been estimated at about 2600 tonnes per annum in total, which illustrates that for all potential that the basic elements of feedstock costs, desirability of the product, low toxicity, and excellent biodegradability represent, the fundamental physical processing or product cleanup costs continue to make this type of product very expensive in comparison with simple ethoxylated products. This results in the low volume of product sold and the high-value application areas (such as food) into which they go.

Less pure products are still finding uses in some market sectors. Mixed sucrose/glyceride ester materials have been claimed as having attractive properties for plant protection and for detergent formulations [48].

VII. POLYOXYALKYLENE POLYOL ESTERS

The preceding sections have described some of the chemistry associated with a number of different polyol moieties. These have included glycerol, polyglycerol, sorbitan, and sucrose materials. In the main, reaction of these materials with fatty acids results in relatively oily (low HLB) esters being produced, which does limit their value as useful surfactants. However, residual hydroxyl functionality in these compounds can be used as the basis for reaction with ethylene oxide to form much more hydrophilic polyoxyethylene chains.

This can be most readily exemplified by considering the polyoxyalkylene derivatives of sorbitan esters, commonly known as polysorbates. For the basic sorbitan ester products themselves, there is a limited range of HLB variation. By varying the length and number of fatty acid chains, the HLB of the ester can be altered from 8.6 (sorbitan monolaurate) to 4.3 (sorbitan monooleate) to 1.8 (sorbitan trioleate). This range of values (1.8–8.6) is still confined to the low end of the HLB scale (1–20). Addition of ethylene oxide, however, can raise the HLB of the resultant ester products to regions that greatly increase their utility. Oxyethylation of sorbitan monolaurate with 4 moles of ethylene oxide raises the HLB of the product from 8.3 to 13.3, and a reaction with a total of 20 moles increases this to 16.7. The common ranges of sorbitan and polysorbate esters and their HLBs are given in Table 4.

It should be mentioned at this stage that continual adjustment of oxyethylene content in order to generate a multitude of products is not really required, as it is possible to combine materials of different HLB values in a straightforward linear mathematical relationship to achieve intermediate HLB values. In fact, the HLB scale was devised on the basis of using sorbitan and polysorbate esters as nonionic emulsifiers in a number of different applications. Perhaps following from this original generation of a spectrum of surfactant emulsifiers capability, now combined with a long history of application utility, sorbitan and poly-

TABLE 4 HLB Values of Commercially Available Sorbitan Esters and Their Derived Polysorbates

Sorbitan esters			Polysorbate products degree of ethoxylation	
			4–5 EO	20 EO
Monoesters	Laurate	8.6	13.3	16.7
	Palmitate	6.7	—	15.6
	Stearate	4.7	9.6	14.9
	Oleate	2.1	10.0	15.0
Triesters	Stearate	4.3	—	10.5
	Oleate	1.8	—	11.0

sorbate ester materials continue to be a common starting point for many new applications.

Oxyethylation of sorbitan esters is generally carried out using procedures that could be equally applied to most other feedstocks (see Chapter 5). Reaction under base-catalyzed conditions (sodium or potassium hydroxide, or the alkali metal salt of the appropriate fatty acid) at temperatures of 130–170°C results in the smooth addition of ethylene oxide. However, just as ester interchange has been previously described as a common feature of base-catalysed processes, so this is probably taking place in these oxyethylation reactions. Although fatty acid groups may originally be attached to the sorbitan ring systems in the feedstock ester, these can easily rearrange to join onto the end of any growing oxyethylene chain [49]. The net effect of this ester interchange would appear to be as if the ethylene oxide inserts itself between the sorbitan ring and the fatty acid moiety. In this way, the overall shape of the final ester is not as simple as the earlier idealized version of a surfactant, i.e., with hydrophilic headgroup and oleophilic tail. It is closer to a core sorbitan ring, with radiating oxyethylene chains, which are capped to a lesser or greater extent by the fatty acid ester functions.

By analogy with the polysorbate ester reaction, similar surfactant ester feedstocks that feature free hydroxyl functions can also undergo oxyethylation reactions to give materials of higher HLB. These include glycerol, polyglycerol, and sucrose esters. In addition, many naturally occurring fats and oils are also similarly reacted to raise their water solubility.

Although many other oxyethylated materials have been studied, such as those derived from reaction of sucrose, sorbitol, or methylglucoside first with ethylene oxide and second with fatty acids, the major commercial ranges of polyoxyalkylene polyol esters are limited. Materials derived from anhydrosorbitol, glycerol (fats and oils), and simply polyoxyethylene make up about 90% of the volume of such products. More specialized esters can be easily tailored, but for the majority of needs, a blend of low- and high-HLB esters, such as the sorbitan and polysorbate materials, will perform efficiently.

VIII. ENZYME-CATALYZED ESTERIFICATION

The above discussion has focused on the relatively nonspecific ways in which commercially available products are made. Although it is possible to synthesize a wide range of chemically pure entities if these species need to be studied, such syntheses tend to be too expensive to operate on a large manufacturing scale for industrial applications. More recently, interest has turned to the possible use of enzyme catalysis to add a degree of product specificity not previously available. This technology is still emerging, and what will be presented here are simply some of the trends that are currently emerging in this field.

Enzymes are natural catalysts that mediate biological processes. For example, various yeasts are able to turn sugars into alcohol, a very well-studied process! However, since the early 1960s, when commercialization of enzymes in detergent formulations was first achieved, interest in their potential organic synthetic ability has expanded rapidly. Although the original commercialization was for relatively impure and expensive actives, development of the technology has led to the production of purer and less costly enzymes. At present, several thousand enzymes have been isolated and characterized, and several hundred are commercially available.

Enzymes themselves are produced by living systems in order catalyze biological processes [50,51]. They are protein-based and are typically polymers of about 250–400 amino acid units. Frequently they contain about 2–15% by weight of sugar residues, but the function of these residues is not absolutely clear.

As the amide functions of the protein chains are very polar, and as each amino acid residue has its own substitution pattern and shape, is it possible for these molecules to fold up and hydrogen-bond in very intricate ways (something like a ball of string). However, most will fold up into one particular shape or primary structure. Once this has formed, the enzymes will possess a small area on the outer surface that will catalyze particular reactions. This is known as the active site, and this area has a degree of flexibility that allows reactant molecules to be accommodated and for product molecules to be released.

However, as the primary structure of any enzyme is dependent on only the relatively weak hydrogen bonding of the amide functions, these materials are not stable to high temperatures. Most cannot be heated above 50°C, although some may continue to operate up to 70°C. Above the appropriate critical temperature, the enzyme structure will unravel (denature), and all catalytic effect is lost. A very slow decrease in temperature may result in the reformation of the original shape, but the phenomenon of denaturation does impose a definite operating limit on any temperatures of organic reactions for which these proteins are used.

There is a great range of potential catalysis that enzymes can offer to synthetic chemists. As they operate at relatively low temperatures and cover a wide range of stereo- and regioselective reactions, this makes them very attractive to use in the synthesis of natural products or pharmaceutical actives, where a high number of selective transformations may be required to define a useful synthetic pathway. This attractive ability for the efficient conversion of raw materials to a desired product is currently offset by the general high cost of these products. The isolation and stabilization of sufficient quantities of enzymes from natural sources remains an expensive activity. However, as these high-value activities, particularly in the pharmaceutical field, continue to stimulate the technology around the production of enzymes, the possible application of these catalysts to moderate-value processes is becoming more realistic.

Enzymes are classified into six main groupings, depending on their mode of action (Table 5). This section concentrates on the chemistry that derives from only one of these, lipases, which are catalysts built by living systems in order to hydrolyze fats and oils. For mammals, such lipases are made and released from the pancreas. The pancreatic lipase is released into the small intestine, where it attacks emulsified triglycerides in preference to those that occur in the walls of the pancreas itself. This has led to the hypothesis that the lipase works best at an oil–water interface. Because of this, most lipases would appear to need a mixture of oil and water to perform at their best. Lipases are also produced by microorganisms such as bacteria as a mechanism for the cleavage of triglycerides. Again, the cleavage reaction is a mechanism for the organism to gain a source of carbon for growth.

Although lipases are used by biological systems in an aqueous environment to promote hydrolysis of esters, it should be remembered that they are essentially performing the function of any other catalyst—they are promoting the

TABLE 5 Enzyme Classifications and Functions

Enzyme classification	Function
Hydrolases	Catalyze the hydrolytic cleavage of C–O, C–N, and C–C bonds. This category includes lipases, proteases, esterases, and glycosidases.
Isomerases	Catalyze molecular isomerizations, and includes racemases and isomerases.
Ligases	Catalyze the formation of a covalent bond between two molecules.
Lyases	Catalyze the cleavage of C-O, C-N, and C-C by elimination.
Oxidoreductases	Catalyze all redox reactions.
Transferases	Catalyze the transfer of a specific group.

speed at which an equilibrium is attained between reactants turning into products. The same basic mass action law discussed above (Sec. 2) still holds. Similarly, in a water-poor environment, lipases will act to esterify an alcohol and an acid, i.e., the mass action law can still be driven in the direction required.

A. Glycerol Esterification Using Enzymes

Of course, given that enzymes have a natural capability for reacting with triglycerides, it follows that esters of glycerol have received much study. Linfield looked at the hydrolysis of tallow, coconut, and olive oils by *Candida rugosa* [52]. He found that the reaction approximated to a first-order kinetics model and was not affected by a temperature change over the range 26–46°C. To hydrolyze all three substrates, however, required above 72 h of reaction time. Lipase from *Aspergillus niger* and *Rhizopus arrhizus* were also studied, with the latter giving a considerably slower reaction rate. Esterification was also studied by stirring a mixture of the appropriate enzyme, glycerol, and an excess of fatty acid for at least 24 h, prior to distilling some water out of the mixture via reduced pressure. No external heat was applied. The mixture was again stirred at ambient temperature and water stripped out at regular intervals. In this way, an almost 80% conversion to product was achieved over a 6-week reaction time. Thin-layer chromatography at this stage showed that the product consisted of 1.4% monoglycerides, 19.8% diglycerides, 43.6% triglycerides, and 35.3% free fatty acid. *Aspergillus niger* and *C. rugosa* produced comparable results. Attempts to esterify other polyols using this technique proved difficult, with sorbitol not reacting with oleic acid, and ethylene and diethylene glycol reacting only very slowly. An interesting observation made during this work was that if vacuum dehydration was too severe, the enzyme would stop working. It could then be reactivated if a small amount of water was added back into the mixture. This observation supports the need for a lipase to have an oil–water interface to operate correctly.

The possibility of converting such a method for glyceride synthesis into a practical process was outlined by Yamane [53]. He reported that use of a microporous membrane reactor could be used to give continuous production of glycerides from separate glycerol and fatty acid streams. On one side of the membrane in the reactor, he recycled a mixture of lipase (*Chromobacterium viscosum*) in glycerol. This circulating loop featured a dehydration chamber, for the removal of the byproduct water, and a fresh glycerol source. On the other side of the membrane, fatty acid was fed in, and glyceride products were produced.

The two most effective methods used for the dehydration section were either molecular sieves or a vacuum dehydrator operating at a pressure of 5 mm Hg. The best conversion (90%) occurred when the glycerol solution contained about

3–4% water. The reaction produced mainly mono- and diglycerides in almost equimolar amounts.

More recently, a full review of immobilized lipase reactors for effecting hydrolysis, esterification, and transesterification was published [54].

As mentioned above, the most desirable glyceride product is normally the monoester. Alcoholysis of an oil or fat to achieve high yields of this material without the need for molecular distillation is technically attractive and should consume much less energy than the conventional process.

Holmberg has shown that it is possible to obtain monoglycerides in yields of up to 80% by simple hydrolysis of a triglyceride with a lipase that was specifically active at the 1,3 positions of glycerol [55]. The byproduct of this reaction was 2 moles of fatty acid. The process was carried out by the formation of an oil-rich microemulsion with isooctane as the hydrocarbon component, and an aqueous phase buffered to pH 7. Use of the 1,3-specific lipase *Rhizopus delemar* to carry out such a specific reaction does have a practical limitation, as the migration of an acyl group from the 2 position to the more thermodynamically stable 1 or 3 position will eventually result in total glyceride hydrolysis. Optimal reaction time for this system was found to be 3 h at 35°C.

This work was extended to show that glycerolysis of palm oil could be carried out in the presence of 2 moles of glycerol to give an impure monoester product [56]. The need to maintain an emulsified system, together with a certain concentration of water to allow the lipase to function, led to a mixture of desired monoglyceride and free fatty acids. This product was obtained in spite of the use of four different enzyme types. It should be noted that although 1,3-specific and positionally nonspecific lipases are known, there has so far been no report of an enzyme that operates only at the 2 position. This has been described as a way of determining the difference between enzymically and chemically catalyzed interesterifications [57] and has also been used to produce 2-monoesters by selective hydrolysis [58].

Yamane reported [59] that the glycerolysis of beef tallow in the solid phase using an enzyme can increase the yield of monoglyceride (up to 70%) over that obtained at temperatures above the melting point of the fat. This has been explained by the observation that monoglycerides have higher melting points than the corresponding triglycerides, and this lower solubility results in the preferential precipitation of the desired product. This effective removal of this product from the reaction medium provides a mechanism for driving the equilibrium away from the typical proportion of monoester that would be achieved for a simple homogeneous reaction mixture. The highest yields obtained with commercially available lipases were given by *Pseudomonas fluorescens* or *Chromobacterium viscosum*. Using the best conditions (temperature 44°C, water content 2%, 2 moles of glycerol per mole of fat), approximately 60% of monoester was formed within 5 h, slowly rising to 70% after a total of 25 h reaction time.

The work was extended by further screening of commercially available lipases, both singly and in combination, together with a stepped reaction temperature profile [60]. Lipases from *Pseudomonas sp.* were highly active, as previously reported. *Geotrichum candidum* was virtually inactive to the fats studied, an observation that was explained away by the known preference of this enzyme for unsaturated fatty acids. *Penicillum camembertii* was also inactive, as this enzyme is most active toward mono- and diglycerides, rather than triglycerides. Combinations of various lipases did not give an increased yield of monoesters above 70%, which may simply be a limiting value for this particular methodology. A product containing 90% monoester was obtained, however, using *Pseudomonas* and olive oil at the relatively low temperature of 10°C. This may be due to more monoglyceride precipitating from the cooler reaction mixture. *Mucor meihei* also showed this effect but gave a slightly lower yield (80%). While this level of monoester is approaching that of a molecularly distilled product, the major hindrance to the use of this technology is the time required to achieve the desired monoester type. An 80% monoester may require 24 h, whereas 90% may require up to 5 days of reaction time.

In contrast, the use of a thermostable lipase and a solvent can considerably hasten reaction time. Lion Corporation reported some studies on 17 different lipases, which showed that *Candida antarctica* and *M. meihei* were among the most thermally stable (withstanding 70°C for 30 min) and active (at 50°C) for the synthesis of glycerol esters [61]. A typical procedure employed immobilized *C. antarctica* catalyzing the reaction between oleic acid and glycerol at 50°C in tertiary butyl alcohol over molecular sieves as a drying agent. After 6 h, the reaction product contained 87% monoester and 9% diester. The molar ratio of oleic acid to glycerol, however, was 1:3. The excess glycerol does speed the reaction to products, but again complicates vessel utilization and process operability for full-scale manufacture.

The Lion patent [61] exemplified reactions using either fatty acids or their lower alkyl esters, and demonstrates that polyglycerols or polypropylene glycol can be similarly converted to high-monoester product. The work was extended to ester interchange reactions, where glycerolysis of oils and fats was carried out. Conversion to products containing mono- and diester content of 80–90% and 10–15% respectively, was obtained for numerous triglycerides. Finally, continuous use of the enzyme over many reaction cycles was demonstrated with a reaction between glycerol and olive oil over *C. antarctica*. The monoester content of the product was shown to remain at about 85% monoester, even after 1500 h of use at 50°C. This continuous use of the enzyme remains a very desirable feature, as recycling over many reaction periods brings down the high cost of enzyme use per unit of product made to a much commercially competitive level.

Schneider and coworkers also looked at the possibility of producing mono-esters of glycerol [62,63]. Due to the inherent mismatch of polarities of glycerol and fatty acids or derivatives, a method was developed whereby preadsorption of glycerol on a silica support was used. Reaction in an aprotic solvent over *Chromobacterium viscosum, Rhizopus delemar,* or *Rhizomucor meihei* was then carried out. However, this methodology did not give rise to exceptionally high yields of monoester, and so a separate purification step was devised. By using a combination of diethyl ether and hexane as solvent, and by cooling the mixture down to between -2 and 10°C, 80–91% yields of monoester could be obtained. All other byproducts were simply returned to the reaction stage, from which the immobilized enzyme and silica support had been filtered. This initial methodology was followed up by work showing its wider applicability. Reaction times unfortunately remained relatively long, at between 24 and 96 h.

In addition to the development processes around monoglyceride synthesis using enzymes, several other facets of this glycerol ester chemistry have been studied. Simple hydrolysis of fats and oils has been reviewed by Sonntag [64], who pointed out that the economic equation for making enzymic fat splitting a favorable alternative to current methods requires that the overall cost be lower than the cost of the heat and electrical energy normally utilized. It should also be remembered that heat recovery is usually designed in many modern plants, making the equation even more difficult to satisfy. In addition, as most lipases interact best at the 1 and 3 positions of glycerol, the critical factor influencing triglyceride hydrolysis is the rate at which the ester at the 2 position moves to the end hydroxyls. Although enzyme hydrolysis can be effective at 40–60°C, temperatures of 100°C may be required to promote this rearrangement at a sufficient speed to allow a fully competitive process to be reached. The existence of an enzyme process capable of operating at such a temperature has yet to be proven.

Lipases have also demonstrated a selectivity for promoting reactions of some fatty acid or ester moieties over others. A review by Macrea [51] concluded that most microbial lipases show little specificity, although that from *Geotrichum candidum* can show a difference in the rate of hydrolysis of fatty acid methyl esters of up to 100 times. Oleate or linoleate alkyl chains are much faster than myristate or stearate. A possible theory to rationalize this effect is that the active site contains a functionality that allows some interaction with π-bond orbitals, or perhaps olefinic bonds.

Alternatively, Jones postulated that the shape of the active site is why some substrates fit in easily, and converted more quickly, than those that do not fit as well [65]. More recently, Mukherjee [66] reported that selective esterification of a hydroxyl-containing substrate is possible such that specific enrichment of the free fatty acid or ester products can be achieved. For the two lipases studied, from either seedlings of rape (*Brassica napus* cv Ceres) or from *Mucor meihei,* γ-linoleic acid is converted very slowly through to ester and so builds up in the free fatty acid portion of the product mixture. When starring from fatty acids derived from evening primrose oil, containing 9.5% of γ-linoleic acid, enrichment of up to 85% has been achieved by converting other acids to their butyl ester. Although this method presents an opportunity to make enriched, unsaturated feedstocks for the production of pharmaceutically interesting materials, it also demonstrates that awareness of this phenomenon is necessary when trying to make esters of fatty acid materials with both saturated and unsaturated components.

In contrast to the work of Mukherjee, Ward [67] has shown that for polyunsaturated fatty acids such as eicosapentaenoic and docosahexaenoic acid, *Mucor meihei* and *Pseudomonas* sp are actually the best enzymes for achieving glycerides from glycerol in a water–hexane mixture at 30°C. Clearly, much more characterization work is needed before fatty acid selectivity and reactivity can be fully defined.

B. Other Carbohydrate Esterification Using Enzymes

Following the interest and extensive reporting on the enzyme-catalyzed synthesis of glycerol esters, other carbohydrate feedstocks have be investigated. These include materials such as sorbitol, glucose, sucrose, and simple glucosides.

Heino published some work in 1984, aimed at using enzymes with this type of raw material [68]. Due to the low temperatures at which these reactions could be carried out, the implication was that much cleaner and purer versions of the chemically catalyzed products could be obtained, as much less decomposition of the feedstock would take place. The reaction procedure consisted of mixing the enzymes studied with fatty acids and sucrose, fructose, sorbitol, and glucose at 40°C in an aqueous buffered solution for 24–95 h. From the eight enzymes studied, *Candida cylindracea* was the most active. A semioptimized process showed that the best reaction conditions occurred when the concentrations of sucrose and oleic acid were 0.05 mol/L and 0.2 mol/L, respectively, after 72 h. Conversion to ester was more than 60% for sucrose, fructose, and sorbitol, but for glucose it was only about 28%.

Klibanov published a number of papers investigating the esterification of similar carbohydrate hydrophiles. Instead of an aqueous environment, he concentrated on the use of dry, aprotic solvents such as pyridine and DMF. However, while these solvents are ideal for the carbohydrate reactant, not many lipases are active in such an anhydrous environment, or with di- or oligosaccharides. In 1988, he reported work with proteases, specially *Bacillus subtilis* [69]. He also utilized a highly activated form (a 2,2,2-trichloroethanol ester) of the fatty acids he studied in order to drive the reaction to completion. Although the conversions to the desired esters were fairly successful (up to 96% based on gas chromatographic disappearance of the carbohydrate), reaction times were still of the order of 2–7 days at 45°C. The acyl group of the ester was predominantly found on one of the primary hydroxyl groups of the carbohydrate substrate. Reactions with salicin, riboflavin, adenosine, and uridine were also studied.

While this work has shown that the esterification reaction is not confined to lipases alone, it is evident that the major problems that need to be dealt with still involve elimination of the use of a toxic solvent and a speed of reaction that makes such a synthetic pathway worthwhile. Klibanov has also studied some transesterification reactions at about the same time [70]. He studied model reactions between triolein (glycerol trioleate) and sorbitol in pyridine. Only porcine pancreatic lipase and, to a lesser extent, *Chromobacterium viscosum* were active. In a typical reaction, sorbitol was reacted with a slight molar excess of triolein in dry pyridine at 25°C. After 3 days, the enzyme was filtered off and the product purified by column chromatography. The pancreatic enzyme gave a monoester compound, as an almost equimolar mixture of the 1 and 6 derivative. The overall yield was very poor, however. The work was extended to natural triglycerides, which demonstrated the generality of the reaction. Surfactant properties of the derived materials were then examined and shown to be quite different from those derived by more conventional chemical means. However, these comparisons were being done between a sorbitan ester (and its relatively few free hydroxyl groups) and a sorbitol ester (where many more hydroxyls are giving rise to a greater water solubility).

Lion Corporation disclosed some details of work on saccharide fatty acid monoesters in a 1990 patent [71]. The process involved use of a fatty acyl donor (acid, lower ester. or triglyceride) with a saccharide or disaccharide compound in a secondary or tertiary alcohol solvent in the presence of a thermostable lipase. As for the glyceride work mentioned above, *Candida antarctica* again featured as the most thermally stable and active enzyme found. The fatty acyl groups included both saturated and unsaturated materials; the saccharides included glucose, fructose, sucrose, and various short chain glucosides (methyl to butyl); and typical solvents included tertiary butanol and diacetone alcohol. Although most of the reactions featured an excess of one of the reactants in order to force the reaction to completion, the reactions with the glucosides appeared to be the easiest to catalyze. This may be due to the extra solubility of these materials in the relatively low-polarity solvent. The reaction temperature was 50–60°C, and the time of reaction varied between 8 and 24 h. The purity of the monoesters produced, however, was claimed as high as 98%.

This work was extended in a further patent, which explored different solvent systems. These included acetylacetone, ethylene, and propylene carbonate, γ-butyrolactone, and various substituted pyridine derivatives [72]. The reaction times were subsequently reduced from 24 to about 10 h, with a conversion that remains high, at about 90%. In addition, byproduct formation of the diester compounds is suppressed. *Candida antarctica* and *Mucor meihei* were the favored lipases used. Reactions with most of the saccharide compounds studied still featured a large excess (usually 4, but up to 20 times) of the acyl-donating material, which again complicates the overall industrial utility of the method. Most of the reactions employed methylglucoside as the saccharide substrate, although the wider applicability of the method was also exemplified.

Some further work [73] on saccharide esters was also patented by Novo, a company that is heavily involved in the production and immobilization of enzymes for commercial sales. Methylglucoside esters were prepared from methy glucose and the appropriate acyl donor material without the need for solvent using *Candida antarctica* at 80°C under reduced pressure (0.01 bar). Reaction time varied between 17 and 36 h, and greater than 90% conversion to ester could be achieved, although these products did sometimes contain significant amounts of diester (up to 15%).

The work was extended in a later patent to other saccharides [74]. *Candida antarctica* was again employed as the appropriate catalyst, and reaction times were relatively long (up to 24 h for high product yields). Although no solvent was used, an emulsifier (normally soap, at up to 20% by weight of the reaction mixture) was required to allow the saccharide and acyl donor to mix effectively. Saccharide substrates studied included ethylglucoside, xylitol, sorbitol, and mannose.

As can be seen from this brief glimpse of practical work that is currently being done in this area, the fundamental problem of physical incompatibility of the feedstock materials remains a large barrier to efficient exploitation of enzyme esterification. In addition, the limited temperatures that commercially available enzymes can withstand may make for more selective reactions, but at the expense of reaction rate. This, coupled with the expense of the enzymes themselves, makes the possibility of industrial utilization of limited attractiveness at present.

IX. SUMMARY

The purpose of this chapter has been to highlight the chemistry, product composition, and technology limits of polyol fatty acid ester surfactants. Although the creation of surfactant products from readily available polyol and fatty acid feedstocks can be easily carried out, this ease of reaction gives rise to little structural discrimination. A fundamental problem is therefore inherent in the chemistry. While an idealized ester surfactant structure may be for a polyol with only one fatty acid group attached, reaction of 1 mole of polyol with 1 mole of fatty acid will not give 1 mole of monoester. It will ordinarily give a mixture of a free polyol, mono- and polyesters. It should also be remembered that the fatty acids usually employed are also mixtures and not single-isomer feedstocks.

In addition to the fundamental random nature of the reaction described above, further complicating factors are evident for the families of surfactants described in the previous sections.

1. Glycerol features both primary and secondary hydroxyl groups, and esterification of the former is much greater than the latter. This difference in esterification ability holds true for all polyol substrates with a mixture of such hydroxyls.
2. Polymerization of glycerol to give the polyglycerol polyol does not proceed in an entirely linear manner, and some cyclic species are evident. The polyol again features primary and secondary hydroxyls.
3. Sorbitan esters feature a wide variety of ester types, based on a number of internally dehydrated sorbitol structures. Full characterization has so far proved very difficult.
4. The conversion of sucrose into fatty acid esters remains an area that has promised much, but delivered little to date, due to physical processing problems.

In contrast to the existing technologies described above, the possibility of using enzyme catalysts to form much more well-defined ester structures does offer an interesting field of study. Although so far processing has remained time consuming, difficult, and costly, it will become progressively cheaper and more commercially attractive to carry out over time. However, given the cost competitiveness of the established polyol ester surfactants ranges, this newer technology may only prove to be of utility in specialist application areas. Overcoming some of the inherent physical processing problems for some substrates has also not been completely achieved.

It should be remembered that the commercial ranges of product, although perhaps multicomponent mixtures, are nevertheless made in well-controlled ways in order to give reproducible application results. The lack of simple structural discrimination in the ester surfactant chemistry should not be seen as negating either the usual good standard of manufacture, or the proven value in use of these materials.

In essence, the major commercial ranges of polyol fatty acid ester surfactants are still those based on glycerol and sorbitol. Their limiting low HLB values easily be overcome by the addition of ethylene oxide. These ranges therefore remain the backbone of this area of surfactant chemistry. Although newer technology and product ranges may be developed over time, they will need to be competitive with this well-established and cost-effective baseline.

REFERENCES

1. J. J. Lewis, in *Preservation of Surfactant Formulations* (F. F. Morpeth, ed.), Blackie, Glasgow, 1995, Chapter 4.

2. E. Fritz, in *Fatty Acids in Industry* (R. W. Johnson and E. Fritz, eds.), Marcel Dekker, New York, 1989, Chapter 3.

3. M. Aslam, G. D. Torrence, and E. G. Zey, in *Kirk-Othmer Encyclopedia of Chemical Technology*, Vol. 9, 4th ed. Wiley, New York, 1994, pp. 758–759. 1994.

4. J. D. Brandner and R. L. Birkmeier J. Am. Oil Chemists Soc. *41*:367 (1964).

5. K. D. Tau, V. Elango, and J. A. McDonough, *Kirk-Othmer Encyclopedia of Chemical Technology*, Vol. 9, 4th ed. Wiley, New York, 1994, pp. 798–799.

6. A. Cahn and J. L. Lynn, in *Kirk-Othmer Encyclopedia of Chemical Technology*, Vol. 22, 3rd ed. Wiley, New York, 1983, pp. 367–373, 1983.

7. M. Ash and I. Ash, *Condensed Encyclopedia of Surfactants*, Chemical Publishing Co., USA, 1989.

8. Ref. 3, pp. 760–761.

9. T. P. Hilditch and J. G. Rigg, J. Chem. Soc. 1774 (1935).

10. H. Saudinger and H. Schwalenstoecker, Chem. Ber. *68*:727 (1935).

11. D. Stern (ed.), *Bailey's Industrial Oil and Fat Products*, Vol. 2, 4th ed. Wiley, New York, 1982, pp. 97–113.

12. Ref. 11, pp. 113–127.

13. N. O. V. Sonntag, J. Am. Oil Chemists Soc. *59*:795A (1982).

14. N. H. Kuhrt, E. A. Welch, and F. J. Kovarik, J. Am. Oil Chemists Soc. 310 (1950).

15. F. R. Benson, in *Nonionic Surfactants* (M. J. Schick, ed.), Surfactant Science Series, Vol. 1, Marcel Dekker, New York, 1967, Chapter 9.

16. R. T. McIntyre, J. Am. Oil Chemists Soc. *56*:835A (1979).

17. N. Garti, A. Aserin and B. Zaidman, J. Am. Oil Chemists Soc. *58*:878 (1981).

18. P. Seiden and J. B. Martin, U.S. Patent 3,968, 169 to the Procter and Gamble Company (1975).

19. M. R. Sahasrabudhe, J. Am. Oil Chemists Soc. *44*:376 (1967).

20. T. N. Kumar, Y. S. R. Sastry, and G. Lakshminarayana, J. Chromatogr. *298*:360 (1984).

21. H. Sebag, GB Patent 2, 168, 348 to L'Oreal (1984).

22. K. S. Dobson, K. D. Williams, and C. J. Boriack, J. Am. Oil Chemists Soc. *70*: 1089 (1993).

23. K. Bock, C. Pederen and H. Thogersen, Acta Chem. Scand. *B35*:441 (1981).

24. P. F. Vlad and N. D. Ungur, Synthesis 217 (1983).

25. Z. Cekovic, J. Serb. Chemists Soc. *51*:205 (1986).

26. R. Barker, J. Org. Chem. *35*:461 (1970).

27. K. R. Brown, U.S. Patent 2,322,820 to Atlas Powder Company (1939).

28. S. Ropuszynski and E. Sczesna, Tenside Surf. Det. *22*:190 (1985).

29. S. Ropuszynski and E. Sczesna, Tenside Surf. Det. *27*:350 (1990).

30. Z. Wang and M. Fingas, J. High Resolut. Chromatogr.*17*:15 (1994).

31. Z. Wang and M. Fingas, J. High Resolut. Chromatogr. *17*:85 (1994).

32. H. Stuehler, Ger. patent DE 3,240,892 to Hoechst (1982).

33. P. Bickert, Ger. patent DE 3,623,371 to Huels (1986).

34. G. D. Gruetzmacher, J. Raggan, and B. Wlodecki, WO Patent 93/00016 to Pfizer (1992).

35. G. Fleche and M. Huchette, Starch *38*:26 (1986).

36. P. Stoss, P. Merrath, and G. Schlueter, Synthesis 174 (1987).

37. R. U. Lemieux and A. G. McInnes, Can. J. Chem. *88*:136 (160).

38. W. N. Haworth and L. F. Wiggins, GB Patent 600, 870 to Edgbaston University (1946).
39. J. B. Martin, U.S. Patent 2,831,855 to Procter and Gamble (1958).
40. N. B. Tucker, U.S. Patent 2,831,856 to Procter and Gamble (1958).
41. L. I. Osipow and W. Rosenblatt, U.S. Patent 3,480,616 to University of Nebraska (1969).
42. L. I. Osipow and W. Rosenblatt, U.S. Patent 3,644,333 to University of Nebraska (1972).
43. K. J. Parker, R. A. Khan, and K. S. Muffi, GB Patent 1,399,053 to Tate and Lyle (1973).
44. H. J. W. Nieuwenhuis and G. M. Viomen, EPA 190, 779 to Cooperatieve Vereniging Suiker Unie(1986).
45. UK Patent 1,332,190 to Dai-Ichi Kogyo Seiyaku (1970).
46. F. H. Mattson and R. A. Volpenheim, U.S. Patent 3,600,186 to Procter and Gamble (1971).
47. S. Matsumoto, Y. Hatakawa and A. Nakajima, U.S. Patent 5,144,022 to Dai-Ichi Kogyo Seiyaku (1990).
48. R. Khan and H. F. Jones, in *Chemistry and Processing of Sugarbeet and Sugarcane* (M. A. Clarke and M. A. Godshall, eds.), Elsevier, Amsterdam, 1988, Chapter 23.
49. Ref. 15, p 270.
50. A. R. Macrae, in *Microbial Enzymes and Biotechnology* (W. N. Fogarty, ed.), Applied Science, Ann Arbor, 1983, Chapter 5.
51. P. H. Nielsen, H. Malmos, Ture Damhus, B. Diderichsen, H. K. Nielsen, M. Simonsen, H. E. Schiff, A. Oestergaard, H. S. Olsen, P. Eigtved, and T. K. Nielsen, in *Kirk-Othmer Encyclopedia of Chemical Technology*, Vol. 9, 4th ed., Wiley, New York, 1994 pp. 567–620.
52. W. M. Linfield, R. A. Barauskas, L. Silvieri, S. Serota, and R. W. Stevenson, J. Am. Oil Chemists Soc. *61*:191 (1984).
53. M. M. Hoq, T. Yamane, and S. Shimizu, J. Am. Oil Chemists Soc. *61*:776 (1984).
54. F. X. Malcata, H. R. Reyes, H. S. Garcia, C. G. Hill, and C. H. Amundsen, J. Am. Oil Chemists Soc. *67*:890 (1990).
55. K. Holmberg and E. Osterberg, J. Am. Oil Chemists Soc. *65*:1544 (1988).
56. K. Holmberg, B. Lasser, and M.-B. Stark, J. Am. Oil Chemists Soc. *66*:1796 (1989).
57. S. S. Roy and D. K. Bhattacharyya, J. Am. Oil Chemists Soc. *70*:1293 (1993).
58. A. Zaks and A. T. Gross, U.S. Patent 5,316,927 to Opta Food Ingredients (1992).
59. G. P. McNeil, S. Shimizu, and T. Yamane, J. Am. Oil Chemists Soc. *67*:779 (1990).
60. G. P. McNeil, and T. Yamane, J. Am. Oil Chemists Soc. *68*:6 (1991).
61. K. Kitano, R. Iwasaki, N. Mori, H. Sasamoto, and T. Akamatsu, EPA 407, 959 to Lion Corporation (1989).
62. M. Berger, K. Laumen, and M. P. Schneider, J. Am. Oil Chemists Soc. *69*:955 (1992).
63. M. Berger, and M. P. Schneider, J. Am. Oil Chemists Soc. *69*:961 (1992).
64. N. O. V. Sonntag, in *Fatty Acids in Industry* (R. W. Johnson and E. Fritz, eds.), Marcel Dekker, New York, 1989, Chapter 2.
65. J. B. Jones, Aldrichemica Acta *26*:105 (1993).
66. M. J. Hills, I. Kiewitt, and K. D. Mukherjee, J. Am. Oil Chemists Soc. *67*:561 (1990).

67. Z-Y Li and O. P. Ward, J. Am. Oil Chemists Soc. *70*:745 (1993).
68. H. Heino, T. Uchibori, T. Nishitani, and S. Inamasu, J. Am. Oil Chemists Soc. *61*:1761 (1984).
69. S. Riva, J. Chopineau, A. P. G. Kieboom, and A. M. Klibanov, J. Am. Oil Chem. Soc. *110*:584 (1988).
70. J. Chopineau, F. D. McLafferty, M. Therisod, and A. M. Klibanov, Biotechnol Bioeng *31*:208 (1988).
71. H. Miyake, M. Kurahashi, H. Toda, K. Kitano, and T. Mori, EPA 413, 307 to Lion Corporation (1990).
72. H. Andoh, M. Tamiya, and R. Iwasaki, EPA 507, 323 to Lion Corporation (1992).
73. O. Kirk, F. Bjorkling, and S. Godtfredsen, WO 90/09451 to Novo-Nordisk (1989).
74. M. Barfoed and J. P. Skagerlind, WO 94/12651 to Novo-Nordisk (1992).

9

Nonionics as Intermediates for Ionic Surfactants

ANSGAR BEHLER, KARLHEINZ HILL, ANDREAS KUSCH,[*] **STEFAN PODUBRIN, HANS-CHRISTIAN RATHS, and GÜNTER UPHUES** Henkel KGaA, Düsseldorf, Germany

[*] *Current affiliation*: Boehringer Ingelheim KG, Ingelheim au Rhein, Germany

I. INTRODUCTION

The surfactant properties of nonionic surfactants (see previous chapters) are modified or further improved with regard to certain applications by introduction of an ionic group. This derivatization is usually carried out by reaction at the terminal hydroxyl group of the nonionic surfactant, typically being a poly-oxyethylene or a glycerol derivative, to build up sulfates, phosphates, and carboxylates as the anionic products [Eq. (1)] [1,2]. The most important feed-stocks for these reactions are the polyoxyethylene alcohols or alkyl polyoxy-ethylene alcohols. It is estimated that the annual growth rate of polyoxyethylene alcohols as the typical nonionic surfactant is 4% [3]. The major reason is the gradual substitution of the polyoxyethylene alkylphenols, which are still used in certain regions of the world. This development, of course, affects the use of the corresponding ionic derivatives as well.

$$RO(C_2H_4O)_xH \quad \rightarrow \quad \rightarrow \quad RO(C_2H_4O)_xZNa \tag{1}$$
$$Z = SO_3, PO_3, CO_2$$

In the case of sulfated products, nowadays the polyoxyethylene alkyl sulfates are by far the most important group of nonionic derivatives. They are mainly used in dishwashing liquids and cosmetics formulations, such as shampoos and shower baths, in combination with other mild surfactants. On the other hand, the sulfated polyoxyethylene alkylphenols lost their important role of 20 years ago, mainly because of less favorable ecological and toxilogic properties compared to the sulfated products based on nonaromatic straight alkyl chains [2]. Phos-phates and carboxylates as nonionic derivatives are produced on the basis of polyoxyethylene alcohols as well. They are mainly used in technical applica-tions, such as the textile and fiber industries.

Another method of producing an anionic surfactant is the sulfation of unsatu-rated nonionic species. Under certain conditions, the sulfating agent reacts with the double bond in the alkyl chain of the starting molecule to form sulfonates as the final products. Examples are sulfonated polyoxyethylene alcohols and sulfosuccinates.

Cationic surfactants or quaternaries based on nonionics are formed by a quaternization reaction of a polyoxyethylene amine with a quaternization reagent R^3X [(Eq. 2)]; however, further modifications of this reaction are known and are discussed in the following in detail. Depending on their structure, these products cover a broad range of applications, mainly in the textile, fiber, and cosmetic industries [4].

$$R^1R^2N(C_2H_4O)_xH + R^3X \quad \rightarrow \quad R^1R^2R^3N^+(C_2H_4O)_xX^- \tag{2}$$

Derivatives from nonionic surfactants based on carbohydrates, such as sorbitane esters, alkyl polyglycosides, and fatty acid glucamides are not part of the following review.

A list of some of the major trade names and their respective manufacturers/suppliers is included in the Appendix. It must be emphasized that this list is not complete and that the reader is encouraged to consult the literature for more complete summaries.

II. SULFATES

A. Sulfated Alkyl Polyoxyethylene Alcohols (Alkyl Ether Sulfates)

1. Introduction

Through conversion of alkyl polyoxyethylene alcohols with sulfation agents, alcohol ether sulfates are obtained, presumably the most significant derivatives of the alkyl polyoxyethylene alcohols. Approximately 34% of alkyl polyoxyethylene alcohols are processed into alcohol ether sulfates. The estimated annual world production of alcohol ether sulfates (AES) in 1993 was about 750,000 t [5].

This product group has been known since the 1930s and was intensively used by the former IG Farben, which, however, mainly utilized alkylphenol or alkylnaphthol as the alcohol. The importance of these phenol-based products for applications in detergents has sharply declined due to increasing ecological awareness and is only dealt with marginally in this chapter. Emphasis is placed on products that are based on aliphatic alcohols.

2. Sulfur Trioxide as Sulfation Agent

On a large technical scale ether sulfates are mainly produced in a continuous process through mild conversation with sulfur trioxide in a multitube falling film reactor [6]. Usually sulfur trioxide (SO_3) is obtained by controlled combustion of sulfur to sulfur dioxide (SO_2) and subsequent conversion into SO_3 at the vanadium pentoxide or cesium oxide contact. An alternative that can be used in small pilot plants is the evaporation of SO_3 from oleum, where sulfuric acid results as a residue, or even from liquid SO_3. Sulfur trioxide is highly reactive, so that conversion with SO_3/air takes place in a dilution of 3–5 vol%. The air should be well dried (dew point $<-60°C$) in order to avoid the formation of sulfuric acid. The reaction with SO_3 primarily leads to a pyrosulfate:

$$RO(C_2H_4O)_xH + 2SO_3(g) \rightarrow RO(C_2H_4O)_xSO_2OSO_3H \qquad (3)$$

The primarily obtained pyrosulfate is metastable and decomposes very rapidly in the presence of further alcohol into the desired sulfuric acid half-ester:

$$RO(C_2H_4O)_xSO_2OSO_3H + RO(C_2H_4O)_xH \rightarrow 2RO(C_2H_4O)_xSO_3H \qquad (4)$$

With $\Delta H = -150$ kJ/mole the reaction enthalpy is very high. Usually in the following step the sulfuric acid half-ester $RO(C_2H_4O)_xSO_3H$ is converted with an aqueous solution of the respective base, e.g., NaOH, KOH, NH_4OH, or with an organic amine into the desired salt:

$$RO(C_2H_4O)_xSO_3H + NaOH \rightarrow RO(C_2H_4O)_xSO_3Na \qquad (5)$$

The process is continuous and the short residence time between formation of the sulfuric acid half-ester and neutralization contributes to the high conversion rate up to 98%. Since sulfur trioxide is oxidative, under unfavorable conditions and with qualitatively inadequate raw materials discolorations (yellow and brown) of the product may occur; they can be eliminated in a bleaching step (preferably with H_2O_2) after neutralization.

Ether sulfates are stable in the alkaline range but are hydrolyzed in an acid medium (autocatalytic reaction). Often phosphate or citrate buffers are added to the finished products for pH stabilization. Further additives used in technical products are EDTA for complexing of traces of iron; preservation with formaldehyde and adjustment of the viscosity through electrolyte, polyethylene glycols, or alcohols.

The increasing consumer awareness with regard to product safety and ecological compatibility leads to a continuous optimization of the product qualities, e.g., reduction of unwanted byproducts [7].

One byproduct during sulfation of alkyl polyoxyethylene alcohols is dioxane; in 1986, the known carcinogenic effect was discussed heatedly. The dioxane quantity can be reduced to <10 ppm, referring to 100% product, by means of technical measures such as conscientious process optimization and aftertreatment. These residual concentrations do not involve any health risk for the consumer [8].

Ether sulfates are usually obtained as aqueous pastes with concentrations of approximately 30 wt% or approximately 70 wt%. This is due to the viscosity behavior of the aqueous products, which is represented in Fig. 1 dependent on the concentration [9].

Processing is in principle possible below 30 wt% and in the lamellar liquid-crystal range around 70 wt%. The viscosity maximum in the hexagonal-liquid crystalline range is approximately at 40–50 wt%, so that such products are solid at room temperature. This phase behavior can be explained as follows: ether sulfates form spherical micelles at low concentrations; from >20–30 wt% they change into rod-like micelles then aggregate into compact hexagonal phases. At higher concentrations (approximately 70 wt%) a lamellar structure develops and the viscosity decreases.

3. Other Sulfating Agents

In addition to sulfur trioxide sulfuric acid (H_2SO_4), chlorosulfonic acid ($ClSO_3H$), and amidosulfonic acid (H_2NSO_3H, mainly for alkylphenol ether sulfates) can also be used on a technical scale. On a smaller scale (mainly in the laboratory) complexes of the SO_3 with bases are used; here the reactivity of the

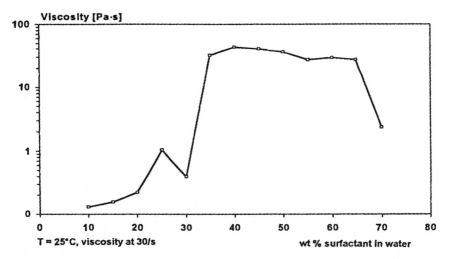

FIG. 1 Viscosities of aqueous solutions of dodecyl/tetradecyl polyoxyethylene [2] sodium sulfate.

SO_3 is reduced. For example, complexes with pyridine, trimethylamine, dimethylformamide, or dioxane can be utilized. In this context chlorosulfonic acid is probably the most important reagent. During the reaction hydrochloric acid (HCl) is produced and removed from the reaction mixture efficiently by degassing:

$$RO(C_2H_4O)_xH + ClSO_3H \rightarrow RO(C_2H_4O)_xSO_3H + HCl \tag{6}$$

The advantage of chlorosulfonic acid sulfation is that the technical apparatus needed is relatively simple. Disadvantages are the high corrosiveness of the resulting HCl gas and the content of chloride in the neutralized product, which can have an influence on the viscosity. Chlorosulfonic acid is particularly suitable for sample preparation in the laboratory. In the case of high-melting products or for conversion under very mild conditions, inert solvents such as chloroform, melthylene chloride, tetrachloroethylene, dichloroethane, and, providing they are soluble, aliphatic hydrocarbons may also be used.

4. Properties

The sulfation of polyoxyethylene alcohols results in increased hydrophilicity of these products, which corresponds to an additional six or seven oxyethylene (EO) units in the starting molecule. The products show a characteristic anionic character, i.e., their foaming behavior is better, they are highly soluble in hot water, the electrolyte content has a strong influence on the phase behavior, and the processability in powder products is improved.

The most commonly used ether sulfates are based on C_{12}/C_{14} or C_{12}/C_{18} primary native fatty alcohols or on similar synthetic alcohols (Ziegler or oxosynthesis). The properties of these products are not dependent on the raw material source (oleochemical or petrochemical) [10]. Alkyl ether sulfates in this range of carbon chain length with degrees of ethoxylation of 1–4 display an especially distinct foaming power. Unlikely many other anionic surfactants, these products are clearly insensitive to water hardness. The foam characteristics of these products can be influenced favorably by addition of alkanolamides, alkylbetaines, and alkylamidobetaines.

From the dermatologic point of view the skin compatibility of alkyl ether sulfates is better than that of the alkyl sulfates, the optimum skin compatibility being approximately 4 EO. Most technical products have EO degrees between 2 and 3 due to a compromise between foam performance, solubility, and thickening, on the one hand, and optimum skin compatibility, on the other hand.

The counterion also influences the skin compatibility; the skin compatibility of ammonium salts and salts with other amines is more favorable. Magnesium salts additionally show advantages with regard to cleaning performance and foam [11,12].

The interfacial activity of the alkyl ether sulfates can be controlled through variation of the carbon chain length and the degree of ethoxylation. The CMC (critical micelle formation concentration) decreases with an increasing EO degree and with an increase in the alkyl chain length. The solubility is improved by introduction of an additional EO group. Usually the solubility of the ether sulfates is characterized by the Krafft temperature which, according to the International Organization of Standardization (ISO), is the temperature at which the solubility of an ionic surfactant suddenly increases. At this temperature the solubility reaches the CMC. The consistency of the aqueous products can be adjusted through addition of an electrolyte. Sodium chloride, for instance, is more efficient than sodium sulfate. The temperature dependence of the viscosity is highly characteristic.

5. Fields of Application

On account of their good foaming power alkyl ether sulfates are preferentially used in foam baths, shampoos, and manual dishwashing detergents; in particular, in combination with sulfosuccinates, amphoteric surfactants, and amine oxides, they have synergistic effects with regard to skin compatibility.

Products with alkyl chain lengths of C_{12}–C_{18} and one to four EO units are emulsifiers for oils and fragrances. Apart from the excellent foaming power of these low-ethoxylated products, they are highly efficient in the reduction of surface tension; their thickening behavior and the cleaning performance is good [13].

The combination with fatty acid monoalkanolamides has a foam stabilizing effect.

Through addition of fatty acid diethanolamide and sodium chloride the viscosity of a diluted ether sulfate solution can be increased. Alternatively, fatty alkyl polyoxyethylene alcohols with narrow-range homolog distribution (NRE) can be utilized (e.g., C_{12}/C_{14} + 2.5EO NRE) [14]. The viscosity of alkyl ether sulfates can also be increased by addition of protein hydrolyzates [15].

Ether sulfates based on alkyl polyoxyethylene alcohols with narrow-range homolog distribution show a lower content of alkyl sulfate and have advantages with regard to their thickening ability [16–19]. The improved skin compatibility, which can be expected because of the lower alkyl sulfate content, has so far not been detected.

Further areas of application are light-duty and liquid detergents and foam cleaners.

Short chain products can be used as hydrotropes, in particular in alkaline cleaners. Products with branched alkyl chains are sometimes advantageous with regard to wetting and washing properties; however, their biodegradability is poor.

6. Alkylphenol Ether Sulfates

Products on the basis of nonylphenyl and dodecylphenyl polyoxyethylene (4, 6) alcohols are of particular technical importance in this context. Preparation is in principle also possible with sulfur trioxide and chlorosulfonic acid [20]; however, usually amidosulfonic acid is used in order to avoid sulfonation of the aromatic ring [Eq. (7)]:

$$C_{12}H_{25}C_6H_4O(C_2H_4O)_6H + H_2NSO_3H \rightarrow C_{12}H_{25}C_6H_4O(C_2H_4O)_6SO_3NH_4 \quad (7)$$

Sulfation with sulfuric acid also takes place without formation of ring sulfonates [21,22]. Typical areas of application are light-duty detergents and dishwashing detergents. However, they are increasingly losing importance on account of their relatively weak biodegradibility.

7. Analysis of Alkyl Ether Sulfates

The combination of oxyethylene and sulfate groups in one molecule makes the complete analysis of these products more complex than, say, the analysis of alcohol sulfates. With regard to the proportion of oxyethylated alcohol, the analysis after acid hydrolysis of the sulfate groups established for alkyl polyoxyethylene alcohols is feasible. If the hydrolysis is carried out with hydrogen iodide (HI), the polyoxyethylene chain can be degraded into diiodoethane, which is thermally unstable and splits off iodine; the iodine can be back-titrated with sodium thiosulfate and provides information about the number of EO units. This is a standard method for the determination of the EO content of nonionic surfactants [23]. Anionic active matter, unsulfated matter, sulfate, and chloride content as well as the water content are the typical specifications of alcohol ether sulfates.

Dioxane traces can be determined via GC head space. The complete analysis of this product class is described in standard methods of the ISO [24] or the Standard Methoden der Deutschen Gesellschaft für Fettwissenschaft e.V. (German Society for Fat Science DGF) [25].

The most important method is that for the determination of the anionic active matter, which will be described in detail. This method was developed by and named after Epton [26]. The Epton titration is also known as two-phase titration in which the anionic surfactant in diluted sulfuric acid solution is titrated with cetylpyridinium bromide, a cationic surfactant, in the presence of an indicator, e.g., methylene blue. The principle is based on the ion pair formation between the anion and the cationic surfactant, which is soluble in an organic medium such as chloroform. The anionic surfactant and methylene blue form a blue complex that is perceived visually as a coloring of the organic phase. In the course of titration methylene blue is displaced by the cationic surfactant and leads to a coloring of the aqueous phase. When the endpoint is reached, both the aqueous and the organic phases show the same blue color.

This standard method has been subjected to many modifications because there are a large variety of cationic surfactants, dyes, and solvents. Often hyamine is used as a cationic surfactant and a mixed system of disulfine blue and dimidium bromide [(DGF-H-III 10 (1992), standard method of the DGF)] is the indicator [27].

In particular because of solvents such as chloroform there is a strong tendency in this area to develop new methods, e.g., potentiometric titration in combination with ion-selective electrodes [28–30] or polarographic methods.

Thin-layer chromatography (TLC) and high-pressure liquid chromatography (HPLC) in combination with ion exchanger columns [31] and ion pair chromatography are also frequently used [32,33]. The determination of the unsulfated matter via simple extraction of the unsulfated proportion from an acid solution with one organic solvent usually fails in the case of ether sulfates because of the high hydrophilicity of the unsulfated alkyl polyoxyethylene alcohols.

1,4-Dioxane, resulting as a byproduct, is usually detected by means of GC head space analysis [34]. Instrumental analysis such as nuclear magnetic resonance (NMR) or infrared spectroscopy is only used in very specific cases.

8. Environmental Aspects

(a) Biodegradability. Linear alkyl ether sulfates are surfactants with good degradability, under both aerobic and anaerobic conditions [35,36]. Although the structure of ether sulfates resembles that of primary alkyl sulfates, ether sulfates are not cracked by sulfatases. The first catabolic step is ether hydrolysis [37,38]. Starting from the unsulfated end, the hydrophilic fractions are shortened successively under separation of glycolic acid molecules until finally the sulfate is released.

One distinguishes primary degradation with a loss of surfactant properties and total decomposition, i.e., final mineralization. The loss of surfactant properties can be determined via the decomposition of the methylene blue active substance (MBAS; see Epton titration); the ether sulfates in the above mentioned range, which is especially interesting for application, achieve values of >96%. For final degradation there are individual screening tests, e.g., determination of the biological oxygen demand (BOD) and also sum parameter analysis methods such as the coupled units test, where ether sulfates reach >60% for final degradation, usually even >70%. Typically, branching in the alkyl chain leads to slower degradability rates compared to the linear molecules.

(b) *Toxicity.* With LD_{50}- values of 4000–10,000 mg/kg the toxicity for mammals is low. The toxicity declines with an increasing EO degree.

The aquatic toxicity also decreases with an increasing EO degree, with LC_{50} = 1.4–20 mg/L for fish and EC_{50} = 1–50 mg/L for daphniae [39].

B. Sulfated Fatty Acid Monoglycerides

In view of a growing environmental awareness of the consumer, the development of surfactants on the basis of renewable raw materials is constantly gaining in importance. Monoglyceride sulfates (Fig. 2) are anionic surfactants that, in terms of their organic backbone, can be obtained exclusively on the basis of oleochemical raw materials. A large variety of processes are available for the preparation of monoglyceride sulfates. As early as 1935 Colgate Palmolive Peet Co. patented a manufacturing process for monoglyceride sulfate [40]. In the same year, Colgate launched a soap-free shampoo formulation under the brand name "Halo" making use of monoglyceride sulfates as a surfactant component. Accordingly, monoglyceride sulfates are among the first synthetic surfactants used in cosmetics. In 1954 Colgate-Palmolive claimed a continuous process for the preparation of monoglyceride sulfates [41]. In this case, in a first reaction step glycerol is converted with oleum to the corresponding acid glycerol sulfuric acid half-ester. This intermediate product is then transesterified with a triglyceride, usually coconut oil, thus leading to monoglyceride

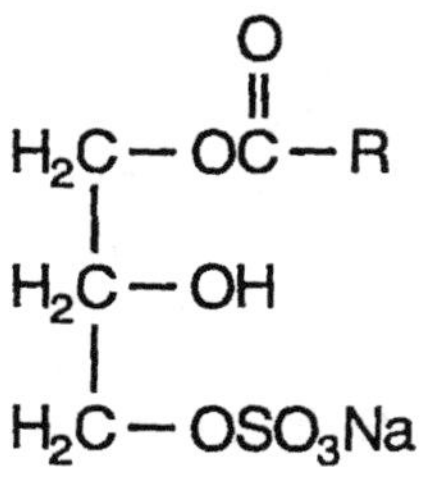

$$
\begin{array}{c}
\quad\quad\quad O \\
\quad\quad\quad \| \\
H_2C - OC - R \\
\quad | \\
H_2C - OH \\
\quad | \\
H_2C - OSO_3Na
\end{array}
$$

FIG. 2 Molecular formula of monoglyceride sulfate sodium salt (R = alkyl).

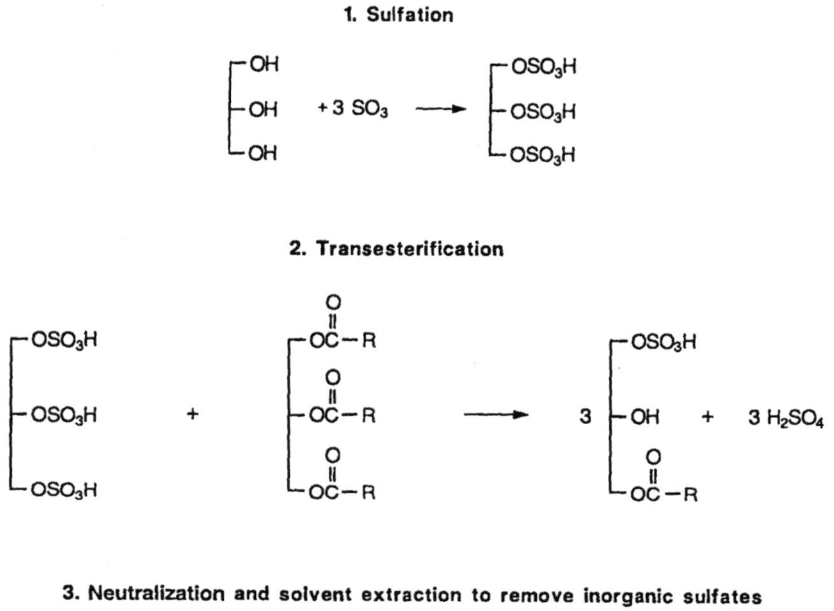

1. Sulfation

$$
\begin{array}{l}
\text{—OH} \\
\text{—OH} \\
\text{—OH}
\end{array}
\quad + 3\ SO_3 \quad \longrightarrow \quad
\begin{array}{l}
\text{—OSO}_3\text{H} \\
\text{—OSO}_3\text{H} \\
\text{—OSO}_3\text{H}
\end{array}
$$

2. Transesterification

$$
\begin{array}{l}
\text{—OSO}_3\text{H} \\
\\
\text{—OSO}_3\text{H} \\
\\
\text{—OSO}_3\text{H}
\end{array}
\quad + \quad
\begin{array}{l}
\overset{O}{\overset{\|}{\text{—OC}}}\text{—R} \\
\overset{O}{\overset{\|}{\text{—OC}}}\text{—R} \\
\overset{O}{\overset{\|}{\text{—OC}}}\text{—R}
\end{array}
\quad \longrightarrow \quad 3
\begin{array}{l}
\text{—OSO}_3\text{H} \\
\\
\text{—OH} \\
\overset{O}{\overset{\|}{\text{—OC}}}\text{—R}
\end{array}
\quad + \quad 3\ H_2SO_4
$$

3. Neutralization and solvent extraction to remove inorganic sulfates

FIG. 3 Reaction pathway to fatty acid monoglyceride sulfates.

sulfate (the mechanism of this conversion is described below). The resulting products were, for example, applied in the household cleaner "Vel," which was marketed by Colgate in the 1950s and 1960s. The industrial production process is a multistep process that uses 20% oleum as a sulfation agent. In the first reaction step, glycerol is converted with oleum in such a way that all three hydroxyl groups of the glycerol are sulfated, thus forming a glycerol trisulfuric acid half-ester (Fig. 3). In a second reaction step this glycerol trisulfuric acid half-ester is converted with a triglyceride, usually hardened coconut oil (molar ratio 2:1).

Similar to a transesterification reaction, sulfuric acid half-ester functions are now exchanged for a fatty acid residue, so that from two molecules of glycerol trisulfuric acid half-ester and one triglyceride molecule three molecules of monoglyceride-disulfuric acid half-esters are obtained. According to Colgate the sulfuric acid half-ester function in the β-position with the fatty acid ester function is unstable. Therefore, after neutralization, e.g., with ammonia or caustic soda solution, a 1,3-fatty acid monoglyceride sulfate is obtained in a highly selective process. Due to the high oleum excess during sulfation of the glycerol, large quantities of sodium sulfate are formed in the neutralization step. The unwanted salts are removed from the monoglyceride sulfate by means of extraction process as follows: By adding alcohol with a low boiling point, e.g., ethanol, to the aqueous, neutralized surfactant solution, two liquid phases are formed—a heavier aqueous phase that is saturated with sodium sulfate and an alcohol phase that contains the monoglyceride sulfate. The salt-containing aqueous phase is separated and the surfactant is obtained by evaporation of the ethanol. The resulting product yields the desired monoglyceride sulfate in a purity of approximately 80%; byproducts are partial glycerides and fatty acid.

Patent literature describes further manufacturing processes for monoglyceride sulfates: Monsanto claims a process for the sulfation of glycerol monolaurate with sulfur trioxide (SO_3) in liquid sulfur dioxide (SO_2) [42]. Colgate also describes a further development of the original manufacturing process; according to the new process, glycerol is initially sulfated with 2 moles SO_3 and subsequently converted with oleum to obtain the glycerol trisulfuric acid half-ester. This mixture is then transesterified with coconut oil and neutralized [43]. The literature [44] provides an overview of the state of the art for the preparation of monoglyceride sulfates until the beginning of the 1970s. In more recent times, attempts were also made to further develop the process originally invented by Colgate. As an example, Wako Pure Chemical Ind. in Japan claimed a process in 1975 for the preparation of pure monoglyceride sulfates via oleum sulfation with monoglyceride and subsequent extraction with isobutanol [45]. In 1978 Riken Vitamin Oil claimed the conversion of glycerol with chlorosulfonic acid and subsequent transesterification with triglycerides [46]. In EP 0267518 Hoechst describes a process for the sulfation of monoglycerides with SO_3 in triethylamine as a solvent [47]. In 1988 Colgate described a new process for the preparation of monoglyceride sulfates via the sulfation of glycerol with chlorosulfonic acid in chloroform as a solvent and subsequent transesterification with fatty acid or fatty acid ester [48]. In the early 1990s, Henkel developed a continuous process for the preparation of monoglyceride sulfates [49]. In this process, technical grade monoglycerides are converted into the corresponding monoglyceride sulfates with gaseous SO_3 in a continuous falling film reactor.

Monoglyceride sulfates, in particular those based on coconut oil, are soluble, high-foaming anionic surfactants. They are distinguished by an excellent skin compatibility, which is comparable to mild anionic surfactants such as sulfosuccinate or ether sulfate [50]. On account of these properties, coconut monoglyceride sulfate was used in a hair shampoo as early as 1935, as mentioned above. In the 1950s and 1960s, the coconut-based monoglyceride sulfates, which Colgate produced on an industrial scale for sale under the brand names Arctic Syntex L and M, and Monad G, were applied in many household products, e.g., a household cleaner with abrasive additives [51] or in the household cleaner Vel. In the United States monoglyceride sulfates are still used as mild surfactants in syndet soaps [52]. In addition to lauryl sulfate, monoglyceride sulfates are described as surfactants for toothpastes/dental care products in combination with specific active substances [53–58]. The excellent skin compatibility of monoglyceride sulfates predestines these products for application in personal care products. A large variety of combinations of monoglyceride sulfates with other mild surfactants has been described for this field of application; e.g., combination of monoglyceride sulfate with phosphoric acid esters [59], with succinic acid [60], with an aminophosphate surfactant [61] for skin-cleansing agents and with combinations with amino acids and amphoteric surfactants for hair shampoos [62].

Detergent mixtures of alkyl polyglycosides with monoglyceride sulfates show synergistic effects with regard to the washing, rinsing, foaming, and cleaning power as well as its skin compatibility [63]. In this combination, high-performance and especially mild shaving preparations [64] or toothpastes [65] can be obtained.

C. Sulfated Alkanolamides

1. Preparation

The synthesis of sulfated alkanolamides has been reviewed in a previous volume of this series [66] comprising literature up to the early 1970s. Concerning the basic preparation steps for sulfated alkanolamides, the reference is still up to date [67], so that recent developments focus on the fields of application and new starting materials.

To produce amide ether sulfates, alkanolamides may be sulfated directly or first oxyalkylated and then sulfated, yielding amide sulfates (1) or amide poly-oxyethylene sulfates (2) after subsequent neutralization with a base as shown in Fig. 4.

The most common alkanolamine basis for sulfated alkanolamides is defi-nitely monoethanolamine [68], although N-alkyl-substituted as well as branched alkanolamines such as isopropanolamine have also been used [69]. Apart from

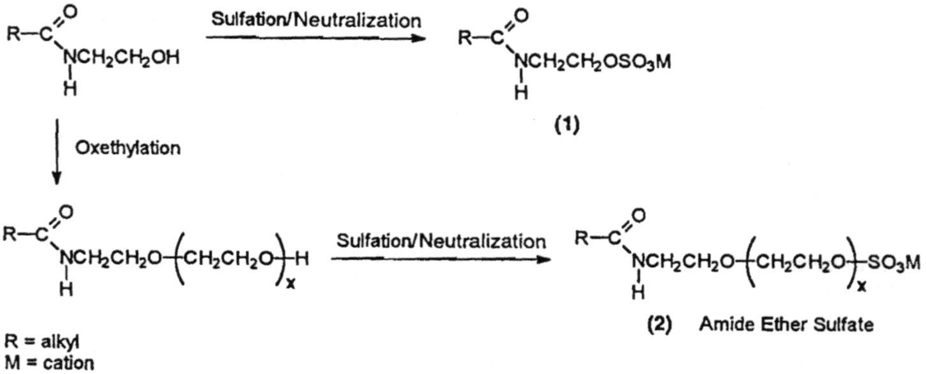

R = alkyl
M = cation

FIG. 4 Reaction paths to amide sulfates and amide polyoxyethylene sulfate.

these monoalkanolamides, polyhydroxyalkanolamides such as diethanolamides or 2,3-hydroxypropylamides have been prepared too [70].

The corresponding alkanolamides are derived from (saturated or unsaturated) C_2- to C_{22}-carboxylic acids or hydroxycarboxylic acids [71], mainly from coconut- or tallow-based feedstocks. A main drawback of the sulfation process of alkanolamides is the high viscosity of the sulfation mixture, which may be overcome by means of a cosulfation with lower molecular weight alcohols [72], alkanolamines [73], fatty alcohols [74], or oxyethylated fatty alcohols [75]. Due to a lower sulfation temperature the products obtained by this route have an improved color. The choice of cations comprises ammonium (including alkyl- and alkanolammonium), alkali, and earth alkali metals.

2. Properties and Application

Sulfated alkanolamides are excellent foaming surfactants with good detergency [76]. Their hydrolytic stability as well as physicochemical data have been compiled elsewhere [66]. Sulfated alkanolamides are used almost exclusively as cosurfactants together with anionic, nonionic, and sometimes cationic components.

Cosmetics (body, hair, and baby care) is the main field of application according to numerous patents (during the last 5 years predominantly Japanese companies have dealt with this issue) because of the low skin irritancy of alkanolamide sulfates [77a], which had already been noticed during the late 1960s [77b]. The same reason applies for an increasing use of alkanolamide sulfates in manual dishwashing formulations [78].

Alkanolamide sulfates are good lime soap dispersants and have thus been used in detergent compositions suitable for hard water applications [79].

Technical applications concerning emulsion polymerization of ethylenically unsaturated monomers [80] or leather preparation [81] relate to the favorable emulsifying properties of sulfated alkanolamides. Sulfated alkanolamides have also been used as mold release [82] or antiadhesive reagents for rubber [83]. Together with cationic surfactants alkanolamide sulfates may serve as dehydration promotion for the production of granular slag [84].

D. Miscellaneous

1. Hydroxyalkyl Polyoxyethylene Sulfates

Alkyl polyoxyethylene-2-hydroxyalkyl ethers (2), which themselves present an interesting class of low-foaming nonionics, are raw materials for the corresponding alkyl polyoxyethylene-2-hydroxyalkyl ether sulfates (hydroxyether sulfates) (3) (Fig. 5). These products are available by acid- or base-catalyzed ring opening of an α-epoxide (1) with an oxyethylated alcohol. The resulting secondary alcohol (2) can subsequently be sulfated in the usual manner to ether sulfate (3). These products show a good biodegradability and can be adjusted

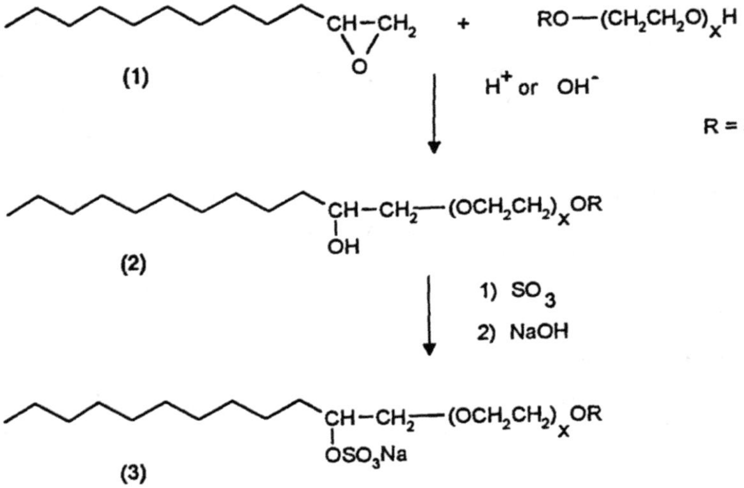

FIG. 5 Reaction scheme for the synthesis of alkyl polyoxyethylene-2-hydroxyalkyl ether sulfates.

concerning their application profile by variation of the starting α-epoxide or the polyoxyethylene alcohol in a broad range. No commercial application is known so far [85–87].

2. Glycerol Ether Sulfates

Glycerol, which is mainly obtained as a "coupled" product in the conversion of fats and oils into fatty acid methyl esters, can be converted to glycerol ethers by

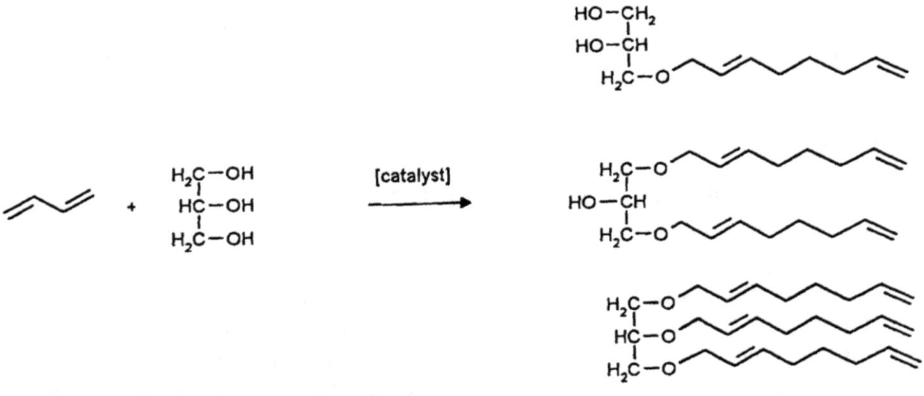

FIG. 6 Telomerization reaction of glycerol with butadiene.

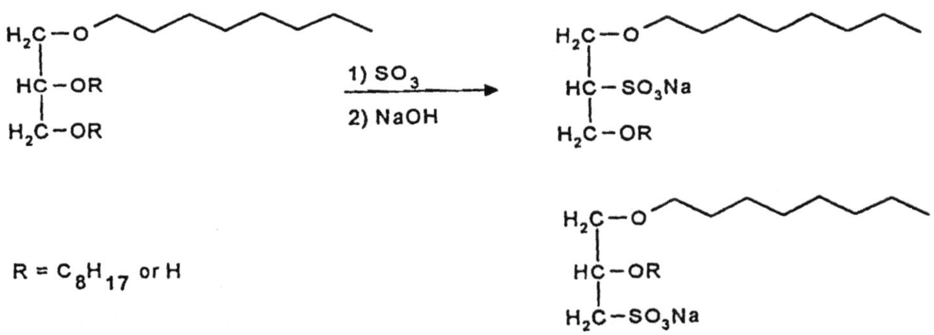

FIG. 7 Sulfation of glycerol octyl ether.

telomerization with butadiene in the presence of a palladium/triphenylphosphine complex as catalyst [88], yielding a mixture of glycerol mono-, di-, and trioctadienyl ethers (Fig. 6). The saturated ethers are available by hydrogenating the resulting mixture or the purified mono- and diethers in presence of palladium-supported charcoal (Pd/C) as catalyst [89].

The glycerol monoether is structurally similar to a monoglyceride, which is an interesting nonionic emulsifiers and also a raw material for sulfation (Sec II.B). The glycerol mono- and dioctyl can be converted to the corresponding glycerol octyl ether sulfates by sulfation with SO_3/air (Fig. 7).

An alternative route to prepare glycerol ether sulfates starts with the reaction of fatty alcohol or fatty alcohol ethoxylate with epichlorohydrin, followed by sulfation [90]. Glycerol ether sulfates are described as high foaming and good biodegradable surface active materials. They are not yet produced on a commercial scale.

III. SULFONATES

A. Sulfosuccinates

Sulfosuccinates (Sulfosuccinic acid esters) are anionic surfactants that are accessible on the basis of maleic anhydride. One distinguishes mono- and dialkyl esters of the sulfosuccinic acid (Fig. 8). Both mono- and diesters are

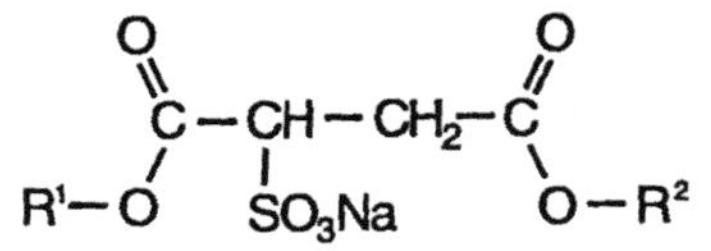

FIG. 8 Structure of sulfosuccinic acid esters. R^1, R^2 = H, alkyl, POE-alkyl.

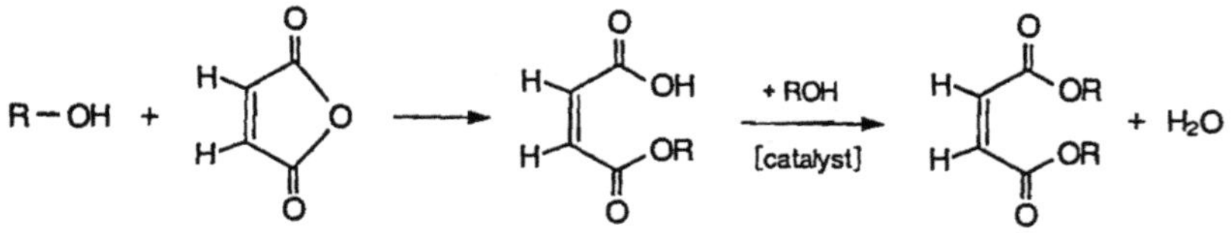

FIG. 9 Reaction scheme for the synthesis of maleic acid mono- and dialkyl esters. R = alkyl, POE-alkyl.

obtained in a two-step process. In the first reaction step, maleic acid anhydride is esterified with compounds containing hydroxyl groups to the mono- or diester (Fig. 9). While diesters are mainly produced with alcohols, many different raw materials with hydroxyl groups are used in the case of monoesters. Fatty alcohols, fatty acid alkanolamides and its oxethylates are most commonly used [91]. Usual esterification catalysts such as p-toluenesulfonic acid are suitable as catalysts for diester production.

In the second reaction step, the maleic acid ester is sulfated with an aqueous sodium sulfite solution to obtain the corresponding sulfosuccinate (Fig. 10). In the case of the sulfosuccinic acid monoester, two regioisomeric sulfosuccinates are possible (Fig. 11). It was detected by ^{1}H NMR analysis that the β position is preferred during sulfation. The ratio β/α is approximately 4:1 [92].

Sulfosuccinates are used in many different fields of application. Comprehensive overviews are given in [91, 93–95]. Sulfosuccinic acid dialkyl esters are weakly foaming surfactants with good wetting power. In particular, products on the basis of octanol or 2-ethyl hexanol are distinguished by their outstanding wetting properties. Even at low concentration they can cause a considerable reduction in the surface tension of aqueous solutions [94]. Sulfosuccinic acid dialkyl esters on the basis of alcohols with fewer than nine carbon atoms are water-soluble. Branched alkyl groups increase the solubility [96]. Because of their good wetting properties, sulfosuccinic acid dialkyl esters are applied as "rapid wetting agents" in the textile industries [97]. In fiber technology these products are used in spinning oils for nylon production. Furthermore, they are used in agriculture for pesticides as well as in paint formulations and in the leather industry. With regard to household products the application of sulfosuc-

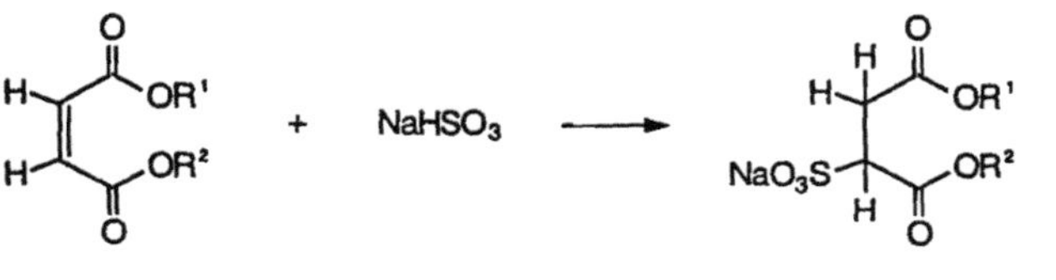

FIG. 10 Reaction scheme to sulfosuccinates. R^1, R^2 = H, alkyl, POE-alkyl.

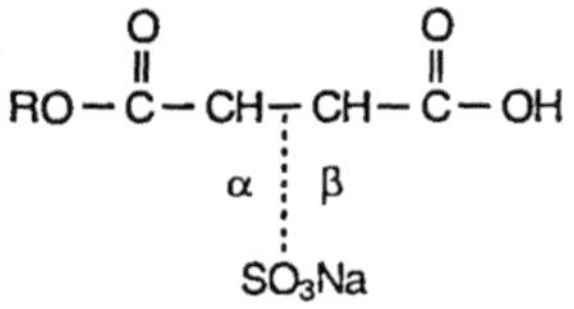

$$RO - \overset{\overset{\displaystyle O}{\|}}{C} - CH \underset{\alpha}{-} CH - \overset{\overset{\displaystyle O}{\|}}{C} - OH$$

$$\underset{\substack{| \\ SO_3Na}}{\overset{\alpha \quad \beta}{}}$$

FIG. 11 Regiomeric isomers of sulfosuccinic acid monoalkyl ester. R = alkyl.

cinic acid dialkyl esters is restricted to specific glass cleaners, e.g., for spectacle lenses or windscreens as well as carpet shampoos.

In contrast to sulfosuccinic acid dialkyl esters, sulfosuccinic acid monoalkyl esters are good-foaming surfactants. Especially products on the basis of oxethylated fatty alcohols, e.g., lauryl/myristyl polyoxyethylene (3) alcohol, exhibit an outstanding skin compatibility [98]. Due to their mildness to skin, large quantities of sulfosuccinic acid monoalkyl esters are used in personal care products such as shower gels, shampoos, and skin-cleaning agents. In particular, they are utilized in mild products such as baby shampoos or shampoos for sensitive skin. Their compatibility is very good, not only with regard to sensitive skin but even on diseased skin [99]. In combinations with anionic surfactants, e.g., lauryl sulfate or lauryl ether sulfate, sulfosuccinic acid monoalkyl esters reduce the skin-irritating effect of these anionic surfactants while maintaining the foaming power [95]. Sulfosuccinic acid monoalkyl esters are very soluble in water and have a good hard water resistance with a low tendency to form calcium soaps. They exhibit a high detergency that is synergistically enhanced in combinations with other surfactants [100]. On account of the hydrolysis-sensitive ester bond their application is limited to a pH range of 6–8. In the industrial sector sulfosuccinic acid monoalkyl esters are, for example, used as emulsifiers for emulsion polymerization. Both the mono-and the dialkyl esters are readily biodegradable and show low toxicity [91].

B. Sulfonated Polyoxyethylene Alkenols

1. Oleyl Polyoxyethylene Alcohol Sulfonates

In the sulfation of unsaturated polyoxyethylated fatty alcohol, reaction occurs mainly at the terminal free hydroxyl group under formation of the corresponding ether sulfate. Internal sulfonates, which form if the reaction takes place on the C–C double bond, are only obtained subordinately. This compound class of internal ether sulfonates is accessible by introducing a protective group.

The corresponding oxethylated fatty alcohol acetic acid esters (1) are formed by esterification of oxyethylated fatty alcohols with acetic acid and/or acetic anhydride as shown in Fig. 12. If oxyethylated unsaturated alcohols, such as oleyl alcohol, are used, the internal C–C double bond can be sulfonated with

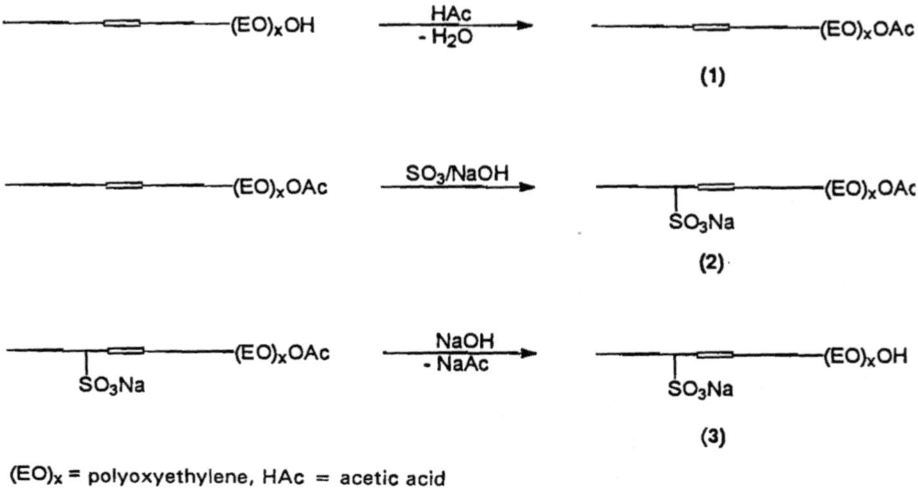

————————————————(EO)ₓOH $\xrightarrow[-\text{H}_2\text{O}]{\text{HAc}}$ ————————————————(EO)ₓOAc

$$(1)$$

————————————————(EO)ₓOAc $\xrightarrow{\text{SO}_3/\text{NaOH}}$ ————————————————(EO)ₓOAc
$\qquad\qquad\qquad\qquad\qquad\qquad\qquad\qquad\qquad\quad$ SO₃Na

$$(2)$$

————————————————(EO)ₓOAc $\xrightarrow[-\text{NaAc}]{\text{NaOH}}$ ————————————————(EO)ₓOH
SO₃Na $\qquad\qquad\qquad\qquad\qquad\qquad\qquad\qquad\qquad\quad$ SO₃Na

$$(3)$$

(EO)ₓ = polyoxyethylene, HAc = acetic acid

FIG. 12 Reaction scheme to oleyl ether sulfonates.

gaseous SO_3. This leads to oxyethylated oleyl alcohol acetates with an internal sulfonate group (**2**). The required internal oleyl ether sulfonates (**3**) are obtained by alkaline saponification.

These products combine the features of anionic and nonionic surfactants. In particular, at higher degrees of oxethylation, the products have a good foaming

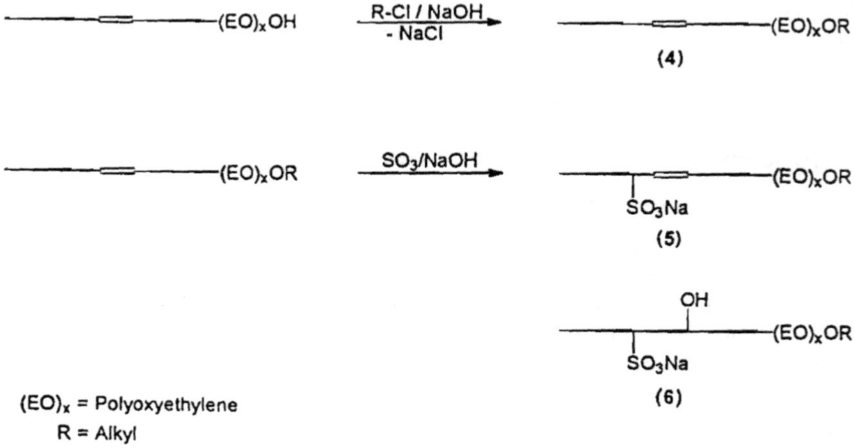

————————————————(EO)ₓOH $\xrightarrow[-\text{NaCl}]{\text{R-Cl / NaOH}}$ ————————————————(EO)ₓOR

$$(4)$$

————————————————(EO)ₓOR $\xrightarrow{\text{SO}_3/\text{NaOH}}$ ————————————————(EO)ₓOR
$\qquad\qquad\qquad\qquad\qquad\qquad\qquad\qquad\qquad\quad$ SO₃Na

$$(5)$$

$\qquad\qquad\qquad\qquad\qquad\qquad\qquad\qquad\qquad\quad$ OH
$\qquad\qquad\qquad\qquad\qquad\qquad$ ————————————————(EO)ₓOR
$\qquad\qquad\qquad\qquad\qquad\qquad\qquad\qquad\qquad\quad$ SO₃Na

(EO)ₓ = Polyoxyethylene
$\qquad$ R = Alkyl $\qquad\qquad\qquad\qquad\qquad\qquad\qquad\qquad$ $$(6)$$

FIG. 13 Reaction scheme for the sulfation of alkenyl polyoxyethylene alkyl ethers.

power and detergency. They are readily biodegradable and show a good skin compatibility. However, no commercial use is known so far [101].

2. Alkenyl Polyoxyethylene Ether Sulfonates

Alkyl polyoxyethylene alkyl ethers (4) are accessible by conversion of alkyl polyoxyethylene alcohol with alkyl halides according to Williamson (Fig. 13) [102]. If unsaturated polyoxyethylene alcohols on the basis of oleyl alcohol are used, the products can be sulfonated with gaseous SO_3. In this case, products with an internal sulfonate group result, i.e., alkene sulfonates (5) and hydroxy-alkanesulfonates (6). Through variation of the alkyl group and the degree of ethoxylation, products with different property profile are accessible [102, 103].

IV. PHOSPHATES

A. Structures and Preparation Methods

Phosphates deriving from nonionic surfactants have long been known. First references trace back to Schöller and Wittwer [104] as well as Steindorff et al. [105]. The classical methods for preparing phosphoric acid esters, i.e., by using phosphorus pentoxide [106], polyphosphoric acid [107], or phosphorus oxychloride, were applied to the phosphatation of polyoxalkylated alcohols. Contrary to the half-sulfuric esters the nonionic-based phosphates are more complex product mixtures, composed mainly of mono- and diesters and triesters as minor components (Fig. 14). In the figure, R denotes residues of alcohols, alkylphenols, and others. In addition, the phosphatation of oxyethylated fatty acids is also known [108]. It was Nüsslein [109] who reported early on the wide range of

monoester
$$R(OC_2H_4)_x - O - \overset{\overset{\displaystyle O}{\|}}{\underset{\underset{\displaystyle OH}{|}}{P}} - OH$$

diester
$$R(OC_2H_4)_x - O - \overset{\overset{\displaystyle O}{\|}}{\underset{\underset{\displaystyle OH}{|}}{P}} - O - (C_2H_4O)_xR$$

triester
$$R(OC_2H_4)_x - O - \overset{\overset{\displaystyle O}{\|}}{\underset{\underset{\displaystyle O - (C_2H_4O)_xR}{|}}{P}} - O - (C_2H_4O)_xR$$

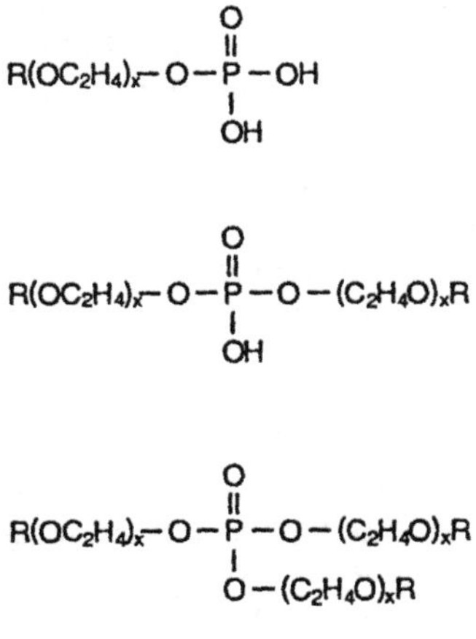

FIG. 14 Structures of phosphoric acid esters.

TABLE 1 Various Examples of Phosphate Ester Compounds [109]

$O=P$ —fatty alcohol / fatty alcohol \ fatty alcohol	Primary, secondary, or tertiary esters; same alcohol or mixed-chain-length alcohols
$O=P$ —glycol, polyglycols, etc. / glycol, polyglycols, etc. — $P=O$ \ glycol, polyglycols, etc.	Esters with polyvalent alcohols
$O=P$ —glycol-fatty acid / glycol-fatty acid \ glycol-fatty acid	made by forming the fatty acid ester last
glycol-fatty acid / $O=P$ — glycol — $P=O$ (OH, OH) \ glycol-fatty acid	Polyesterification

variations for phosphoric acid esters using the different synthesis methods. They are summarized in Table 1.

The large variety of synthesis procedures have been expressed by a huge number of publications and patents. Accordingly, only the fundamental reaction types using the above-mentioned phosphorylating agents are referred to. For technical and economic reasons phosphorus oxychloride is used in nearly all cases for the preparation of triesters [Eq. 8]. In order to prevent the formation of alkyl chlorides [110], the generated hydrochloric gas must be removed from the reaction mixture immediately. Tertiary amines are used preferentially as absorbents [111]; amine salts formed must be removed by filtration. Phosphorus oxychloride ($POCl_3$) has also been suggested for the preparation of monoesters [112,113]. The reaction only succeeds in the presence of large quantities of nonpolar solvents at low temperatures. The production of diesters is described by a patent of Kao Corporation [114], according to which the alcohol component can be reacted with $POCl_3$ using substoichiometric amounts. To destroy the mono- and triesters the resulted mixture is treated with alkali hydroxide. The diesters remain stable.

$$3\,R(OC_2H_4)_xOH \ + \ POCl_3 \longrightarrow \left(R(OC_2H_4)_xO\right)_3P=O \ + \ 3\,HCl \tag{8}$$

In industrial and laboratory practice, phosphatation by means of phosphorus pentoxide is the most important route. The highly purified reagent exists as a cyclic structure having the sum formula P_4O_{10}. The phosphatation reaction is very complex and may occur according to Eq. (9). As suggested by König and Langley [115], pure phosphorus pentoxide reacts spontaneously with 3 moles of alcohol by forming the triester of a tetrameric phosphoric acid in the first step. In general, it is recommended to work at a temperature of 50°C and with efficient agitation because of the strong exothermic reaction. The remaining P–O–P anhydride bonds are then split off successively by reacting with an additional 3 moles of alcohol in subsequent reaction steps.

$$P_4O_{10} + 3\ RO(C_2H_4O)_xH \longrightarrow$$

$$\underset{\text{(I)}}{HO-\overset{\overset{O}{\|}}{\underset{\underset{OH}{|}}{P}}-O-\overset{\overset{O}{\|}}{\underset{\underset{OR'}{|}}{P}}-O-\overset{\overset{O}{\|}}{\underset{\underset{OR'}{|}}{P}}-O-\overset{\overset{O}{\|}}{\underset{\underset{OR'}{|}}{P}}-OH}$$

$$(I) + RO(C_2H_4O)_xH \longrightarrow HO-\overset{\overset{O}{\|}}{\underset{\underset{OH}{|}}{P}}-OR' + \underset{\text{(II)}}{HO-\overset{\overset{O}{\|}}{\underset{\underset{OR'}{|}}{P}}-O-\overset{\overset{O}{\|}}{\underset{\underset{OR'}{|}}{P}}-O-\overset{\overset{O}{\|}}{\underset{\underset{OR'}{|}}{P}}-OH}$$

$$(II) + RO(C_2H_4O)_xH \longrightarrow HO-\overset{\overset{O}{\|}}{\underset{\underset{OH}{|}}{P}}-OR' + \underset{\text{(III)}}{HO-\overset{\overset{O}{\|}}{\underset{\underset{OR'}{|}}{P}}-O-\overset{\overset{O}{\|}}{\underset{\underset{OR'}{|}}{P}}-OH} \qquad (9)$$

$$(III) + RO(C_2H_4O)_xH \longrightarrow HO-\overset{\overset{O}{\|}}{\underset{\underset{OH}{|}}{P}}-OR' + HO-\overset{\overset{O}{\|}}{\underset{\underset{OR'}{|}}{P}}-OH$$

$$P_4O_{10} + 6\ RO(C_2H_4O)_xH \longrightarrow 2\ HO-\overset{\overset{O}{\|}}{\underset{\underset{OR'}{|}}{P}}-OR' + 2\ HO-\overset{\overset{O}{\|}}{\underset{\underset{OR'}{|}}{P}}-OH$$

$$R' = -(C_2H_4O)_xR$$

In practice it is not possible to obtain a molar ratio of 1:1 between mono- and diester as it is impossible to get 100% pure phosphorus pentoxide and to charge it unchanged into the reaction vessel. Traces of water in the alcohol and the air cause an undesirable splitting of the ring system into tetrameric phosphoric acid, so that the content of monoesters is larger than the quantity of diesters. Additionally, technical phosphorus pentoxide mostly contains polymers [116], which also influences the molar ester ratio. In case the reaction has not been carried out under optimum conditions pyrophosphate esters may be formed as byproducts, which may change the properties of the desired product.

Furthermore, phosphoric acid esters can be synthesized by reacting nonionic surfactants with polyphosphoric acid [108,117,118]. The concentration of the commercial quality is typically 84 % on the basis of phosphorus pentoxide. According to Clark and Lions [119], polyphosphoric acid has the composition shown in Table 2. Polyphosphoric acid can also be made in situ by mixing o-phosphoric acid, water, and phosphorus pentoxide. The composition then depends on the mixing ratio of the single components [120,121]. The

TABLE 2 Composition of a Technical Grade Polyphosphoric Acid

n	1	2	3	4	5	6	7
%	5	17	16.5	14.2	12	8.7	2.2

$$HO \left[\begin{matrix} O \\ \| \\ P-O \\ | \\ OH \end{matrix} \right]_n H$$

alcoholysis of the P–O–P anhydride bonds and formation of the final product occur in subsequent steps as shown in Eq. (10).

$$R' = (C_2H_4O)_x R$$

Tetrameric phosphoric acid is reacted with a maximum of 3 moles of alcohol. During this procedure only monoesters are formed along with free phosphoric acid. It should be pointed out that the alcoholysis reaction is relatively slow and requires elevated temperatures of 80–100°C. In case of the direct esterification of o-phosphoric acid, olefins may be produced [122]; consequently, its use as a phosphorylating agent was not considered for a long time. By adding small quantities of basic substances, especially tertiary amines, the formation of olefins can be prevented virtually completely [123]. According to the procedure of azeotropic dehydration at >160°C mixtures of mono-, di-, and triesters are obtained. The final composition is determined by the molar ratio of the educts.

In addition to the aforementioned possibilities of synthesizing phosphoric acid esters of oxyalkylated alcohols, the literature offers certain more specific procedures. Two interesting routes are described in the following paragraphs.

As mentioned before, a successful production of triester can only be guaranteed by the phosphatation with phosphorus oxychloride. For a series of special applications it has been suggested that the ester mixtures, resulting from the reaction with phosphorus pentoxide, be reacted with oxyalkylenes (EO;PO) in order to obtain triesters [Eq. (11)].

$$
\underset{\substack{|\\ OR}}{RO - \overset{\overset{\displaystyle O}{\|}}{P} - OH} \; + \; H_2\overset{\displaystyle O}{\overset{\displaystyle \diagup\!\!\!\diagdown}{C} - CH_2} \longrightarrow \underset{\substack{|\\ OR}}{RO - \overset{\overset{\displaystyle O}{\|}}{P} - O - CH_2 - CH_2 - OH} \qquad (11)
$$

Another very interesting synthesis route is the preparation of highly concentrated diesters [124], starting with triesters and phosphorus pentoxide [Eq. (12)]. Reaction conditions are given as an example: Within 1 h 0.5 mole of P_4O_{10} is added at 80°C to 2 moles of triphosphate by stirring for further 2 h. Two moles of alcohol is then added and kept at a temperature of 100°C for 5 h. The tetraalkyl diphosphate having been prepared in addition to the 2 moles of diester will also be hydrolyzed to diester with 1 mole of water at 85°C within 3 h. The final product should contain more than 85 mol% of the diester. Additionally, this route makes it possible to synthesize mixed esters.

$$
2\;\underset{\substack{|\\ OR}}{RO - \overset{\overset{\displaystyle O}{\|}}{P} - OR} \; + \; 0.5\;P_4O_{10} \longrightarrow \underset{\substack{|\\ OR}}{RO - \overset{\overset{\displaystyle O}{\|}}{P} - O -}\underset{\substack{|\\ OR}}{\overset{\overset{\displaystyle O}{\|}}{P} - O -}\underset{\substack{|\\ OR}}{\overset{\overset{\displaystyle O}{\|}}{P} - O -}\underset{\substack{|\\ OR}}{\overset{\overset{\displaystyle O}{\|}}{P} - OR}
$$

(I)

$$
(I) \; + \; 2\;ROH \longrightarrow 2\;\underset{\substack{|\\ OR}}{RO - \overset{\overset{\displaystyle O}{\|}}{P} - OH} \; + \; \underset{\substack{|\\ OR}}{RO - \overset{\overset{\displaystyle O}{\|}}{P} - O -}\underset{\substack{|\\ OR}}{\overset{\overset{\displaystyle O}{\|}}{P} - OR} \qquad (12)
$$

(II)

$$
(II) \; + \; H_2O \longrightarrow 2\;\underset{\substack{|\\ OR}}{RO - \overset{\overset{\displaystyle O}{\|}}{P} - OH}
$$

B. Analytical Methods

Finding reliable methods regarding the quantitative detection and characterization of all potential components, above all alkyl polyoxyalkylene phosphates, was problematic in the past. A simple potentiometrical titration procedure has been known for some time. Its use, however, is limited to specific applications, especially when quantitative results are needed [125]. A solution of the sample in ethanol in the presence of small amounts of sodium chloride and water is titrated with 0.1 or 0.5 M sodium hydroxide solution. As soon as a second inflection point appears, 10 mL of a 10% aqueous calcium chloride solution is added. Titration is continued until a third inflection point is indicated as shown in Fig. 15. The first inflection point comprises one proton of the monoester and

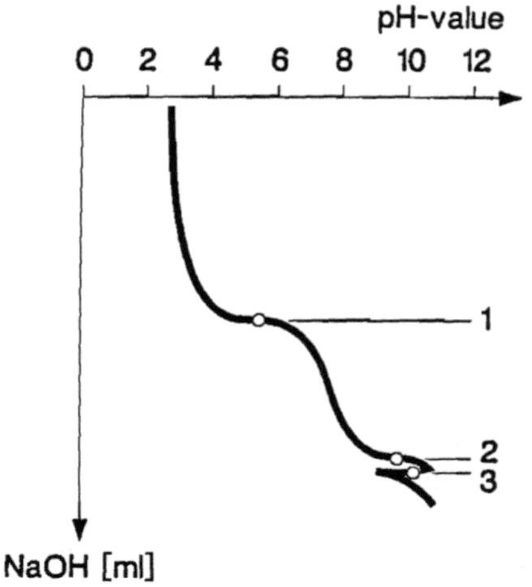

pH-value

$$\% \text{ monoester} = \frac{(2\ V_2 - V_1 - V_3)\ m\ M}{w\ 10}$$

$$\% \text{ diester} = \frac{(2\ V_1 - V_2)\ m\ M}{w\ 10}$$

$$\% \text{ H}_3\text{PO}_4 = \frac{(V_3 - V_2)\ m\ M}{w\ 10}$$

V_1 = ml NaOH (inflection point 1)
V_2 = ml NaOH (inflection point 2)
V_3 = ml NaOH (inflection point 3)
m = molarity of the sodium hydroxide solution
M = molecular weight of sample
w = weight of the sample

FIG. 15 Titration of phosphate esters.

the free phosphoric acid as well as the proton of the diester. The second inflection point indicates the second proton whereas the third inflection point stands for the third proton of the free phosphoric acid. The composition of the ester mixture is calculated according to the equations indicated in Fig. 15. For ester mixtures not containing any pyrophosphates the results are extremely accurate. In the presence of P–O–P anhydride bonds the method can provide qualitative data, i.e., more than three inflection points or a molar mono-/diester quotient below 1 indicates the presence of pyrophosphates. Quantitative data in this case are obtained by means of ^{31}P NMR analysis [126].

C. Properties and Applications

Regarding their properties phosphoric acid esters based on nonionics differ distinctly from the classical phosphates. Depending on the length of the polyoxyethylene chain the stability to hard water is increased considerably. Furthermore, the dissolution in highly alkaline media is improved. The antistatic effectivity is made possible by the polyether chain. The compatibility in hydrophilic systems is based on the same fact. As expected, there is a loss of compatibility in hydrophobic systems. The surface active properties such as foaming, wetting power, depression of the surface tension, and emulsifying ability depend on the kind of alkyl residues, the mono-/diester ratio, the polyether chain, as well as on the degree of neutralization (pH value). The distinct stability to hydrolysis reactions over a wide range of pH values is comparable with the classical phosphates. The ester bonds are cleaved only under extremely acid or alkaline conditions, most of all caused by increased temperatures. The high-temperature resistance of the esters is also remarkable. The improved solubility in water accelerates the biological degradability.

The nonionic-based phosphates are used according to their wide-ranging properties. Various applications are described in the literature, mainly in patents. Especially in the cosmetic sector alkyl polyoxyethylene ether phosphates are of high value because of their mildness to hair and skin in a number of detergent compositions [127]. They have even been recommended for use in toothpastes [128]. Triesters of oxethylated fatty alcohols (e.g., Hostaphat, Hoechst AG) are excellent emulsifiers for ointments, creams, and lotions [129]. Shampoos containing mono-/diphosphates based on lauryl polyoxyethylene (3) alcohol make hair easily combable, shiny, and antistatic [130]. Transparent perfume gels can be obtained by combining with suitable nonionics, paraffin, and perfume oils [131]. The nonfading properties of antibacteriologic soaps are improved by the addition of partial ether phosphates [132].

In the metal industry, there are also a large number of patent applications. Generally speaking, partial esters of phosphoric acid are used in lubricants as extreme pressure (EP) and antiwear additives with corrosion inhibiting effects [133]. They are also used in acid [134] and alkaline [135] cleaners for aluminum. In metal flotation polyoxyalkylene phosphates are used for the separation of fluorides [136].

In the plastics industry, phosphates of oxalkylated polysiloxanes can be used as lubricants and antistats [137]. Alkyl polyoxyethylene phosphates act as polymerization emulsifiers for olefins. The dispersions remain homogeneous and stable [138] while also having a better light stability [139].

For textile fiber applications, alkyl polyoxyethylene phosphates are distinguished regarding their multifunctional properties. Depending on their composition they can be used as wetting agents or emulsifiers, in view of their thermal stability as antistats [140], spinning aids [141], or lubricants. EO/PO block C_9/C_{13} alkyl phosphates that were prepared by reaction with polyphosphoric acid are used as wetting agents for the textile pretreatment in alkaline media [117]. Several proposals exist for the use of alkyl polyoxyethylene phosphate detergent compositions for washing and cleaning agents [118,142,143]. Large quantities of special phosphates are delivered to the agricultural industry as emulsifiers for pesticides. Last but not least, it is noteworthy that alkyl polyoxyethylene phosphates included in phospholipids are described as having pharmaceutical effects, i.e., inhibitory activities against HIV-1 and HIV-2 [144], malate dehydrogenase activation analogous to *Mycobacterium smegmatis* [145], or antileukemic effects [146].

V. CARBOXYLATES

In recent years, alkyl polyoxyethylene carboxylic acid or alkyl(poly-1-oxapropen)oxaalkene carboxylic acid, also known as ether carboxylic acids, have gained in importance. The combination of advantageous ecological and toxicologic properties, outstanding dermatologic behavior, and excellent emulsifying, wetting, and cleaning properties implies that this class of surfactants has a broad field of application. Ether carboxylic acids are valuable components in cosmetic formulations, washing and cleansing agents, as well as auxiliaries in the textile, printing, paper, plastics, and metalworking industries. Due to their structure, the physicochemical properties of ether carboxylic acids can be adjusted to the application requirements by modification of the hydrophobic alkyl or alkylaryl chain, the number of ethylene oxide units, and the degree of carboxylation.

A. Preparation

The conventional methods of preparation of ether carboxylic acids start with fatty alcohols, which are then primarily ethoxylated [147,148]. Conversion to ether carboxylic acid may be afforded, in principle, by several routes (Fig. 16).

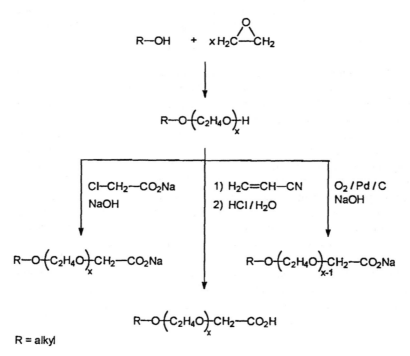

FIG. 16 Synthesis of ether carboxylic acids.

The alkyl polyoxyethylene alcohols can be carboxymethylated by reaction with monochloroacetic acid in the presence of sodium hydroxide or through terminal oxidation of the fatty alcohol ethoxylate. The ether carboxylic acid can also be synthesized by the addition of a vinylic system, i.e., acrylonitrile, to an oxyethylated fatty alcohol and subsequent hydrolysis.

In analogy to Williamson's ether synthesis, conversion of the alkyl polyoxyethylene alcohols with sodium chloroacetate and sodium hydroxide into the sodium salt of the ether carboxylic acid is accomplished. In the first reaction step, the alcoholate is formed [Eq. (13)]. This reacts subsequently with sodium chloroacetate, resulting in the ether carboxylate [Eq. (14)]:

$$RO-(CH_2CH_2O)_x-H + NaOH \rightarrow RO-(CH_2CH_2O)_x^- \ Na^+ + H_2O \qquad (13)$$

$$RO-(CH_2CH_2O)_x-H + ClCH_2COONa \rightarrow$$
$$RO-(CH_2CH_2O)_x-CH_2COONa + NaCl \qquad (14)$$

The inductive effect of the oxygen atoms in the polyglycol ether chain has an acidifying effect on the terminal hydroxyl group. Therefore, the reactivity of the alkyl polyoxyethylene alcohol increases with an increasing degree of ethoxylation [149,150], as does the dissociation constant of the resulting ether carboxylic acid. Thus, in a partial carboxymethylation, the homologous distribution of the ether carboxylic acid will be shifted to products with a higher degree of ethoxylation, whereas the nonconverted alkyl polyoxyethylene alcohol contains more low-oxyethylated fractions. Furthermore, the reactivity and the realizable conversion level also depends on the structure of the alkyl component. In the case of ethylene oxide- and propylene oxide-containing ether carboxylic acids, the proportion of propylene glycol in the polyglycol chain is also important.

Residual, nonreacted alkyl polyoxyethylene alcohol, glycolic acid, water, and sodium chloride result as byproducts. The free ether carboxylic acids can be isolated by adding sulfuric acid to the product mixture and separating the resulting phases. Using this method, salt contents in ether carboxylic acids of <10 ppm can be achieved without any problem. The disadvantage, however, is the high wastewater load due to the process-related salt formation. High degrees of carboxymethylation and product purity can be achieved by use of bromoethyl acetate and subsequent saponification of the resulting ester [151]. It is also possible to carboxymethylate polyhydroxy compounds, polyalkylene oxides [152], acetylene derivatives, and amino- or amidooxyethylates with this method.

The Michael addition of primary or secondary alcohols to acrylonitrile allows the synthesis of cyanoethyl ethers [153] [Eq. (15)]. These can be hydrolyzed to form the corresponding ether carboxylic acids [154] [Eq. (16)]. The cyanoethylation of aliphatic alcohols conveniently takes place at room temperature when strong bases such as sodium, sodium hydroxide, or ion exchange resins [155] are employed as catalysts:

$$RO-(CH_2CH_2O)_x-H + CH_2=CH-CN \rightarrow RO-(CH_2CH_2O)_x-CH_2CH_2-CN \quad (15)$$

$$RO-(CH_2CH_2O)_x-CH_2CH_2-CN + H_2O/H^+ \rightarrow$$
$$RO-(CH_2CH_2O)_x-CH_2CH_2-COOH \quad (16)$$

This method can also be applied to convert diols and polyols to cyanoethyl ether. Other functionalities, such as dialkylamino, halogen, or olefin groups, do not affect the reaction. Cycloaliphatic alcohols, (poly)glycols, even high molecular weight hydroxy compounds and polyvinyl alcohol, can easily be converted with acrylonitrile. However, in the case of phenols, higher temperatures are required.

After cyanoethylation, the alkaline catalyst must be neutralized; otherwise the dissociation of the product into the initial alcohol and polymerization of the acrylonitrile will take place. At a pH of 4–6, however, the ether nitriles are stable and storable. Nevertheless, polyacrylonitrile was sometimes found as a byproduct in this reaction. These nitriles are important as intermediates for plastics, synthetic fibers, pharmaceutical products and insecticides, but rarely for surfactants.

An advantage of this synthesis is the fact that the entire conversion of the fatty alcohol to the ether carboxylic acid can be carried out as a one-pot reaction. In most cases, the catalyst used for the oxyethylation of alcohol is also sufficient for the catalysis of the Michael addition. The subsequent hydrolysis can be undertaken with alkali, but often concentrated hydrochloric acid is used. Ammonium chloride is formed as a byproduct.

An elegant process for the preparation of ether carboxylic acids is the catalytic oxidation of the terminal hydroxyl groups of an alkyl polyoxyethylene alcohol [156]. Although alcohol ethylene oxide adducts can be oxidized with inorganic oxidizing agents such as chromic acid, potassium permanganate, or peroxides, the oxidation with oxygen in the presence of metal catalysts is more selective [Eq. (17)]. This reaction is carried out with a mixture of the alkyl polyoxyethylene alcohol in an aqueous solution of sodium hydroxide with palladium/charcoal in an autoclave at 60–90°C. Bismuth, lead, cadmium, and cobalt also serve as cocatalysts or catalyst activators. For technical reasons the oxidation takes place in a dilute, approximately 10–20% solution of the alkyl polyoxyethylene alcohol in water. Although the catalytic oxidation proceeds with high selectivity and yield and is also advantageous from an ecological point of view, it has not gained importance on a technical production scale. Difficulties in connection with the workup of the reaction mixture and the low concentration of the ether carboxylate in the solution inhibit large-scale production.

$$RO-(CH_2CH_2O)_x-H + O_2 + NaOH \rightarrow$$
$$R-(OCH_2CH_2)_n-OCH_2COONa + H_2O \quad (17)$$

B. Analytical Methods

For the qualitative and quantitative determination of ether carboxylic acids, a variety of analytical methods are suitable [157]. The qualitative evaluation of the reaction products and reaction monitoring is done by thin-layer chromatography [158,159]. The separation is carried out according to functionality and not to polymer size. The eluent system chloroform/methanol 20:80 has proven to be particularly suitable. Reverse phase HPLC allows identification of the ether carboxylates in the form of tetramethylammonium salts [160]. In this way, the tetramethylammonium salts of the ether carboxylic acids can be separated from other anionic or nonionic surfactants.

Ether carboxylic acids can be separated by gas chromatography into the individual polymer homologs and subsequently determined with the aid of mass spectroscopy. In the case of nonylphenyl polyether carboxylates it is indeed possible to distinguish the ortho and para isomers [157].

Infrared spectroscopy is very helpful for reaction control in the preparation of ether carboxylic acids according to the nitrile route [Eq. (15)]. The characteristic absorptions for the oxyethylates [v(OH) 3400 cm^{-1}], ether nitriles [v(CN) 2250 cm^{-1}], and ether carboxylic acids [v(CO) 1710 cm^{-1}] enable the identification of the individual components of a reaction mixture [161].

The quantitative evaluation of ether carboxylic acids is also achievable through potentiometric titration. In this case, the determination of free ether carboxylic acids is carried out by titration of the ether carboxylic acid with sodium hydroxide. The salts of ether carboxylic acid are titrated against hydrochloric acid. Solubility problems and insufficiently defined potential steps can be avoided by using a 95% solution of the ether carboxylic acid in methanol. For technical products, the quantitative content of ether carboxylic acid is calculated on the basis of the mean molecular weights of the corresponding ethoxylates. The pK_a values of ether carboxylic acids of the type R–(EO)$_x$–COOH are between $pK_a = 4.1$ ($x = 12$) and $pK_a = 5.6$ ($x = 3$). The acidity increases with an increasing degree of oxyethylation [4]. Two-phase titration, which is established as a standard method [26] for quantitative determination of anionic surfactants, shows difficulties in the determination of ether carboxylic acids depending on the structure and composition of the product. The titration of ether carboxylates with other cationics is also described in the literature [162].

For fast or routine analysis of technical products, where usually both polymer distribution and nonconverted educt and byproducts prove to be problematic, NMR spectroscopy is not recommendable. On account of the overlap of resonances of the polyethylene oxide unit with the signal of the methylene protons adjacent to the carboxyl group, ^{1}H NMR spectroscopy allows conclusions only in optimal conditions because the resonances of these protons depend strongly on the concentration and pH value of the solution. For the estimation of the yield or degree of conversion, ^{13}C NMR spectroscopy is more suitable because the resonance of the carboxyl carbon atom is sufficiently shifted to low fields [163].

The literature also discusses ion exchange chromatography [157], oleochemical indices, and elemental analysis as methods for characterization of ether carboxylic acids.

C. Properties and Applications

The physicochemical properties of ether carboxylic acids are particularly determined by the hydrophobic alkyl chain, the degree of ethoxylation, and the degree of carboxymethylation, as well as the type of neutralization and the pH of the solution. Ether carboxylic acids are temperature-stable and resistant to alkali and hydrolysis, even under strongly acidic or alkaline conditions.

Through variation of the length and structure of the hydrophobic carbon chain, above all the wetting and foaming behavior of the ether carboxylic acids can be influenced. The foam volume increases with an increasing chain length. Unlike many nonionic surfactants, which are only low foaming at the cloud point, short chain ether carboxylic acids keep their low foaming properties over a wide temperature and pH range [164].

On account of its complexing capacity for cations, the hydrophilic polyglycol group is responsible for the hard water stability and the lime soap dispersing power of the ether carboxylic acids [165]. Both properties improve with increased ethoxylation. Therefore, ether carboxylic acids are outstandingly suitable for surfactant applications in hard water. And especially, as stated above, the foaming properties of the ether carboxylic acids can be controlled by the degree of ethoxylation.

The carboxyl group enables the ether carboxylic acids to perform a double function as nonionic and anionic surfactant. In an acidic medium, ether carboxylic acids act as nonionic surfactants, whereas in an alkaline medium they gain the properties of an anionic surfactant. In aqueous solutions the ether carboxylic acids, as weakly dissociating acids, are present in nonionic form. In aqueuos solutions, the formation of micelles starts at the same concentration as in the case of the corresponding nonionic surfactant [164]. The salts of ether carboxylic acids form micelles at higher concentrations; therefore the surface tension is reduced at low pH values.

The carboxymethylation level of the ether carboxylic acids also influences the properties of technical ether carboxylic acids because, even in an optimized reaction, in practice there will always remain a certain proportion of alkyl polyoxyethylene alcohol that is not carboxymethylized. This combination of ether carboxylic acid and alkyl polyoxyethylene alcohol, for example, has a positive effect on the emulsifying capacity. Synergistic effects can also be observed with regard to wetting power and foaming behavior. The oxidation and alkali stability of the ether carboxylic acids is increased with a higher conversion level.

The pH influences the viscosity of an ether carboxylic acid solution. Generally, a reduction of the pH lowers the viscosity of the ether carboxylic acid solution.

On account of their advantageous ecological, toxicologic, and physico-chemical properties and the good compatibility with representatives of all surfactant classes, ether carboxylic acids can be applied effectively in many fields. Ether carboxylic acids, for example, are used in washing and cleaning agents as well as cosmetics. They are utilized as emulsifying and auxiliary agents in the textile, printing, paper, plastics, metalworking, and pharmaceutical industries [166].

The salts of ether carboxylic acids with high degree of ethoxylation are considered to be very mild and skin-compatible surfactants. Therefore, they are particularly suitable for applications in cosmetics [77a]. The good dermatologic and ecotoxicologic properties such as low defatting effect, nontoxicity, easy biodegradability, as well as outstanding cleaning and emulsifying effect are exploited in numerous cosmetic formulations. In particular, for medical cleansing agents, the disinfecting effect of salts of the ether carboxylic acids with rare earths is important.

Because ether carboxylic acids show very good hard water stability and lime soap dispersing power, they are interesting as surfactants for phosphate-free washing agents and dishwashing detergents. Furthermore, they are distinguished by low-foaming behavior and high redeposition-inhibiting capacity. Through the variation of the hydrophobic carbon chain and hydrophilic polyglycol chain as well as the degree of carboxymethylation, the detergency of ether carboxylic acids can be optimized [167]. Ether carboxylic acids are also used for manual dishwashing detergents, carpet cleaners, and other household products [168].

Due to their good wetting properties and their outstanding alkali and oxidation resistance, ether carboxylic acids are predestined for applications in formulations based on sodium hypochlorite or hydrogen peroxide. They are frequently used in textile-bleaching agents, toilet cleaners, and high-pressure bottle and metal cleaners. Thanks to their beneficial properties, members of this surfactant group are applied in the pretreatment of wool and cotton, as dyeing and printing auxiliaries, as additives for paints and inks, and for the de-inking of paper. In the plastics industry ether carboxylic acids are employed as auxiliary agents for emulsion polymerization and as antistatics.

Because they exert a good corrosion inhibiting effect, ether carboxylic acids are also used as emulsifiers in drilling, rolling, and cutting oil emulsions and cooling lubricants [169]. Ether carboxylates show a high long-term stability, even under extreme chemical and thermal conditions. As they are compatible with brine and crude oil, they are also effective surfactants for recovering oil by micellar polymer flooding from reservoirs of high salinity [170]. Based on their structure, alkyl ether carboxylates can be tailored to the needs of the particular reservoir.

Due to their high alkali and electrolyte stability, ether carboxylic acids can also be utilized as hydrotropes. They allow formulation of stable products even under high salt or alkaline conditions. Apart from this, synergistic effects regarding foaming properties, detergency, wetting, and cleaning power are frequently observed when ether carboxylic acids are combined with other surfactants. Conversion products of ether carboxylic acids with monoethanolamide can serve as thickeners for surfactant solutions [171].

D. Environmental and Toxicologic Aspects

Alkyl ether carboxylic acids are considered to be a toxicologically safe surfactant class. The acute toxicity as well as skin and mucous membrane compatibility were examined. With an LD_{50} of over 2000 mg/kg (C_{12}/C_{14}–4.5EO-carboxylate) alkyl ether carboxylates are assessed as nontoxic [164]. The Zein test, which allows statements about the skin compatibility of anionic surfactants and tests on rabbit skin and rabbit eye mucous membranes, identify alkyl ether carboxylic acids as a skin-compatible substance class. The skin mildness of ether carboxylic acids increases with a higher degree of ethoxylation.

The biodegradability of ether carboxylic acids was proven in the OECD test [164]. Above all, ether carboxylic acids based on linear fatty alcohols were proven to be easily biodegradable in the OECD confirmatory test with a degradation rate of 95% [167]. The degradation of fatty alkyl polyoxyethylene alcohols, and therefore indirectly that of ether carboxylic acids, was examined on the system *Pseudomonas*. It showed that both etherases and alcohol and aldehyde dehydrogenases are involved in the degradation process [172]. Ether carboxylic acids, which correspond to terminally oxidized fatty alkyl polyoxyethylene alcohols, were found as metabolites. The degradation products, as alcohol, aldehyde, and carboxyl group-containing metabolites, were detected, but none accumulated under the test conditions. The structure of the hydrophobic alkyl chain has a great influence on the biodegradability of ether carboxylic acids.

Tests on fish toxicity show that the aquatic toxicity depends on the degree of carboxymethylation. With a increasing carboxymethylation level, the fish toxicity of the ether carboxylic acid and alcohol ethoxylate mixture decreases [167]. As compared to other, conventional anionic and nonionic surfactants, certain alkyl ether carboxylic acids have been shown to be less toxic. However, in the ecologic evaluation of a substance the biodegradability is more important than the aquatic toxicity because surfactants usually lose their aquatoxic activity during the degradation process.

VI. QUATERNARY AMMONIUM COMPOUNDS

The practically used quaternary surfactants are based exclusively on a tetra-coordinated nitrogen atom. Tetraalkylammonium salts containing at least one longer alkyl chain represent the group of cationic surfactants. Often the analogous amine salts are described as quaternaries as well. But contrary to the true quaternary surfactants, they are formed by neutralization with acids. Therefore they consist of protonated amines and should be named as pseudocationics (Fig. 17). The substantial difference between both classes of cationics is that the pseudocationics only show surfactant properties at a pH value lower than 7. In this section "true quaternaries" based on nonionics will be discussed.

Polyoxyethylene ether quaternaries represent an extremely complex class of substances that allow a large number of tailor-made products due to the high potential of variations. On the other hand, the chemical diversity makes a systematic classification regarding only a few main types more difficult. The structure differences in this section could be very useful. The examination of the comprehensive literature allows the classification of nonionic surfactant based quaternaries into the following groups:

1. Substances prepared by quaternization of polyoxyalkylene amines
2. Substances in which the polyether chain is not directly bound to the nitrogen atom
3. Substances containing a polyether chain that is introduced by the quaternization agent

A fourth group which, strictly speaking, does not correspond to the main title "nonionic intermediates" should also be presented:

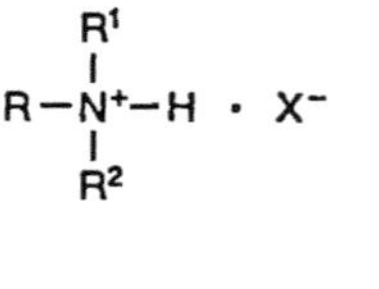

$$
\begin{array}{c}
R^1 \\
| \\
R-N^+-R^3 \cdot X^- \\
| \\
R^2
\end{array}
$$

quaternary ammonium salt

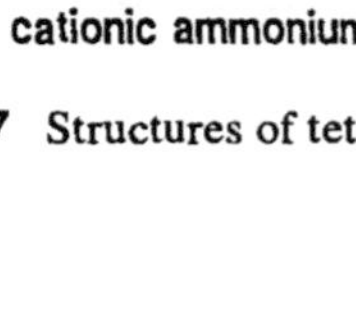

$$
\begin{array}{c}
R^1 \\
| \\
R-N^+-H \cdot X^- \\
| \\
R^2
\end{array}
$$

pseudo cationic ammonium salt

FIG. 17 Structures of tetra-coordinated ammonium salts.

4. Substances in which the polyoxyalkylene moiety is introduced during or after the quaternization step

On the basis of characteristic examples an introduction into the chemistry of the individual groups is given. In group 1, oxyalkylated, mostly oxyethylated amines serve as raw materials. In a simple manner the oxyethylated fatty amines will be quaternized analogous to the ordinary tertiary amines by means of methyl chloride, dimethyl sulfate, benzyl chloride, etc., as alkylation agents [173] [Eq.(18)].

$$
\begin{array}{ccc}
(C_2H_4O)_xH & & (C_2H_4O)_xH \\
| & & | \\
R-N & + \quad CH_3Cl \longrightarrow & R-N^+-CH_3 \quad \cdot \quad Cl^- \\
| & & | \\
(C_2H_4O)_xH & & (C_2H_4O)_xH
\end{array}
\qquad (18)
$$

Sometimes the addition of potassium iodide is recommended as a catalyst. Nevertheless it is difficult to reach a conversion to quaternaries of more than 90%. Probably a steric hindrance is responsible. By increasing the polyether chain length the reactivity drops continuously. Quaternaries based on partial fatty amides of polyvalent amines [174], as described in Eq. (19), have shown to be of practical use as less hydrophobic textile softeners.

$$
H_2N-C_2H_4-NH-C_2H_4-NH_2 + 2R\ COOH \xrightarrow{-2 H_2O} R\ CO-NH-C_2H_4-NH-C_2H_4-NH-OC\ R
$$
$$
(I)
$$

$$
(I) + x H_2\overset{O}{\overset{\diagup\diagdown}{C-CH_2}} \longrightarrow R\ CO-NH-C_2H_4-\underset{\underset{(C_2H_4O)_xH}{|}}{N}-C_2H_4-NH-OC\ R
$$
$$
(II)
$$

$$
(II) + (CH_3)_2SO_4 \longrightarrow R\ CO-NH-C_2H_4-\underset{\underset{(C_2H_4O)_xH}{|}}{\overset{\overset{CH_3}{|}}{N^+}}-C_2H_4-NH-OC\ R \cdot CH_3SO_4^-
$$

Very interesting is the synthesis shown in Eq. (20), where in the first reaction step oxyalkylated triethanolamine is reacted with long-chain epoxides followed by a quaternization reaction [175].

$$
\begin{array}{ccc}
(C_2H_4O)_xH & & (C_2H_4O)_x CH_2-CH(OH)-R \\
| & O & | \\
N+C_2H_4O)_x H + 3 H_2\overset{\diagup\diagdown}{C-CH_2-R} \longrightarrow & N+C_2H_4O)_x CH_2-CH(OH)-R \\
| & & | \\
(C_2H_4O)_x H & & (C_2H_4O)_x CH_2-CH(OH)-R
\end{array}
$$
$$
(I) \qquad\qquad (20)
$$

$$(I) \quad + CH_3Cl \longrightarrow \begin{array}{c} (C_2H_4O)_{\overline{x}}CH_2-CH(OH)-R \\ | \\ H_3C-N^+(C_2H_4O)_{\overline{x}}CH_2-CH(OH)-R \cdot Cl^- \\ | \\ (C_2H_4O)_{\overline{x}}CH_2-CH(OH)-R \end{array}$$

In a completely different route the quaternization can also be done with alkylene oxide in the presence of water [176, 177] as demonstrated in Eq. (21).

$$R-N-(CH_3)_2 \quad + \quad x\,H_2C\overset{O}{\overset{\triangle}{-}}CH_2 \quad + \quad H_2O \longrightarrow R-\overset{\overset{\displaystyle CH_3}{|}}{\underset{\underset{\displaystyle CH_3}{|}}{N^+}}-(C_2H_4O)_xH \cdot OH^- \quad (21)$$

Primary and secondary amines can also be used in this reaction and form quaternaries if the amount of epoxide is appropriate. Additional hydroxyl functions will be formed [178] that can react further with the alkylene oxide (see group 4 products). The thermal instability of the formed quaternary ammonium bases is real disadvantage for the procedure. As a consequence, the resulting dealkylation reaction decreases the yield of quaternaries. But fortunately, the neutralization of the amine before the addition of alkylene oxide seems to be a successful alternative [179]. However, for the dissociation of the amine salt the presence of water is necessary and glycols will be formed as byproducts. Table 3 shows results of the quaternization reaction with ethylene oxide depending on the presence of acid. The respective reaction temperature was adjusted to 80°C. The concentration of water in the final product was 50% [180].

Polyoxyethylated amines can also be quaternized in such a manner that, as shown in Eq. (22), unusual acid compounds, i.e., partial phosphates, form the anion [178,181]:

$$\begin{array}{c} (C_2H_4O)_xH \\ | \\ R-N \\ | \\ (C_2H_4O)_{x'}H \end{array} + \begin{array}{c} O \\ \| \\ HO-P-OR \\ | \\ OR \end{array} + x' H_2C\overset{O}{\overset{\triangle}{-}}CH_2 \overset{H_2O}{\longrightarrow} \begin{array}{c} (C_2H_4O)_xH \\ | \\ R-N^+-(C_2H_4O)_{x'}H \\ | \\ (C_2H_4O)_{x'}H \end{array} \cdot \begin{array}{c} O \\ \| \\ RO-P-O^- \\ | \\ OR \end{array}$$

$$(22)$$

TABLE 3 Quaternization with Ethylene Oxide

Amines	Acid	EO-supply [mole]	Quaternization in the presence of acid	Degree of quaternization	Glycols
				Analytical results (%)	
Tallow amine	H_3PO_4	5	No	22	8.6
Tallow amine	H_3PO_4	5	Yes	91	3.1
Tallow amine·5EO	H_3PO_4	5	No	40	11.1
Tallow amine·5EO	H_3PO_4	5	Yes	91	6.5
Tallow amine·5EO	Lactic acid	5	Yes	98	13.7

Following Eq. (23), the preparation of polyquaternary cationics is possible, with so-called polyester tertiary amines being produced by polycondensation reaction of polyoxyethylated fatty amines with dicarboxylic acids like adipic acid [182] as well will be quaternized with ethylene oxide [183].

$$n\ H(OH_4C_2)_x-N-(C_2H_4O)_x\cdot H\ +\ n\ HOOC-(CH_2)_4-COOH\ \xrightarrow{-2n\ H_2O}\ \left[CO-(CH_2)_4-CO-(OC_2H_4)_x-N-(C_2H_4O)_x\right]_n$$

$$\overset{|}{R}$$

(I)

(23)

$$(I)\ +\ n\ CH_3COOH\ +\ x\cdot H_2\overset{O}{\underset{}{C}}-CH_2 \longrightarrow \left[CO-(CH_2)_4-CO-(OC_2H_4)_x-\overset{(C_2H_4O)_x\cdot H}{\underset{R\cdot CH_3COO^-}{N^+-(C_2H_4O)_x}}\right]_n$$

Basically, other epoxides are also applicable, i.e., propylene or butylene oxide. Even long chain oxiranes are mentioned in the literature [184]. It should also be mentioned that the use of strong acids during the quaternization reaction with ethylene oxide generates dioxane.

The references for product types of group 2, the polyoxy alkylene chain of which is bonded to basic nitrogen atoms by a spacer, are less numerous. Methods to obtain such substances have been published by R. J. Stenberg [185] and E. I. Carpenter [186] who added acrylonitrile to oxyethylated alcohols. The resulting propionitriles were converted to primary amines by hydrogenation followed by an alkylation reaction before a quaternization could be applied. The reaction sequences are shown in Eq. (24).

$$RO(C_2H_4O)_xH\ +\ CH_2=CH-CN\ \xrightarrow{base}\ RO(C_2H_4O)_x-CH_2-CH_2-CN$$

(I)

$$(I)\ +\ 2\ H_2\ \longrightarrow\ RO(C_2H_4O)_x-CH_2-CH_2-CH_2-NH_2$$

(II)

(24)

$$(II)\ +\ 2\ CH_3Cl\ +\ 2\ NaOH\ \xrightarrow[-2\ H_2O]{-2\ NaCl}\ RO(C_2H_4O)_x-CH_2-CH_2-CH_2-\overset{CH_3}{\underset{CH_3}{N}}$$

(III)

$$(III)\ +\ CH_3Cl\ \longrightarrow\ RO(C_2H_4O)_x-CH_2-CH_2-CH_2-\overset{CH_3}{\underset{CH_3}{N^+}}-CH_3\ \cdot\ Cl^-$$

Another, simpler route [180] is represented in Eq. (25) according to which polyether fatty carboxylic acids are reacted with dimethylaminopropylamine to amidoamines, which can easily be quaternized leading to the cationic product in high yields.

$$RO(C_2H_4O)_x-CH_2-COOH \ + \ H_2N-(CH_2)_3-N(CH_3)_2 \ \xrightarrow{-H_2O} \ RO(C_2H_4O)_x-CH_2-CO-NH-(CH_2)_3-N(CH_3)_2$$
$$(I)$$

$$(25)$$

$$(I) \quad + \ CH_3Cl \ \longrightarrow \ RO(C_2H_4O)_x-CH_2-CO-NH-(CH_2)_3-\overset{\overset{\displaystyle CH_3}{|}}{\underset{\underset{\displaystyle CH_3}{|}}{N^+}}-CH_3 \cdot Cl^-$$

At this point it is interesting to mention that amphoteric surfactants such as betaines, which also contain a tetra-coordinated nitrogen atom, can be synthesized by reaction of the aforementioned amidoamines with chloroacetic acid salts [Eq. (26)]:

$$RO(C_2H_4O)_x-CH_2-CO-NH-(CH_2)_3-N(CH_3)_2 \quad + \quad Cl-CH_2-COONa$$

$$\downarrow$$

$$(26)$$

$$RO(C_2H_4O)_x-CH_2-CO-NH-(CH_2)_3-\overset{\overset{\displaystyle CH_3}{|}}{\underset{\underset{\displaystyle CH_3}{|}}{N^+}}-CH_2-COO^- \quad + \quad NaCl$$

Quaternaries of group 3, where the polyether chain is introduced by the quaternization agent, are represented by numerous examples in the literature. Here, typical syntheses are described. Often the use of chloroacetic acid esters of polyoxyethylated alcohols is proposed for the quaternization of tertiary fatty amines [187,188], as follows:

$$R-N(CH_3)_2 \ + \ Cl-CH_2-CO(OC_2H_4)_x-OR' \ \longrightarrow \ R-\overset{\overset{\displaystyle CH_3}{|}}{\underset{\underset{\displaystyle CH_3}{|}}{N^+}}-CH_2-CO(OC_2H_4)_x-OR' \cdot Cl^-$$
$$(27)$$

The product from Eq. (27), the so-called betaine ester, can be hydrolyzed easily to betaines [Eq. (28)]:

$$R-\overset{\overset{\displaystyle CH_3}{|}}{\underset{\underset{\displaystyle CH_3}{|}}{N^+}}-CH_2-CO(OC_2H_4)_x-OR' \cdot Cl^- \ + \ NaOH \ \longrightarrow \ R-\overset{\overset{\displaystyle CH_3}{|}}{\underset{\underset{\displaystyle CH_3}{|}}{N^+}}-CH_2-COO^- \ + \ H(OC_2H_4)_x-OR' \ + \ NaCl$$
$$(28)$$

In order to obtain cationic surfactants that are stable toward hydrolysis, the quaternization reaction is carried out with chlorohydrin ethers of nonionic surfactants.

Polyquaternary products are obtained by reaction of bischlorohydrin ethers with tetramethyldiaminoethane as shown in Eq. (29) [180]:

$$n \; Cl \; CH_2-\overset{\displaystyle OH}{\overset{|}{CH}}-CH_2-O+C_2H_4O\tfrac{1}{x}CH_2-\overset{\displaystyle OH}{\overset{|}{CH}}-CH_2Cl \quad + \quad n \; \overset{CH_3}{\underset{CH_3}{\overset{|}{N}}}-C_2H_4-\overset{CH_3}{\underset{CH_3}{\overset{|}{N}}}$$

$$(29)$$

$$\left[+CH_2-\overset{\displaystyle OH}{\overset{|}{CH}}-CH_2-O+C_2H_4O\tfrac{1}{x}CH_2-\overset{\displaystyle OH}{\overset{|}{CH}}-CH_2-\overset{CH_3}{\underset{CH_3}{\overset{|}{N^+}}}-C_2H_4-\overset{CH_3}{\underset{CH_3}{\overset{|}{N^+}}}+ \right]_n \cdot 2nCl^-$$

Bisquaternary ammonium bromides with polyoxyethylene ether chains [189] were synthesized in analogy to Eq. (30) and the physical-chemical properties were studied.

$$3 \; HO+C_2H_4O\tfrac{1}{x}C_2H_4OH \; + \; 2 \; PBr_3 \; \longrightarrow \; 3 \; Br+C_2H_4O\tfrac{1}{x}C_2H_4Br$$
$$(I)$$

$$(30)$$

$$(I) \quad + \quad 6 \; R-N(CH_3)_2 \; \longrightarrow \; 3 \; R-\overset{CH_3}{\underset{CH_3}{\overset{|}{N^+}}}+C_2H_4O\tfrac{1}{x}C_2H_4-\overset{CH_3}{\underset{CH_3}{\overset{|}{N^+}}}-R \; + \; 6 \; Br^-$$

While the last synthesis route only has academic interest, chloroacetic acid esters and chlorohydrin ethers act as simply available quaternization agents with highly variable possibilities.

Cationic ammonium salts that were obtained by quaternization with alkylene oxide (see group 1) and by a further oxyalkylation reaction are mentioned here as examples for group 4 products [178].

The oxyalkylation of hydroxy groups containing quaternaries in a separate reaction step fails as the high reaction temperature causes a dealkylation reaction of the quaternary nitrogen atom. Finally the cationic modification of nonionics with glycidyltrimethylammonium chloride as suggested by the manufacturers (i.e., Degussa, Shell) is mentioned [Eq. (31)]:

$$RO+C_2H_4O\tfrac{1}{x}H \quad + \quad H_2C-\overset{\displaystyle O}{\overset{\diagup\diagdown}{CH}}-CH_2-N(CH_2)_3 \; \cdot \; Cl^-$$

$$(31)$$

$$RO+C_2H_4O\tfrac{1}{x}CH_2-\overset{\displaystyle OH}{\overset{|}{CH}}-CH_2-N^+(CH_3)_3 \quad \cdot \quad Cl^-$$

The above-described quaternary ammonium compounds based on nonionic surfactants possess the general characteristics of classical quats. Most of all, the affinity to anionic surfaces may be the significant property. Depending on length and kind of the polyoxyalkylene chains, however, additional effects will be obtained. For example, the water solubility will be improved as well as the compatibility with anionic surfactants or the wetting power.

Typical applications are given in the textile and fiber industry where they act as softening, antistatic, fixing, or emulsifying agents [175,190–193]. It was also suggested that they be used as spin finishing for fibers and yarns including lubricating effects. Accordingly, their compatibility with anionic surfactants make them applicable for hair care and hair rinse products in which they impart antistatic properties and an improved combability [194–196]. Cationics that are prepared by quaternization with propylene oxide find application as demulsifiers for mineral oil–contaminated wastewater [197].

APPENDIX: TRADE NAMES AND SUPPLIERS

Alkylether sulfates	Supplier
Perlankrol	Harcros Chemicals
Elfan NS	Akzo Nobel
Empicol	Albright & Wilson
Cosmopon, Rolpon	Auschem Cesalpinia
Chemsalan	Chem-Y
Daclor, Dacpon	DAC International Surfactants
Laural	Elf Atochem
Caflon	Ellis & Everad
Disponil, Euperlan, Standapol, Texapon	Henkel KGaA
Manro, Tensagex	Hickson Manro
Genapol, Geropon	Hoechst AG
Marlinat	Hüls
Emal, Sapanol	Kao Corp.
Maprofix	Millchem
Nikkol	Nikko Chemicals
Rhodapex	Rhone-Poulenc
Montelane, Oronal	Seppic
Dobanol	Shell
Polystep, Steol	Stepan Europe
Surfac	Surfachem
Neopon, Rewopol, Witcolate	Witco

Sulfated fatty acid monoglycerides	Supplier
Plantapon CMGS	Henkel KGaA
Nikkol SGC-80 N	Nikko
POEM-LS-90	Riken Vitamin Oil

Sulfated alkanolamides	
Genapol AMS	Hoechst AG
Genapol AMG	Hoechst AG

Phosphates	
Alkaphos	Alcaril Chem.
Crodafos	Croda Chem.
Gafac	GAF Chem.
Crafol AP, Forlanit, Standapol, Nopcostat	Henkel KGaA
Hostaphat	Hoechst AG
Marlophor	Hüls
Alkawet	Lonza Chem.
Monafac	Mona Ind.
Pecosil	Phoenix Chem.
Triton	Röhm & Haas
Servoxyl	Servo
Rewophat, Ultraphos	Witco Chem.

Carboxylates	
Akypo R-Types	Kao Corp.
Akypo-Soft-Types	Kao Corp.
Akypofoam-Types	Kao Corp.
Marlowet	Hüls
Rewopol C-Types	Rewo
Sandopan	Sandoz
Velsan	Sandoz

Quarternary ammonium compounds	
Rewoquat	Witco Chem.
Mirataine	Rhone Poulenc

REFERENCES

1. H. Stache, *Anionic Surfactants*, Surfactant Science Series, Vol. 56, Marcel Dekker, New York, 1965.
2. J. Falbe, *Surfactants in Consumer Products: Theory, Technology and Application*, Springer-Verlag, Berlin, 1987.
3. P. Hövelmann and B. Brackmann, 21st World Congress and Exhibition of the International Society for Fat Research (ISF), The Hague, 1995.
4. B. R. Bluestein and C. L. Hilton (eds.), *Amphoteric Surfactants*, Surfactant Science Series, Vol. 12, Marcel Dekker, New York, 1982.
5. M. F. Cox, 3rd World Conference and Exhibition on Detergents, Montreux, Sept. 1993.
6. W. H. deGroot, *Sulfonation Technology in the Detergent Industry*, Kluwer Academic, Dordrecht, 1991.
7. W. Skrypzak and O. Szappan, in Kozmet (Sec. Issue), pp. 59–66 (1994).
8. Study of the German "Industrieverband Körperpflege und Waschmittel e.V.," Frankfurt/M., Nov. 1986.
9. T. Förster, H. Hensen, R. Hofmann and B. Salka, Cosmet. Toiletries *110*:29 (1994).
10. B. D. Condon, K. L. Matheson, J. Am. Oil Chemists Soc. *71*(1):53–59(1994).
11. H. Plate, Parfüm. Kosmet. *76*:28(1995).
12. M. E. Spiess, Parfums, Cosmet., Aromes *109*:65 (1993).
13. M. F. Cox, J. Am. Oil Chem. Soc. *66*:1637 (1989).
14. A. Behler, H. Hensen, H.-C. Raths, and H. Tesmann, Seifen Öle Fette Wachse *116*(2):60 (1990).
15. H. P. Welzel, H. P. Neumann, and K. Haage, Tenside Surf., Det. *31*(5):286.
16. H. Hensen, H.-C. Raths, and W. Seipel, Seifen Öle Fette Wachse *117*:592 (1991).
17. G. Crass, Seifen Öle Fette Wachse *118*:921 (1992).
18. M. F. Cox, J. Am. Oil Chemists Soc. *67*(9):599 (1990).
19. D. L. Smith, J. Am. Oil Chemists Soc. *68*(8):629 (1991).
20. Wu Guozhong, Zhang Zhicheng, Huaxue Shiji *16*(1):58 (1994).
21. V. P. Gordnov, V. V. Sorokin, A. A. Petrow, Khim. Tekhnol. Topl. Masel *4*:23 (1974).
22. R. T. Savel'yanova, N. N. Evsees, V. P. Savel'yanov, T. A. Shashkova Zh. Org. Khim. 2253 (1984).
23. J. Cross, in *Nonionic Surfactants* (J. Cross, ed.), Surfactant Science Series, Vol. 19, Marcel Dekker, New York, 1987, pp. 137–224.
24. *Methods of Analysis*, International Organization for Standardization (ISO),1, rue de Varembeè. Case postale 56, CH-1211 Geneve 20.
25. *German Standard Methods*, Abteilung H-Tenside, Deutsche Gesellschaft für Fettwissenschaft e.V., Münster, Wissenschaftliche Verlagsgesellschaft mbH, Stuttgart 1992.
26. S. R. Epton, Trans. Faraday Soc. *44*:226 (1948).
27. V. W. Reid, G. F. Longman, and E. Heinerth, Tenside *4*(2):292 (1967).
28. S. Alegret, J. Alonso, J. Bartroli, J. Baro-Roma, J. Sanchez, and M. de Valle Analyst *119*(11): (1994).
29. R. Schulz, R. Gerhards, Tenside Surf. Det. *32*(1):6 (1995).
30. H. J. Buschmann, U. Denter, and K.-F. Elgert, Melliand Textilberichte, 951, No. 12 (1995).

31. G. Krusche, Tenside Surf. Det. *27*(2):122 (1990).
32. M. Y. Ye, R. G. Walkup, and K. D. Hill, J. Liq. Chromatogr. *17*(19):4087 (1994).
33. A. Stemp, V. A. Boriraj, P. Walling, and P. Neill, J. Am. Oil Chemists Soc. *72*(1):17 (1995).
34. S. Scalia, and G. Frisina, *Proc. Chromatogr. Soc. Int. Symp.* (D. Stevenson and I. D. Wilson, eds.), Plenum Press, New York, 1991, pp. 219–226.
35. P. Schoeberl, Munch. Beitr. Abwasser-, Fisch-, Flussbiol. *44*:367 (1990).
36. J. Steber and H. Berger, in *Biodegradability of Surfactants* (D. R. Karsa and M. R. Porter, eds.), Blackie, London, 1995
37. S. G. Hales et al., Environ. Microbiol. *44*:790 (1982).
38. B. Ziolkowsky, Seifen Öle Fette Wachse *116*:450 (1990).
39. P. Schöberl, K. J. Bock, and L. Huber, Tenside Surf., Det. *25*:86 (1988).
40. B. R. Harris, U.S. Patent 2,023,387 to Colgate-Palmolive-Peet Company (1932).
41. J. Ross, U.S. Patent 2,693,479 to Colgate-Palmolive Company (1950).
42. J. K. Fincke, U.S. Patent 2,634,287 to Monsanto (1948).
43. F. W. Grag, U.S. Patent 2,979,521 to Colgate-Palmolive Company (1958).
44. (a) G. Gawalek in *Tenside*, Akademie- Verlag, Berlin, 1975, p. 171; (b) M. Linfield (ed.), *Anionic Surfactants*, Surfactant Science Series, Vol. 7, Marcel Dekker, New York, 1976, pp. 219–.
45. K. Yamashita et al., JP 50/70322 to Wako Pure Chemical Ind. (1975).
46. M. Shinohara, JP 78/77014 to Riken Vitamin Oil (1975).
47. H. Stühler, EP 267,518 to Hoechst AG (1987).
48. F. Ahmed et al., DE 3,831,446, to Colgate-Palmolive Company (1988).
49. A. Behler et al., DE 4,038,477, to Henkel KGaA (1990).
50. U. Zeidler, J. Soc. Cosmet. Chem. Japan *20*:17 (1986).
51. A. Straw et at., GB 2031455, to Colgate-Palmolive Company (1978).
52. F. Ahmed, EP 318729 to Colgate-Palmolive Company (1989).
53. H. Raaf et al., DE 2440802 to Blendax Werke Schneider (1974).
54. W. J. King et al., U.S. Patent 3956478 to Colgate-Palmolive Co. (1975).
55. R. Zimmerer and R. Berg, EP 18019 to Procter + Gamble Co. (1979).
56. A. Takahashi et al., JP 84/040940 to Lion Corp. (1984).
57. Y. Yamazaki, JP 85/161920 to Lion Corp. (1990).
58. I. D. Hill, U.S. Patent 4942034-A to Hill ID (1990).
59. S. Mitsyama, JP 03,153,796 to Tamanohada Soap Co., Ltd. (1991).
60. H. Takahashi et al., JP 85/269,578 to Sanyo Chem. Ind. Ltd. (1987).
61. K. Tsubone, JP 90/28099, to Kanebo KKK (1990).
62. J. Shiomi, JP 93/95493 to Mandamu KK (1994).
63. A. Behler et al., WO 95/06702 to Henkel KGaA (1993).
64. W. Breitzke and A. Behler, DE 4328355 to Henkel KGaA (1993).
65. W. Breitzke and A. Behler, DE 44067488 to Henkel KGaA (1994).
66. J. K. Weil and A.J. Stirton, in *Anionic Surfactants* (W.M. Linfield, ed.), Surfactant Science Series, Vol.7, Marcel Dekker, New York, 1976, pp. 224–230.
67. R. G. Bistline, Jr., in *Anionic Surfactants* (H. Stache, ed.), Surfactant Science Series, Vol. 56, Marcel Dekker, New York, 1995, pp. 633–634.
68. A. K. Reng, Parf. u. Kosmet. *61*:87 (1980).
69. a) J. K. Weil, N. Parris, and A. J. Stirton, J. Am. Oil Chemists Soc. *47*:91(1970).
    b) M. Fukuda and T. Tamura, JP 62001448 to Lion Corp (1987).
70. A. Fujio et al., JP 7070047 to Kao Corp. (1995).

71. A. El-Sawy, S.A. Essawy, M.M. El-Sukkary, and A.M.F. Eissa, Hung. J. Ind. Chem. 20:25 (1992).

72. R. G. Bistline, Jr., W.R. Noble, F.D. Smith, W. M. Linfield, J. Am. Oil Chemists Soc. 54:371 (1977).

73. M. Yoshio, S. Etsuo, and H. Yasushi, JP 55149243 to Nippon Oils and Fats (1980).

74. M. A. Podustov et al., SU 1051068 to Kharkov Polytechnic Institute (1983).

75. M. A. Podustov et al., SU 1544769 to Kharkov Polytechnic Institute (1990).

76. A. K. Reng and J.M. Quack, Seifen Öle Fette Wachse 102:307 (1976).

77. (a) A. Turowski, W. Skrypzak, A.K. Reng, and P. Jürges, Parfüm. Kosmet. 76:16 (1995); (b) J. Bohunek, G. Täuber, H. Berghausen, DE 1815657 to Farbwerke Hoechst AG (1968).

78. G. Crass, Tenside Surf., Det. 30:408(1993).

79. (a) T. Konekiyo et al., JP 1016898 to Mitsubishi Petrochemical Co. (1989); (b) P. Goffinet, R. Canaguier, EP 189687 to Union Generale de Savonnerie (1986).

80. F. Hironobu, K. Fumitaka, and M. Yoshio, JP 58015501 to Nippon Oils and Fats (1983).

81. H. H. Friese, U. Ploog, and F. Pieper, DE 3419405 to Henkel KGaA (1985).

82. R. Prather, U.S. Patent 4376088 to Upjohn Co.(1983).

83. G. D. Kudinova et al., SU 1151470 to Belorussian Technological Institute (1985).

84. A. Yamamoto, Y. Arimoto, and M. Konno, BE 902736 to Chemical Works Co. (1985).

85. R. Piorr, R. Höfer, H. J. Schlüssler, and K. Schmid, Fette Seifen, Anstrichm. (1987).

86. M. J. Schwuger and R. Piorr, Tenside 24:70 (1987).

87. H. Baumann, Fat Sci. Technol. 92:49–56 (1990).

88. A. Behr, in Aspects of Homogeneous Catalysis Vol. 5 (R. Ugo, ed.) Reidel, Dordrecht, 1984 pp. 5–73.

89. B. Gruber, B. Fabry, B. Giesen, R. Müller, and F. Wangemann, Tenside Surf. Det. 30:422(1993).

90. B. D. Condon, J. Am. Oil Chemists Soc. 71:739 (1994) and 71(7):743 (1994).

91. A. Domsch and B. Irrgang, in Anionic Surfactants (H.W. Stache, ed.), Surfactant Science Series, Vol. 56, Marcel Dekker, New York, 1995, p. 501.

92. Henkel KGaA, unpublished results.

93. J. A. Milm, R. Soc. Chem. (Ind. Appl. Surf.) 77:76(1990).

94. A. J. O'Lenick, Jr., and W.C. Smith, Soap Cosmet. Chem. Spec. 64:36 (1988).

95. T. Schoenberg, Cosmet. Toiletries 105:105(1989).

96. J. Falbe (ed.), Surfactants in consumer products, Springer-Verlag, Berlin, 1987, p. 83.

97. K. Lindner, in Tenside Textilhilfsmittel Waschrohstoffe, Vol. 1, Wissenschaftliche Verlagsgesellschaft, Stuttgart, 1964, p. 747.

98. W. Kästner and P. J. Frosch, Fette Seifen Anstrichm. 83:33 (1981).

99. J. Valeé, Parfum. Cosmet. Savons 4:205 (1961).

100. G. Gawalek, in Tenside, Akademie Verlag, Berlin, 1975, p. 232.

101. B. Fabry, Tenside Surf. Det. 29:320 (1992).

102. R. Piorr and A. Meffert, 40th DGF lecture meeting, Regensburg, 1984.

103. R. Piorr and A. Meffert, DE 3331513 to Henkel KGaA (1983).

104. C. Schoeller and M. Wittwer, U.S. Patent 1,970,578 to IG Farbenindustrie A. G. (1934).

105. A. Steindorff, G. Balle, K. Horst, and R. Michel, U.S. Patent 2,213,477 to General Aniline and Film Corp. (1940).
106. F. Vögeli, Ann. Chem. 69:180 (1849).
107. A. Lassaigne, Ann. Chim. 3:294 (1820).
108. DE 2,658,862 to Hoechst AG (1976).
109. J. Nüsslein, Parfum. Cosmet. Savons 2:554 (1959); CA 54,7074 f(1960).
110. T. Kurosaki and A. Manbu, DE 3,047,378 to Kao Soap Co. (1979).
111. K. Sasse, in Methoden der Organischen Chemie (E. Müller, ed.), Thieme, Stuttgart, 1964, Bd. 12/2, p. 311.
112. G. Imokawa et al., J. Am. Oil Chemists Soc. 55:839 (1978).
113. H. Schlecht and H. Distler, DE 2,350,851 to BASF (1973).
114. T. Kurosaki and N. Nishikawa, DE 3,325,337 to Kao Soap Co. (1982).
115. S. König and J. Langley, Henkel Corp., private information.
116. T. Fay, G. P. Sheridan, and D. R. Karsa, presented by Comite Espanol de la Detergentes, XII. Jornado (1981).
117. A. Wiedemann, DE 3,735,049 to Sandoz (1986).
118. S. C Williamson, U.S. Patent 4,493,782 to Amchem Products (1983).
119. F. B. Clarke and J. W. Lyons, J. Am. Chemists Soc. 88(19): 4401 (1966).
120. T. Kurosaki and A. Manba, DE 3,047,378 to Kao Soap Co. (1979).
121. T. Kurosaki, J. Wakasuki, H. Furugaki, and K. Kojima, DE 3,520,053 to Kao Soap Co. (1984).
122. K. Sasse, in Methoden der Organischen Chemie (E. Müller, ed.), Thieme, Stuttgart, 1940, Bd. 12/2, p. 143.
123. G. Uphues and U. Ploog, EP Patent 0,272,568 to Henkel KGaA (1987).
124. O. Marten, Henkel do Brasil, private information.
125. U. Ploog and G. Uphues, Henkel Ref. 26:63 (1990).
126. G. Jaumann, E. Keck, M. Köhler, G. Uphues, A. Wilsch-Irrgang, and W. Winkle, Fat Sci. Technol. 92:389 (1990).
127. H. Hirota, H. Ogino, H. Ishido, Y. Shibata, S. Igarashi, and C. Fukami, U.S. Patent 4,758,376 to Kao Soap Co. (1985).
128. D. H. Birtwistle, P. Carter, and D. A. Rosser, EP 0,371,801–EP Patent 0,371,804 to Unilever (1988).
129. W. Skrypzak, A. K. Reng, and J. M. Quack, Parfüm. Kosmet.60:317 (1979).
130. I. Homma, DE 2,927,278 to Kao Soap Co. (1980).
131. H. Hoffmann, F.-J. Gohlke, and B. Krönert, FR 2,011,893 to Hoechst AG (1968).
132. H. Watanabe, M. Arisawa, and M. Koike, GB Patent 2,025,451 to Kao Soap Co. (1980).
133. J. P. G. Beiswanger, and A. Nassry, U.S. Patent 3,933,658 to GAF Corp.(1970).
134. J. Kandler, K. Merkenich, G. K. Koehler, and H. Peters, DE 2,249,639 to Knapsack AG (1972).
135. L. Westermann, G. Sorbe, H.-D. Wasel-Nielen, and W. Klose, EP 0,078,918 to Hoechst AG (1981).
136. H. Schranz, DE 1,142,803 to Klöckner-Humboldt-Deutz AG (1957).
137. A. J. O'Lenick, Jr., U.S. Patent 5,070,171 to Siltech Inc. (1991).
138. C. H. Hwa, S. F. Gelman, and P. Kraft, DE 2,035,792 to Stauffer Chemical Co. (1970).
139. D. H. Lorenz and E. P. Williams, DE 2,533,043 to GAF Corp (1974).
140. R. L. Hawkins, U.S. Patent 4,294,883 to Hoechst AG (1979).

141. R. Veitenhansl, and H. Thiel, in *Schriftenreihe Deutsches Wollforschungsinstitut*, Aachen (1984).
142. J. Kamegai, Y. Kajihara, and M. Arisawa, EP 0,324,575 to Kao Soap Co. (1988).
143. J. Kamegai and M. Arisawa, EP 0,324,451 to Kao Soap Co. (1988).
144. T. Calageropoulou, M. Koufaki, A. Tsotinis, J. Balzerini, E. De Clrecq, and A. Makriyannis, Antiviral Chem. Chemother. 6(1):43 (1995).
145. T. Imai and T. Murata, Biochem. Int. 19(16):1277 (1989).
146. Y. Homma et al., Cancer. Chemother. Pharmacol. 11(2):73 (1983).
147. J. G. Aalbers, Ph.D. thesis, Amsterdam, 1964.
148. W. Gerhardt et al., Tenside Surf. Det. 29:169 (1992).
149. J. G. Aalbers, Fette Seifen Anstrichm. 70:174 (1968).
150. S. Ebel and W Parzefall, *Experimentelle Einführung in die Potentiometrie*, Verlag Chemie, Weinheim, 1975, p 150.
151. A. F. Bückmann et al., Makromol. Chem. 182:1379 (1981).
152. M. Leonard et al., Tetrahedron 40:1581 (1984).
153. H. A. Bruson, Org. Reactions 5:79 (1949).
154. R. V. Christian and R. M. Hixon, J. Am. Chem. Soc. 70:1333 (1948).
155. C. S. Hsia Chen, Org. Chem. 27:1920 (1962).
156. K. Heyns and L. Blasejewicz, Tetrahedron 9:67 (1960); K. Heyns and H. Paulsen, Angew. Chem. 69:600 (1957); K. Fiege and Wedemeyer, Angew. Chem. 93:812 (1981).
157. W. Gerhardt et al., Tenside Surf. Det. 29:285 (1992).
158. E. Kunkel, Tenside Surf. Det. 17:10 (1980).
159. E. T. Griffiths et al., Biotechnol. Appl. Biochem. 9:217 (1987).
160. H. König and Strobel, Fresenius Z. Anal. Chem. 728 (1990).
161. P. S. Klose and D. Menz, Erdöl-Erdgas-Z. 97:167 (1981).
162. J. A. Brenkmann, Chem. Weekbl. 63:73 (1967).
163. M. Matsuda et al., Kiyo-Suzuka Kogyo Koto Senmon Gakko 20:51 (1987).
164. N. A. I. van Paassen, Seifen Öle Fette Wachse 109:353 (1983).
165. E. Stroink, Seifen Öle Fette Wachse 115:235 (1989).
166. G. Czichocki et al., Fat Sci. Technol. 94:66 (1992).
167. K. Schulze, Seifen Öle Fette Wachse 101:37 (1975).
168. L. Leusink, C. Broer, Seifen Öle Fette Wachse 119:810 (1993).
169. H. Meijer, Chem. Ind. 58 (1988).
170. D. Balzer, Erdöl-Erdgas-Z. 36:514 (1983).
171. H. Meier, Seifen Öle Fette Wachse 114:159 (1988).
172. E. T. Griffiths et al, Biotechnol. Appl. Biochem. 9:217 (1987).
173. M. E. Chiddix and R. L. Sundberg, U.S. Patent 2,759,975 to General Aniline and Film Corp. (1956).
174. C. Schoeller and E. Ploetz, U.S. Patent 2,214,352 to General Aniline and Film Corp. (1940).
175. H. Rutzen, DE 2,844,451 to Henkel KGaA (1978).
176. E. Ploetz and H. Ulrich, U.S. Patent 2,127,476 to I. G. Farbenindustrie A. G. (1938).
177. E. Ploetz and H. Ulrich, U.S. Patent 2,173,069 to I. G. Farbenindustrie A. G. (1939).
178. G. Demmering, DE 2,052,321 to Henkel KGaA (1970).

179. J. Kolbe, W. Kortmann, and J. Pfeiffer, EP 0,075,770 to Bayer A. G. (1982).
180. G. Uphues, unpublished results.
181. G. Uphues, U. Ploog, W. Becker, and I. Goebel, DE 3,807,069 to Henkel KGaA (1988).
182. U. Ploog and M. Petzold, DE 3,032,216 to Henkel KGaA (1980).
183. M. Hofinger, A.K. Reng, and J. M. Quack, DE 3,345,156 to Hoechst A. G. (1983).
184. H. Rutzen, Fette Seifen Anstrichm. *84*:87 (1982).
185. R. J. Stenberg, U.S. Patent 2,983,738 to Archer-Daniels-Midland Co. (1961).
186. E. L. Carpenter, U.S. Patent 2,372,624 to American Cyanamide Co. (1945).
187. O. Manzke, DE 1,618,026 to Wella A. G. (1967).
188. J. R. Wechsler and M. Lane, U.S. Patent 4,370,272 to Stepan Chemical Comp. (1981).
189. H. C. Parreira, E. R. Lukenbach, and M. K. O. Lindemann, J. Am. Oil Chemists, Soc. *86*:1015 (1979).
190. M. Saiki, Y. Imai, and M. Takagi, EP 0,240,622 to Takemoto Yushi Kabushiki Kaisha (1986).
191. M. Hofinger, R. Kleber, and L. Jäckel, DE 3,325,228 to Hoechst A. G. (1983).
192. M. Tsumadori, K. Shimizu, J. Inokoshi, and M. Murata, EP 0,299,176 to Kao Corp. (1988).
193. W. Wagemann, A. May, and H.-W. Bücking, DE 2,926,772 to Hoechst A. G. (1979).
194. S. Nakamura, H. Kurokowa, and J. Mitamara, DE 3,642,009 to Lion Corp. (1985).
195. J. Faucher, EP 0,153,435 to Union Carbide Corp. (1984).
196. M. E. Spiess, Parfum. Kosmet. 72: 370 (1991).
197. C.-P. Herold, U. Ploog, G. Uphues, B. Spei, and V. Wehle, DE 4,002,472 to Henkel KGaA (1990).

Index

Milton Keynes UK
Ingram Content Group UK Ltd.
UKHW020022071024
449327UK00032B/2890